The Nature of Cities

The Nature of Cities

*Ecological Visions and
the American Urban Professions, 1920–1960*

JENNIFER S. LIGHT

The Johns Hopkins University Press

Baltimore

© 2009 The Johns Hopkins University Press
All rights reserved. Published 2009
Printed in the United States of America on acid-free paper
9 8 7 6 5 4 3 2 1

The Johns Hopkins University Press
2715 North Charles Street
Baltimore, Maryland 21218-4363
www.press.jhu.edu

Library of Congress Cataloging-in-Publication Data
Light, Jennifer S., 1971–
 The nature of cities : ecological visions and the American urban professions, 1920–1960 / Jennifer S. Light.
  p. cm.
 Includes bibliographical references and index.
 ISBN-13: 978-0-8018-9136-6 (hardcover : alk. paper)
 ISBN-10: 0-8018-9136-1 (pbk. : alk. paper)
 1. Urban renewal—United States—History—20th century. 2. Conservation of natural resources—United States—History—20th century. 3. Social sciences—United States—History—20th century. 4. Real estate business—United States—History—20th century. 5. City planning—United States—History—20th century. 6. Chicago (Ill.)—History—20th century. I. Title.
HT175.L54 2009
307.7609773'11—dc22   2008028732

A catalog record for this book is available from the British Library.

Special discounts are available for bulk purchases of this book. For more information, please contact Special Sales at 410-516-6936 or specialsales@press.jhu.edu.

The Johns Hopkins University Press uses environmentally friendly book materials, including recycled text paper that is composed of at least 30 percent post-consumer waste, whenever possible. All of our book papers are acid-free, and our jackets and covers are printed on paper with recycled content.

To Anja and Jon

Contents

Acknowledgments   ix

Introduction: Revisiting American Antiurbanism   1

1 The City Is an Ecological Community   6
2 The City Is a National Resource   36
3 A Life Cycle Plan for Chicago   69
4 From Natural Law to State Law   98
5 A Nation of Renewable Cities   128

Conclusion: From Ecology to System   161

Abbreviations   173
Notes   179
Essay on Sources   287
Index   299

Acknowledgments

Many individuals and institutions supported this project. A grant from the Northwestern University Research Grants Committee funded travel to archives and library collections as well as image reproduction costs. At Cambridge University, a term as Derek Brewer Visiting Fellow provided the necessary critical distance to focus my argument, and Richard Barnes and Lord Richard Wilson of Emmanuel College were genial hosts.

Archivists and librarians at the U.S. National Archives and Records Administration, Chicago Historical Society, DePaul University, Detroit Historical Society, Harvard University Graduate School of Design and Gray Herbarium, and University of Chicago assisted in locating essential source materials. Reminiscences from Donald Foley, who worked under Homer Hoyt and Louis Wirth in the 1940s, were invaluable. In my own distant past, graduate seminars with Everett Mendelsohn and Leo Marx on the history of ecology, environmentalism, and meanings of nature in America sparked my curiosity about ecological metaphors in the urban professions, and the secondary readings from these long-ago courses provided important background for this study. Colleagues in the Rhetoric and Public Culture program at Northwestern University's Department of Communication Studies, especially Angela Ray, suggested sources to refine my ideas about how the circulation of technical language matters as much as the circulation of the technical objects that were the focus of my first book.

John Cloud and Robert Kargon offered sharp feedback on an earlier draft of this manuscript. Portions of the project were presented to the Department of History and Sociology of Science at the University of Pennsylvania; the University of Chicago Workshop on City, Society, and Place; and the Boston Environmental History Seminar, where participants offered many constructive criticisms. At Johns Hopkins University Press, Robert J. Brugger once again was a terrific editor, judicious and wise in his suggestions for revisions. Andre Barnett made many improvements to the final manuscript. I put the finishing touches on the project as a visiting faculty member of the Department of Urban Studies and Planning at the Massachusetts Institute of Technology, and I appreciate in particular the intellectual hospitality of Joseph Ferreira, Robert Fogelson, and Lawrence Vale.

The Nature of Cities

INTRODUCTION

Revisiting American Antiurbanism

At a 1960 campaign speech to an audience of mayors, Democratic presidential candidate John F. Kennedy made the case for increasing federal investment in America's cities. Despite several years of urban renewal demonstrating "what wonders can be worked through Federal-city partnership," a slew of problems remained. In a nation of cities characterized by "slums, traffic jams, crime and delinquency," a "long-term commitment to urban renewal" was urgently needed, the Massachusetts senator insisted, calling for a federal department of urban affairs. "The Department of Agriculture was created 98 years ago to serve rural America," he explained. In an era when "the new urban frontier . . . exists in every city in America . . . it is time the people who live in urban areas receive equal representation."[1]

With its frontier rhetoric, Kennedy's address evoked traditional American attitudes toward urban versus rural areas—as oppositional territories, with favored status for nature in the country's political life.[2] The local government officials whose press releases announced, "More Americans live in slums than on farms," welcomed his call to correct this imbalance.[3] Yet as the renewal policy Kennedy praised exemplified the nation's uneven investments, it simultaneously showcased how Americans' prior approaches to understanding and managing nature had significantly shaped the histories of urban analysis and reform. Renewal emphasized a comprehensive strategy of redevelopment, rehabilitation, and conservation that targeted phases of cities' "life cycles" and relied on collaborations between national technical experts and local grassroots volunteers. The program marked the culmination of four decades of work from academic social scientists, city planners, and real estate appraisers who sought to deliver the prestige and predictability of the natural sciences and scientific management of nature to the development of urban research and policy.

From the 1920s, when America was newly recognized to be predominantly urban, the country's emerging urban professions set out to understand and manage cities with an eye on the nation's obsessive concern for its agrarian past. Drawing on American traditions in the natural sciences, as well as the scientific management of nature, prominent figures working in urban studies, in real estate, and in city planning believed that cities, as ecological communities and national resources, operated according to predictable laws—and by extension that solutions to urban problems should keep the "nature" of cities in mind. This book recounts the story of these figures' enduring efforts to build an American tradition of urban research and policy in the image of the country's approach to the analysis and administration of nature. The irony of their campaigns' delayed successes at a time when the conservation movement that served as their model had lost its public acclaim will also be addressed.[4]

American Antiurbanism

From political observers such as Thomas Jefferson to literary giants such as Ralph Waldo Emerson and landscape architects such as Frederick Law Olmsted, American commentators and critics have long positioned cities in opposition to nature and rural life. Urban and environmental historians' accounts of the conceptual and physical history of U.S. cities have echoed such observations to emphasize how the nation's distinct antiurbanism—the contention that the naturalism of rural living was highly preferable to the artifice of cities—shaped its urban policies and practices. In studies of Garden Cities, public parks, and suburbia, stories abound of individuals and institutions who, by bringing nature to cities, saving nature in cities, or fleeing cities for nature, sought to deliver to property owners a closer relationship with the natural world.[5]

This focus on the persistence of antiurbanism has obscured another side to American attitudes about relationships between cities and nature, however. "If we consider how much we are nature's," wrote Ralph Waldo

Emerson in his essay "Nature" (1844), "we need not be superstitious about towns, as if that terrific or benefic force did not find us there also, and fashion cities. Nature, who made the mason, made the house."[6] Emerson's suggestion that cities were expressions of the same forces as the pastoral existence he preferred typified another enduring strain of thinking about urban areas, whose influences on the emergence of the American urban professions in the early twentieth century has not yet been a focus for historical research.[7] Building on the work of the plant and animal specialists called to service in the country's conservation efforts, and for whom urbanization had become a topic of increasing concern, the urban professionals described in this book became convinced that close correspondences existed between patterns of human life in urban environments and patterns of plant and animal life in the natural world. They argued that the array of interventions organized to apply ecological knowledge to scientifically manage resources, including forests and farms, could be extended to maintain and indeed improve cities. Five chapters trace the slow unfolding of these beliefs from theory to implementation and how, between 1920 and 1960, this alliance of professionals sold their ideas to varied audiences from federal housing agencies to local community organizations.

Why did so many prominent figures in the American social science, city planning, and real estate professions see urban landscapes as "ecological communities" and "national resources" in need of "conservation"? What difference did their analogical thinking make in how the evolution of U.S. cities unfolded? To answer these questions about the sociology of urban knowledge and urban problems, this book draws together sources in environmental and urban history.

Scholars have noted the rise of ecological thinking in early twentieth-century sociology, the influences of these naturalistic theories on real estate appraisal practices and federal housing policies during the New Deal, and the continued relevance of the life cycle template to postwar urban policy.[8] Yet, missing from these accounts is how such episodes in the history of the American urban professions were chapters of a much larger story about the place of science in constructing urban expertise and why this specific scientific tradition gathered momentum and then waned when it did. During the decades that spanned the rise and fall of the nation's conservation movement—with ecologist Frederic Clements's theories of succession and climax as a model for natural resources planning policy — naturalistic visions of cities in general and a belief in the predictable urban life cycle in particular provided a powerful frame for diverse actors working to mobilize federal and local investments in the city improvement cause.[9] These actors saw rhetorical potency not only in nature's symbolic remove from society. Rather, when the mastery of natural resources was widely praised, associations between cities and nature suggested that urban resources might also be scientifically managed and that within the ranks of the urban professions resided the necessary expertise. In the dominance of ecological perspectives at the origins of academic urban studies, in the influences of the conservation movement on the development of the real estate and city planning professions, and in the vision of cities as renewable resources that characterized the first nationwide program for city improvement, the effect of Americans' enduring preference for nature on the history of urban society was significantly broader than scholars have presumed. *The Nature of Cities* offers new insights into ways that knowledge about nature shaped how major figures in U.S. urban history understood their landscapes and themselves.

Chicago is the geographic center of this story, not only because of its status as the archetypal American city of the early twentieth century but also because of the central roles that local actors played on the national stage. Chicago was headquarters to most of the nation's real estate, city planning, and housing organizations; its identity as a paradigmatic center for political corruption did not diminish urban professionals' interest in scientific approaches to urban analysis and reform.[10] Faculty and students at the University of Chicago's departments of Sociology and Geography, seeking a social "science" to make sense of the apparent chaos of swelling urban areas, applied concepts and methods from plant and animal ecology to explain life in cities, creating the "human ecology" research tradition in the United States. Scholars at Northwestern University's Institute for Research in Land Economics and Public Utilities with close ties to industry and government, hoping to build a science of real estate economics, borrowed liberally from human ecology as they set the curriculum for the National Association of Real Estate Boards' Standard Course in Real Estate and the appraisal practices of federal agencies, including the U.S. Home Owners Loan Corporation and the U.S. Federal Housing Administration. Convinced by practitioners of the new urban science that a life cycle guided the unfolding of cities' growth, decay, and change, the association, together with these government agencies, organized an experiment with urban "community conservation" in Chicago's Woodlawn neighborhood to demonstrate the possibilities for reversing predictable urban patterns. The Chicago Plan Commission expanded its vision of urban resources planning in its *Master Plan of Residential Land Use in Chicago*, with collaborations between technical experts and citizen volunteers central components of a long-term, citywide program to plan a future Chicago in line with natural forces of urban growth and decay. A Community Appraisal Study in several South Side neighborhoods enlisted academics, city planners, and members of the real estate industry to engage community organizations in the urban conservation effort, which inspired state legislation to enable life cycle planning in Illinois cities. In each example, and many others, urban professionals in Chicago, the city characterized as "nature's metropolis," pressed to make sense of urban areas in light of knowledge about the natural world.[11] These individuals never aspired to inscrutable fidelity with the traditions from which they borrowed. Their loose analogies became "metaphors we live by" as federal legislation delivered a long-awaited program for the scientific management of urban resources to cities across the United States.[12]

ONE

The City Is an Ecological Community

"The modern city," wrote the esteemed sociologist Robert Park in 1940, "has long since ceased to be what the peasant village was, an agglomeration of individual habitations." For two decades, America had been recognized as an urban nation, and the city, "like the civilization of which it is the center and focus," had become "a vast physical and institutional structure in which men live, like bees in a hive."[1] Countering those contemporaries who pointed to similarities between human and animal life in the chaos of the urban "wilderness" and "jungle," however, Park's analogical reasoning emphasized how urban life was far "more regulated, regimented, and conditioned" than the average resident or observer could see.[2] Scientific studies of bees, and indeed all animals and plants, helped to reveal the hidden order in urban environments, he argued, insisting that, while cities might not be "natural," logical rules guided their operations.

How did Park, like so many early twentieth century social scientists, come to express such confidence that cities were "ecological" communities organized by predictable laws? Faced with population trends swelling urban areas in the 1920s and a subsequent economic depression for cities in the 1930s, these scholars found a grab bag of scientific theories derived from earlier studies of plants and animals valuable tools for making sense of trends in human society. In studies of diverse urban patterns and processes — the formation and disintegration of neighborhoods, the location and migration of industries, and the history and future of land values —

sociologists, geographers, and economists (along with a few political scientists and anthropologists) extended the work of scientists, including Eugenius Warming, Charles Child, and Frederic Clements, to build analogies between the structure, behavior, and evolution of dynamic communities in nature and the structure, behavior, and evolution of nature's apparent antithesis: the rapidly changing American city.[3] Their informal appropriations of ecological thought — the result of a program of scientific self-education rather than of close associations with scientific experts — would dominate urban studies in the early twentieth century.

Stories of the rise of "human ecology" to date have emphasized the sociological aspects of this research tradition.[4] Yet the study of cities has never been the exclusive territory of a single field, and although sociologists led the charge to borrow concepts and methods from ecology, this approach to urban studies had remarkably broad appeal. Human ecology was "a new social science," as geographer George Renner put it, a metadiscipline for making sense of human environments.[5] For the scholars who characterized their work as "ecological," the cross-disciplinary relationships developed around this shared focus would not be without occasional tensions. Yet such disputes were minor distractions from the larger networks that interest in a scientific understanding of U.S. cities helped to build.[6]

Converging Research Traditions in Science and Social Science

The appearance of human ecology did not mark the first convergence between biological and social thought. Analogies between plant, animal, and human societies have existed for centuries.[7] Historians have characterized the late nineteenth and early twentieth centuries as periods when an evolutionary mindset and an acceptance of biological determinism gripped American science and popular culture. In the theories of Social Darwinism that inspired Margaret Sanger's campaign for birth control and the eugenical notions of racial hierarchies that shaped immigration policies, the idea that knowledge of animals and plants might extend to understanding humans was popular, if controversial. Scholars have documented the prominent role of scientists in mediating these public controversies and how their attachment to such beliefs was in decline by the end of the 1920s.[8] Yet, as this generation's interest in linking plants, animals, and humans waned, a new field making similar connections arrived.[9]

The development of ecological approaches to urban analysis in the United States followed the convergence of interests among natural scientists and social scientists independently observing recent national trends in land use and population distribution. According to the U.S. Census Bureau, the frontier officially closed by 1890. Migration from rural to urban areas, together with immigration from Europe, Asia, and Mexico at the end of the nineteenth and beginning of the twentieth centuries, created human densities and diversities never before seen in an American context. The nation whose identity long had been tied to its agricultural power had become, like many European countries, increasingly urbanized. With this "revolution in land use," to use Charles Abrams's phrase, Chicago, for example, a city of less than 70 people in the 1820s had grown to 1,698,575 in 1900; 2,185,283 in 1910; and 2,701,705 by 1920.[10] Human activities were transforming plant and animal communities, and these changes were nowhere as evident as the urban fringe where nature and city met.

For natural scientists whose focus was the American landscape and the plant and animal life within, the rapid urbanization that surrounded many university centers of ecological research became a matter of some consternation.[11] Ecological research had traditionally focused "on native or wild plants and animals and not on cultivated plants and domesticated animals," as North Dakota Agricultural Experiment Station head Herbert Hanson observed, expressing concern about the rise of cities as fundamentally "unnatural" developments with implications for researchers' ability to understand the natural world.[12] Charles Adams, an associate in animal ecology at the University of Illinois and a founder of the Ecological Society of America, was one of the many scientists who bemoaned human modifications of nature, pointing to the difficulty investigators now confronted of studying "disturbed, artificial and pathological conditions" without knowledge of the " 'normal' conditions that preceded them."[13] In the wake of the frontier's closing, conservation had become a prominent political issue, and between 1890 and 1920, many plant and animal experts worked to apply scientific knowledge to protecting natural areas on the outskirts of cities — for example, creating state parks and forest preserves. The Ecological Society of America established a Committee on the Preservation of Natural Conditions for Ecological Study.

Yet, on other occasions, plant and animal ecologists, remarking on trends in urbanization, emphasized not how the forces of urban development stood in opposition to nature but rather how knowledge of plant and animal communities might extend to explaining human and urban affairs. "The wonder has always been," wrote plant scientist Frederic Clements in 1916, "not that there are so many differences in structure between such disparate organisms as insects and man, but that there are so many striking similarities in behavior."[14] Even Charles Adams, in the same *Guide to the Study of Animal Ecology* (1913) where his anxious expression of concern for human disturbances to nature appeared, mused about the potentially broad applicability of ecological ideas. "Human ecology is a part of general animal ecology," he offered. "The response of man, as an animal, to a part or the whole of his environment is strictly ecological. So far as known to the writer, human activities in general have never been fully and comprehensively oriented from the ecological standpoint." Here Adams imagined an expanded territory for the science of communities, able to account for the study of humans as well as plants and animals, and able to give "a new unity to all studies of human relations."[15]

Such comments about parallels between life in plant, animal, and human communities from scientists were oftentimes offhanded remarks or, when more deliberate, intended to rally interest in ecology, itself a comparatively new field.[16] Among a community of listeners in the social sciences, however, more substantial conclusions were drawn. These social scientists, less interested in the implications of urbanization for plants and animals than in achieving intellectual rigor for the emerging social science profession, saw in ecology a new approach to studying humans' adaptation to the changing urban environment.[17] Taking the view, as Yale geographer Ellsworth Huntington put it in 1925, that "plant ecology, animal ecology, and human ecology are all governed by the same great laws; and all are essential to the understanding of how and why the phenomena of the earth's surface are so peculiarly distributed," they concluded that the forces of urbanization disturbing the balance of nature for plants and animals simultaneously were creating "the natural habitat of man."[18] The city had become the nation's newest frontier, these scholars would argue, deserving of research attention.

Ecology was not the obvious first choice among American social scientists seeking a scientific basis for their inquiries into urban affairs. As discussions about future directions for urban research unfolded, many initially expressed hope for developing an experimental tradition in the "laboratory" of the city, with chemists and physicists as their model.[19] Yet, experimentation in urban contexts raised ethical and practical concerns. "There is at least one important difference between the laboratory of the physical scientist and that of the social scientist," University of Chicago sociologist Ernest Burgess reflected in 1929. "In chemistry, physics, and even biology the subjects of study can be brought into the laboratory and studied under controlled condition," an approach "not feasible in the social sciences."[20] Scientific practice in the "natural laboratory" of forests, fields, and other environments studied by plant and animal ecologists, where the behavior of living systems could be observed with minimal disturbance, was a much better fit.[21] With their field study procedures, natural history surveys, and map-making techniques, these plant and animal specialists employed a research tool kit similar to what sociologists, geographers, and economists already used.[22] As Arthur Tansley observed in a presidential address to the British Ecological Society, "there are very few important techniques which are distinctively ecological."[23] Plant and animal ecologists, seeing similar laws structuring a diversity of living systems, had already borrowed numerous concepts from sociology (including the idea that there was a "plant sociology" and "animal sociology"). They had identified among geographers such as Alexander von Humboldt the giants in their field. And they had observed common intellectual origins with economics (as the shared root word "oikos" attests). The interest that plant and animal ecologists displayed in social science soon attracted social scientists' attention; turning to ecology provided researchers with not only a model for future investigations but also an opportunity to reframe aspects of their prior work as "scientific" owing to its adoption in this natural science field. "Ecological methods are surely applicable to men," observed Charles Redway Dryer, head of the Association of American Geographers in a presidential address. "I will read Warming's statement of the problems of ecological plant geography, and as I read, I ask you to substitute men for plants, humanity for vegetation, institutions for species... Can the economic and social geographer devise a better scheme than this?"[24]

From the 1920s, then, not only sociologists but also geographers and economists, working in fields where studies of location and land values gained importance for research, overturned conventional antagonisms between cities and nature to point to features of urban life that most closely resembled plant and animal life. The center of this intellectual tradition was the University of Chicago. Yet, similar modes of inquiry were found at institutions across the United States.[25] Whether their focus was New York or Los Angeles, Minneapolis or Madison, in their efforts to identify the natural laws guiding growth, decay, and change in urban areas, social scientists relied on the template that plant and animal ecologists had provided.[26] Such approaches would provide a model for city investigations across disciplines for decades to come.[27]

Creating the Language of Human Ecology in the 1920s

As with many sciences, the ecological study of plants and animals in the early twentieth century was not a monolithic discipline. Ecology, cobbled together from preexisting traditions that included botany, zoology, and field biology, was a lively enterprise comprising competing theories and schools. Within scientific circles, debates about the extent of competition versus cooperation in living systems, and whether all natural systems achieved stable climax, persisted throughout the period.[28] The social scientists who sought to construct a predictive science of cities from outside these circles paid less attention to the details of such disputes than to the broad explanatory powers of concepts, including natural area, competition, and succession.

Natural Area

One of the foundational units in human ecology was the "natural area" (preferred by sociologists), "natural region" (preferred by geographers), or "natural zone" (preferred by economists). For plant and animal ecologists, "the animals living in a small brook, the littoral zone of a lake, in a colony of breeding gulls, or on the floor of a forest" were "treated as a unit" such that "the entire history of the animals in the habitat" was "considered as a response to the conditions of life."[29] Because city planning and zoning were relatively new phenomena, many social scientists suggested their observations offered nearly unvarnished insights into the forces guiding city formation. "As cities have grown up without zoning laws or comprehensive city planning, a tendency is observed for certain districts to specialize in certain uses," University of Wisconsin economists Richard Ely and Edward Morehouse explained. "This process is called natural zoning to distinguish it from zoning which is controlled by public officials according to a definite public policy."[30] These scholars read slums, ghettos, shopping districts, and neighborhoods defined by a single nationality as manifestations of the laws of nature in the city, each a reflection of more general principles of human community formation and evolution. "The Negro community is one of the most fundamental 'natural areas' in the city," wrote Howard University sociologist William Jones in a 1929 study of Washington, DC, "because it is the result of social attitudes which are generated by racial differentiation." What was true for "the Negro communities in Washington" was "true of communities everywhere," he concluded, arguing that such homogenous spatial arrangements "had their inception in the traits of human nature."[31] "Practically all types of businesses are found to have 'natural habitats' or positions within the business districts, i.e. ecological positions," University of Chicago graduate student Ernest Shideler agreed. "Certain types of business establishments are complementary to each other and these tend to group together."[32]

Making both implicit and explicit references to the findings of plant and animal ecologists, this emerging scholarly community suggested that the physical organization of urban environments expressed the natural order, and by extension that it was possible to read in the neighborhoods of any city the scientific laws of the urban system. They discovered a distinct set of natural areas in most cities: "Every large city tends to have its Greenwich Village just as it has its Wall Street," Robert Park explained citing the work of Eugenius Warming. "Every great city has its racial colonies, like the Chinatowns of San Francisco and New York, the Little Sicilies of Chicago," echoed Columbia University sociologist Nels Anderson.[33] Such findings confirmed the possibility of developing a generalizable scientific model of cities.

Competition, Dominance, Competitive Cooperation, and Symbiosis

Observing the existence of homogenous racial enclaves and their commercial or industrial equivalents was merely a first step."[34] Once they established that chaotic and bustling cities were physical manifestations of natural laws and outlined in general terms the divisions among neighborhood types, these scholars set out to understand how, precisely, such divisions might have occurred. If, as sociologist and research director of the Institute for Social and Religious Research Harlan Paul Douglass observed, "natural selection" determined the locations of individuals and institutions, an explanation could be found in the same forces that guided such separations and segregations in plant and animal communities.[35] Ecologists had identified predictable patterns in nature with reference to "competition" and "dominance"; in their analyses of human behavior, ecologically minded urban analysts quickly followed suit. "Just as there is a plant ecology whereby, in the struggle for existence, like geographical regions become associated with like 'communities' of plants, mutually adapted, and adapted to the area," Ohio Wesleyan sociologist Harvey Zorbaugh explained, "so there is a human ecology whereby, in the competition of the city and according to definable processes, the population of the city is segregated over natural areas and into natural groups."[36] Citing Charles Darwin's theory of the struggle for existence, Zorbaugh highlighted the role of competition among racial and nationality groups in shaping residential neighborhoods in American cities.

Competition had a long history within economic thinking. Now, "the violent and eruptive competition of uses" economists such as Herbert Dorau and Albert Hinman observed "going on in our American cities" took on new scientific associations as ecological studies of urban natural areas concluded that a chain of dominance among population subgroups had developed as a result of competition—hierarchies that mapped onto spatial arrangements within metropolitan areas.[37] University of Washington sociology professor Roderick McKenzie drew upon animal ecologist Charles M. Child's *The Physiological Foundation of Behavior* to underscore the scientific basis of this belief: "Social groups, whether animal or human . . . become integrated in a dynamic relation of dominance and subordination, leaders and followers. This is the pattern of the pack, the herd, the flock of migrating birds, as well as that of all human groups."[38] If "the principle of dominance operates in the human as well as in the plant and animal communities," Robert Park elaborated in a paper that referred to Charles Darwin, Charles Elton, and William Morton Wheeler, then cities' natural areas, "for example, the slum, the rooming-house area, the central shopping section, and the banking center—each and all owe their existence directly to the factor of dominance, and indirectly to competition."[39] For most cities, it appeared that ecological principles could not only explain why specific individuals and institutions were segregated from one another but also why they were located in specific neighborhoods. According to this view, the location of some residences and businesses close to the city center and others toward the periphery mapped onto the natural organization of the city.

Such academic emphasis on principles of competition and dominance implied frequent intergroup conflict. Yet scholars made clear that these were not the only types of relationships that existed among the individuals and institutions in contemporary cities. Plant and animal ecologists had studied a range of mutually beneficial interactions among species, and in his 1928 study of ghetto neighborhoods past and present, in Europe and in the United States, University of Chicago sociology graduate student Louis Wirth applied their notions of "competitive cooperation" and "symbiosis" to his interpretation of the sorting of human groups. Although the "segregation of the population into distinct classes and vocational groups" was "essentially a process of competition," then, it was "akin to the competitive co-operation that underlies the plant community." For different human groups with different desires and needs managed "to live side by side, much like plants and animals, in what is known as symbiosis."[40] Such interpretations, which suggested that the processes of competition might be resolved in a stable way as groups in conflict accommodated one anothers' needs, helped to explain why, as University of Chicago researchers had noted in Chicago and Columbia University economist Robert Murray Haig had observed in New York City, "some of the poorest people live in conveniently located slums on high-priced land . . . Tiffany and Woolworth, cheek by jowl, offer jewels and jimcracks from substantially identical sites." To "the superficial observer," Haig explained, "the assignment of the land to the various uses seems . . . to have been made by the Mad Hatter at Alice's tea party." But "most of the apparent anomalies and paradoxes dissolve into commonplaces when subjected to serious study and detailed examination."[41]

Invasion, Succession, and Sequent Occupance

Theories of natural areas, competition, dominance, competitive cooperation, and symbiosis helped to make sense of the overarching spatial arrangement of cities. Such conceptual frameworks for understanding urban structure, while indicating the dynamic character of interactions among individuals and institutions, could not fully capture the continuous nature of urban change. Many of the era's social scientists had become interested in studying cities on account of the large-scale patterns of migration, immigration, and building they witnessed in their own backyards. The question of how, in any city, interactions among individuals and institutions might give rise to transformations in the urban fabric became a topic of substantial research attention.

Here too the era's urban analysts found explanations in plant and animal ecology. Studies of plant and animal habitats had observed the "invasion" of new species as a leading edge of area change, and this concept was adapted to analyses of the migrations of populations and land use types. "Residential property is always invaded more or less rapidly by industry and business," observed Harlan Paul Douglass in St. Louis, "and the old houses of the well-to-do are being turned over, in progressive stages of disrepair, for the occupance of the poor."[42]

The cascade of events that followed the initial invasion of an area by a new population or land use type came to be referred to as "sere," "sequent occupance," and "succession," also borrowed terms. Plant and animal ecologists, most notably Henry Cowles and Frederic Clements, had used this vocabulary to characterize stages of change in local environments; in their studies of the evolution of urban areas, ecologically minded urban analysts followed suit. (Geographers preferred "sequent occupance" and "sere" and sociologists and economists favored "succession" for the same basic concepts.)[43] "The analogy between sequent occupance in chorology [the science of the distribution of living organisms over the earth's surface] and plant succession in botany will be apparent to all," noted Harvard geographer Derwent Whittlesey in 1929. He juxtaposed examples from agriculture and urbanism to explain the basic transferability of ideas:

> Human occupance of area, like other biotic phenomena, carries within itself the seed of its own transformation. Some examples of this principle are everyday commonplaces. The American farmer, inaugurating a stage of occupance by plowing and planting virgin soil, sets in motion agents which at once begin subtly or grossly to alter the suitability of his land for crops; in extreme cases the ground deteriorates to a point where it must be converted into pasture or forest, or even abandoned; when either of these events occurs, human occupance of that area has entered upon a new stage. To take an example from urban life: the normal increase in population of a wisely located and hence successful urban area in time forces the resident population to remove to a zone some distance from the center, which thenceforth is occupied in a wholly new fashion.[44]

An explosion of studies followed, each an effort to schematize the stages of sequent occupance or succession in urban areas. Some took a macro-level perspective, tracing the evolution of cities from their agricultural origins to the present day. Geographer Charles Redway Dryer, for example, noting how "one of the outstanding discoveries of ecology is the existence of plant succession," the "regular series of associations which follow one another in the same habitat," and known to ecologists as a "sere," suggested a four-stage sere was apparent in human communities as well. "Why should not the successive stages of economic life in the human occupation of an area, as in the Middle West, Indian hunting, white pioneer deforestation, developed agriculture and industrial exploitation, constitute and be known as a sere?"[45] University of Minnesota economic historian Norman Scott Brien Gras, by contrast, offered a five-stage theory of the "rise of the metropolis" to the American Sociological Society's Section on Human Ecology in 1925: (1) the collectional economy, (2) the cultural nomadic economy, (3) the settled village economy, (4) the town economy, and (5) the metropolitan economy as the predictable sequence of categories experienced by many societies.[46] Other scholars interested in sequential processes of change took a micro-level perspective to focus their attention on transformations within cities once urban areas had been fully formed. Following Frederic Clements's four-stage model of transformations in plant communities from succession to climax, Ernest Burgess, for example, delimited a four-stage process of (1) "invasion" — the initial entry of a new group; (2) "reaction" — the resistance of existing residents, (3) "influx" — the accelerated entry of newcomers and departure of old timers, and (4) "climax" — the stability of the area with its new inhabitants.[47] Roderick McKenzie, by contrast, saw only three distinct steps, "namely (a) the initial stage, (b) the secondary or development stage, and (c) the climax stage."[48]

Human Ecology at the University of Chicago

Following the official recognition that America had become an urban nation, then, social scientists grew eager to understand the "nature" of cities. The University of Chicago became the headquarters for the effort to make sense of urban environments in ecological terms. Scholars have documented how the Sociology Department was the site for this work, where, at the behest of faculty members Robert Park and Ernest Burgess, sociology students encountered the ideas of leading figures in plant and animal ecology in course readings, in reviews from the Chicago-edited *American Journal of Sociology*, and in person at the meetings of the graduate Society for Social Research.[49] It was in this department that the first two textbooks in human ecology were written: The "Green Bible" from Park and Burgess, formally known as *Introduction to the Science of Sociology* (1921), juxtaposed readings from plant and animal ecologists, including William Morton Wheeler, Frederick Clements, Eugenius Warming, and J. Arthur Thompson, alongside the work of social thinkers.[50] More explicit in its appropriations from plant and animal science — "a botanist passing through a forest gathers facts concerning plants and trees which are unnoticed by the layman" — department researcher Vivien Palmer's *Field Studies: A Student's Manual* (1928) was created to satisfy the "urgent demand" for "sociological equivalents to laboratory manuals" for learning scientific observational techniques.[51] Although Park and Burgess still sent their students to read original texts in plant and animal ecology, the appearance of these instruction manuals signaled sociology's effort to establish itself as a "scientific" endeavor.[52] In close studies of Chicago neighborhoods, including the Gold Coast, South Chicago, and Hyde

Park, as well as in citywide investigations of suicides, chain stores, and specific nationality groups, students followed their mentors' lead to frame these research subjects in ecological terms.

In fact, even before University of Chicago sociologists expressed interest in human ecology, their colleagues at the institution's geography department — which spanned inquiry in natural and social sciences — had taken initial steps to synthesize these fields. As early as 1908, geographer J. Paul Goode was teaching a course on ontography — the study of plant, animal, and human ecology. (Goode mentioned the term "human ecology" in a presentation that year.[53]) Henry Cowles's courses in field studies and natural history surveys were cross-listed in geography (during his career, Cowles served as president of both the Ecological Society of America and the Association of American Geographers), and the atmosphere Cowles created over his many years at Chicago apparently "exerted a favorable influence in the extension of plant and animal ecology to the still newer field of human ecology."[54] By the time that the university's Department of Geography was encouraging its students to sign up for sociologist Roderick McKenzie's Human Ecology course in the mid-1920s, that program had already received wide exposure to ecological concepts and methods.[55] Even faculty uninterested in urban geography made the link: At the 1923 meeting of the American Association of Geographers, for example, association president and University of Chicago geographer Harlan Barrows proposed in his keynote address that the field of geography had become "human ecology."[56] When the university's new laboratory for social science opened at the close of the decade, it offered shared statistical and mapping facilities to researchers in the expanding field of urban studies, reflecting the multidisciplinary nature of the field, and a physical site for cross-disciplinary communication.[57]

As these scholars published their findings, traveled to other campuses, and graduated students to faculty positions, they built momentum for ecological approaches to urban study at universities around the United States. Robert Park, for example, traveled widely to lecture on the topic. Roderick McKenzie, teaching for several years on the faculty at University of Washington, returned to Chicago to offer three summer-session courses and one at Stanford. Erle Fiske Young, on the social work faculty at the University of Southern California, created a Human Ecology course there. And in an early example of cross-campus collaboration, the Rockefeller

Foundation–sponsored Pacific Coast Race Relations Survey, which ran from 1923 to 1926, drew together sociologists, including Robert Park (University of Chicago), Roderick McKenzie (University of Washington), and Emory Bogardus and William Smith (University of Southern California), with geographers such as Eliot Mears (Stanford), Charles E. Rugh (Clarion State Normal School), and a smattering of faculty and students in disciplines that included psychology (Albert Chandler, Occidental), education (John C. Almack, Stanford), and religion (Isaac Granadosin, University of Southern California). Although the project sputtered because of lack of funding, it created networks for the circulation of ideas and inspired numerous dissertations and subsequent faculty research on the ecological dimensions of life in the cities and regions on the West Coast. Of the project, Bogardus concluded, "The retroactive effects of cooperative research are most noticeable [in how] each research worker tries to be intelligible to the others in related fields."[58] In a testament to these relationships, Park sent the book *North America: Its People and Resources, Development, and Prospects of the Continent as an Agricultural, Industrial, and Commercial Area* to Stanford geographer Mears with a note that read, "This book goes further than anything I have yet found in the way of Human Geography and Human Ecology. I am sure you will find it interesting and valuable."[59]

Analogy and Accuracy in Ecological Thought

As scholarship on the ecology of cities proliferated apart from the scientific community on whose ideas these social scientists relied, the growing community of human ecologists turned to reflect on the limits to analogical reasoning as a means for explaining urban life. A few did not, simply following Charles Redway Dryer's call that every human geographer "avail himself of the methods, formulas, and as far as practicable, the language of the plant and animal ecologist."[60] Yet, from the appearance of the Green Bible, with its discussion questions inviting students to elaborate on similarities and differences between plant, animal, and human society, more moved beyond pointing to the use of social science by natural scientists — and in turn of scientists' own frequent use of analogy as a tool for reasoning in defense of their work — to acknowledge the points at which ecological thinking about cities might be a misstep. As University of Hawaii sociologist Andrew Lind observed in his study of population mobility in Seattle, for example, humans had far greater opportunities for deliberate migration than plants.[61] Harvard geographer Derwent Whittesley, writing in a paper on sequent occupance, noted that studying the phenomenon for plant communities was "less intricate" than attempting similar studies in human communities, where "changes in any of the complex elements of natural environment, and in the equally complex cultural forms" required simultaneous consideration.[62] Scientists "do not assume that the corresponding features revealed by analogical reasoning or observation constitute proofs," sociologists Nels Anderson and Eduard Lindeman observed in an early urban studies textbook, insisting that human ecologists do similar due diligence.[63] Thus, when New York University sociologist Frederic Thrasher investigated zoologist Charles Child's musings on "whether many of the more primitive and evanescent types of integration among human beings at the present day," including crowds, mobs, and gangs, were "to a considerable extent . . . very similar or essentially identical with many social integrations among animals," he concluded that "the processes of growth and differentiation of parts which take place within the community" were only "somewhat analogous in principle to those that occur in the development of an organism or of a plant community."[64] Although most human ecologists did not aspire to complete fidelity with plant and animal science, they recognized that a field that focused on the close correspondences between patterns of human life in urban environments and patterns of plant and animal life in nature needed to recognize the distinctions between these patterns as well.

These distinctions were occasional grounds for pride as the new social science appeared even more verifiable than the plant and animal studies after which it was modeled. Such was the case with "life history" research techniques, Vivien Palmer's textbook explained. While scientists such as Frederic Clements could describe how "the endeavor has been made to have each garden or culture tell its story as completely as possible, partly to provide a definite or permanent mental picture but also to furnish a graphic demonstration," plant and animal ecologists could not interview the subjects of their research.[65] "The social scientist is in the very unique position of being able to interrogate his subject matter concerning the process which is taking place," Palmer noted, "this adding a valuable check to the discovery of his facts and the formulation of his generalizations."[66]

By the 1920s, such discussions about analogical reasoning and its limits had convinced many of the field's contributor of two ideas. First was the basic soundness of such an approach to urban theory development. Although "the biological analogy has its weaknesses and they have been subjected to searching criticism over the past forty years," as University of Pennsylvania sociology graduate student W. Wallace Weaver put it, "that there are many similar processes in the 'social' life of plants or non-human animals and human beings has never been denied."[67] Second was that more systematic standards were required before human ecology could assume its rightful place among the social "sciences." The variety of independent explanations of phenomena such as "sequent occupance" and "succession" that scholars had generated was so great that, without agreement on basic terms, these developments were almost counterproductive. "The importance of classification has long been recognized in the biological and physical sciences and it has been playing a leading role in the rapid development that has characterized these sciences during the last seventy-five or hundred years," observed Northwestern University economists Herbert Dorau and Albert Hinman in 1928, and proposed that "the science of land economics" in turn learn from this work.[68] Like plant and animal scientists' accounts of "the forest," the tundra," and "the animal community," these scholars hoped to move urban studies beyond the many individual stories of "Boston," "Chicago," and "Seattle" toward conclusions about "the American city," past, present, and future. "There is implicit in all these studies," wrote Robert Park the following year, "the notion that the city is a thing with a characteristic organization and a typical life-history, and that individual cities are enough alike so that what one learns about one city may, within limits, be assumed to be true of others."[69] In the context of human ecologists' growing interest in questioning limits to analogical reasoning, setting disciplinary standards, and developing a predictive understanding of urban processes, Ernest Burgess's earlier work to devise a "zonal" model of "the city" would play a starring role. Adopted for studies of cities around the United States, the continuing relevance of this template gave colleagues confidence that a rigorous urban science ultimately would be achieved.

Communicating an Ecological Vision and Its Limits: Burgess's Zonal Model of "The City"

Created at the University of Chicago's Sociology Department, Ernest Burgess's 1925 zonal model remains the best-known graphic representation of "the city" in the human ecology tradition. Supporting his claims with data from faculty and student field investigations of Chicago as well as those of scholars whose focus was other cities, Burgess argued that all urban areas comprise five concentric zones: (1) a business zone, (2) a transitional zone between commercial and residential areas, (3) a zone of workingmen's homes, (4) a zone of apartments, and (5) a zone of single family areas.[70] Transportation was a central factor in these arrangements; zones grew outward in successive rings around the central business district as road networks and rail lines were extended from the central city. "I can think of no better analogy than the web of the common spider," Norman Scott Brien Gras observed to a meeting of the American Sociological Society, "The efficient builder establishes first his radial lines running out in all directions from the center. Then the concentric fasteners are put in. At last the spider, posted at the center, is ready to do business."[71] Burgess applied this analogy to explaining his pattern as well. Its organization was "much like that of a spider's web with lines of transportation represented by threads with subsidiary centers with their radial lines and by the chief center where the main threads converge, dominating the whole web."[72]

Burgess's use of concentric circles was not the first such urban representation. In the early nineteenth century, German economist Johann von Thünen had detailed least-cost location theory for agriculture and industry in concentric terms. At the turn of the century, Ebenezer Howard's Garden City ideal had featured concentric designs. Compared with these earlier representations, however, the significance of Burgess's map was distinct: Created as a depiction within the "ecological" tradition by a sociologist with special interest in building a "scientific" urban studies (for example, it was through interactions with Charles Child that he developed his gradient theory and soon after through the university's Behavior Research Fund that he developed close relations with other campus scientists), this idealized map—with its proposal that the structure of human communities closely resembled those in the plant and animal world—

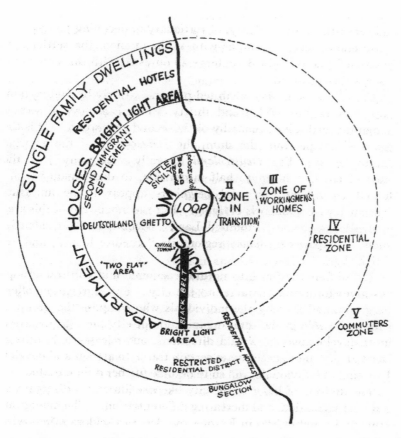

Ernest Burgess's Zonal Model of "The City." From Robert Park, Ernest Burgess, and Roderick McKenzie, eds., *The City* (Chicago: University of Chicago Press, 1925). Courtesy University of Chicago Press.

took on a new identity as a scientific modeling tool.[73] "Zonation" was an ecological concept applied to demarcating areas of climatic or species difference as well as aspects of organism development. Plant zonation was not always a matter for concentric depiction, but many ecologists found this mode of representation relevant to their work. Following University of Chicago ecologist Henry Cowles's observations in local fieldwork how in ecological classification, "great emphasis is placed on border lines or zones of tension, for here, rather than at the center of the society, one can best interpret the changes that are taking place," for example, Cowles

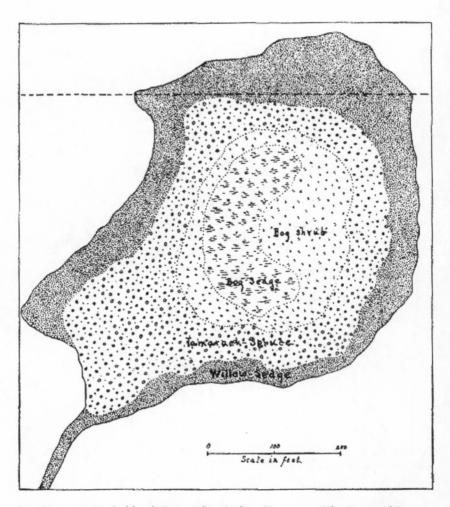

Bog Succession in Oakland. From Edgar Nelson Transeau, "The Bogs and Bog Flora of the Huron River Valley," *Botanical Gazette* 40, no. 6 (1905): 418–448. Courtesy University of Chicago Press.

student Hazel Schmoll had drawn a concentric half circle of the structure of plant communities in the Chicago suburb of Glencoe in the 1910s.[74] A few years earlier, Alma College botanist Edgar Transeau had depicted bog succession in the Huron River Valley in similar, if less stylized, terms. H. W. Graham and L. K. Henry annotated their photographs of the changes to plant life around a lake in the 1920s to highlight the "concentric

Plant Succession around a Lake in the 1920s. From H. W. Graham and L. K. Henry, "Plant Succession at the Borders of a Kettle-Hole Lake," *Bulletin of the Torrey Botanical Club*, 60, no. 4 (Apr. 1933): 301–315. Courtesy Torrey Botanical Society and Library of the Gray Herbarium, Harvard University.

zones about the lake which represent the annual recessions of the lake level" and the "succession of plants in the denuded shore."[75] Burgess applied this template as a shorthand explanation for the overarching urban patterns he observed. Defining each zone by the predominant type of property, he showed how it was a "natural" occurrence to find high-end businesses located next to slums.

The visual rhetoric of this concentric model emphasized the structural similarities between plant, animal, and human communities. Within its outlines also lay the assumption that cities, like communities in nature, were dynamic — and by extension that through the processes of invasion and succession, the location and character of zonation were subject to change. "Recent studies clearly show that the main fact of expansion is found in the tendency of each inner zone to extend its area by the invasion of the next outer zone," Burgess explained. "This aspect of expansion may be called succession, a process which has been studied in detail in plant ecology."[76] Henry Cowles, who in his "physiographic ecology" pointed to how change might be a slow, but ongoing process, and Frederic Clements, with his theories of "invasion-succession" and "climax communities," had been the chief expositors of this view. In two widely cited papers on plant associations of the Chicago area, Cowles had argued there was a geographical dimension to the history of vegetational development that well-trained observers could see.[77] In *Plant Succession* (1916) and in later papers, where he detailed his four-stage model of succession to climax, Clements had documented cyclical patterns in the life of associations in nature.[78] Following from these and other scientists, who interpreted from the structural arrangement of natural landscapes cause-and-effect linkages between space and time, Burgess insisted a similar set of patterned connections could be read in the urban landscape. The zonal model of cities' natural areas, in other words, described not only urban structure but also the processes that generated such organization. Taking these ideas a step further suggested that city neighborhoods and institutions, indeed, entire cities — like plant and animal species and communities — experienced linear and predictable life cycles.[79]

As with many of the ecological concepts that social scientists borrowed from ecology, the "life cycle" had a history in biological and social thought.[80] Sociologists, geographers, and economists were already convinced that cyclical processes operated in a variety of social contexts. Committed to the view that "prediction is the aim of the social sciences as it is of the physical sciences," they made ample use of occupational, business, and industrial cycles as forecasting tools.[81] Burgess's model now added the possibility not only for a scientific explanation of transitions between urban past and present but also for identifying what lay ahead for cities. The unfortunate downside to this discovery was that as these scholars reflected on the limits to their analogical reasoning what they saw in the urban future did not look bright. "Through a cumulative process of natural selection that is continually going on as the more ambitious and energetic keep moving out," Burgess explained, "the unadjusted, the dregs and the outlaws accumulate" in city centers.[82] The continuous wavelike patterns of out-migration that pushed increasingly affluent residents and some businesses toward uninhabited areas away from the city's core suggested the urban future would consist of prosperous distant areas and deteriorated central neighborhoods, with economic and social problems mapping directly onto physical disrepair.[83] Although they followed plant and animal ecologists to conclude that "the human community tends to develop in a cyclic fashion," whose "point of maximum development may be termed the point of culmination or climax, to use the term of the plant ecologist," as Roderick McKenzie explained, they eschewed these scientists' view of climax as a stable stage.[84] In contrast to the communities that plant and animal ecologists studied, which reached equilibrium with maturity, urban communities peaked before declining to stabilize in a decadent state. Climax came earlier in a pattern of events that moved from vacant land to build up to an equilibrium of decay. Burgess's 1930 update to his zonal model, depicting agricultural districts and hinterland on the urban fringe, further emphasized how "the city grows like a tree, taking on new growth at the periphery but dying at the heart," as Nels Anderson and Eduard Lindeman put it that year, citing ecologists such as Charles Child, J. Arthur Thompson, and Eugenius Warming.[85] For these figures, the recognition of gentrification in outlying areas did not mollify anxieties about urban decline, and most interpretations of cities' internal repatternings characterized the direction of change as taking a negative turn.

Created early in the development of human ecology, Burgess's stylized city model gathered strength as scholars tested its reliability in urban areas around the United States, mapping cities and neighborhoods, and comparing them to the concentric ideal. Although maps have not figured centrally in the recent history of urban studies, at the field's origins, cartography was an essential statistical research technique.[86] "From the preliminary survey of the area to the preparation of the final report," wrote W. Wallace Weaver of his investigations in the nation's third largest city, Philadelphia, "one of the most necessary elements of study was map work. Through the examination of recent maps it was possible to delimit and describe the area selected for study, while the use of older maps permitted a more specific reconstruction of historical changes than any data available from printed sources. Outline maps furnished a convenient means of recording observed data, and for displaying summaries of certain findings. In fact, the use of maps was an essential part of almost every portion of the project."[87] With every city operating according to the same "natural laws," human ecologists widely believed that the correct template for understanding urban structure should be easily transferable from city to city—an expectation their research confirmed. "In the growth process of any city there is a definitely marked tendency for the gross structural anatomy to conform to a characteristic pattern which can be ideally represented by a series of zones expanding radially from the downtown business district," explained Calvin Schmid, in a study of Seattle.[88] Weaver, comparing his own cartographic analyses to Burgess's model, saw its explanatory power in a single neighborhood, West Philadelphia. "It has had a historic development," he explained, "which illustrates every stage of growth from exploration to urban congestion, and represents every zone in the assumed concentric arrangement of the city from the central business district to the commuters' suburbs."[89] By the late 1920s, then, as anxieties about disciplinary standards swelled, Burgess's model—and the breadth of its relevance in space and time—assuaged the scholarly community that it was making progress toward a scientific understanding of urban life.

Economic Depression Arrives

The ecological approaches to understanding the dynamism of urban life that blossomed in the 1920s—a boom period for U.S. cities—gained new data points in the early 1930s, as a nationwide depression and trends, including falling birthrates, slowing immigration, and escalating race migration, brought changing fortunes to many urban communities. In the wake of these developments, the federal government pressed academics to inventory and assess the state of the nation's cities. The U.S. Department of Commerce and U.S. Civil Works Administration under Republican president Herbert Hoover, which launched Financial Surveys of Cities and Real Property Inventories, also assembled sociologists and economists, including Roderick McKenzie, Jesse Steiner, Howard Odum, Charles Johnson, Herbert Dorau, Herbert Nelson, and Richard Ely at a conference on homebuilding to debate such topics as "Negro Housing" and slum repair.[90] With support from the Rockefeller Foundation, a presidential research committee on social trends convened Howard Odum, Roderick McKenzie, Ernest Fisher, Robert Park, Charles Newcomb, and Calvin Schmid to offer their views on the status of metropolitan areas.[91] Several years later, Democratic president Franklin Roosevelt formed a Research Committee on Urbanism within his National Resources Planning Board, inviting distinguished social scientists, including Louis Wirth and Warren Thompson, to evaluate urban trends and future prospects.[92] Yet while federal officials recognized the potential for social scientists' predictive models to assist in the creation of new scientific social policies, their commitments to intervention in the depression chiefly focused on rural problems rather than urban concerns.[93] Anxious local governments, private foundations, nonprofit agencies, and academics soon undertook their own survey research and social problems studies to obtain the latest information on conditions in specific neighborhoods and to contemplate possible action for cities.[94]

For theoretically minded scholars who followed Robert Park's call to take the role of "the calm, detached scientist" observing human relations like "the zoologist dissects the potato bug," these studies of the increasingly desperate situation in cities around the United States confirmed that theories of human ecology held up in the new economic context.[95] Individuals and institutions continued to segregate by type, and cooperative as well as competitive relationships among such groupings could still be seen. Through a "reciprocal interchange of services—Zion Town furnishing Weshampton with domestic and personal service, and Westhampton furnishing Zion Town with means of subsistence"—Howard Harlan, the University of Virginia Phelps-Stokes Fellow, noted of wealthy whites and poor African Americans in two Richmond neighborhoods, "the conflict between the two contrary racial and cultural groups is accommodated... this relationship, to borrow terms from plant ecology, we may call commensalism or symbiosis."[96] Patterns of sequent occupance and succession were still apparent. "The literature on urban land economics and of human ecology is full of descriptions of the functional change from the center outward that occurs in any dynamic urban area," wrote economist Roy Burroughs in 1935, having explored these processes in depth in Cleveland.[97] In the nation's central cities, close correspondences persisted between the geographies of social and economic problems and the geographies of urban physical decay. As T. Earl Sullenger of the Municipal University of Omaha suggested, "practically every social problem occurs with other problems as a cause or result."[98] From Rochester and St. Louis to Honolulu and Harlem, the concentric model of zonation generally continued to fit.[99] "The fundamental concept of city growth as described by E. W. Burgess is applicable to the city of Long Beach," wrote Elsa Schneider

Longmoor, of the Los Angeles County Relief Administration, together with University of Southern California social work professor Erle Fiske Young, "although the form of radical [sic] expansion in concentric circles from the center outward, as presented by him, is somewhat distorted, due to the natural contour of the ground and the natural fixed boundaries."[100] In short, despite popular portrayals of cities in crisis, for these scholars, urban areas during the depression, as during the earlier building boom, were orderly in their disorder. "The human community tends to pass through a series of cycles of alteration and growth which historically constitutes an ecological succession," explained George Renner, "analogous to that of an animal or plant community."[101] "From the many studies that have been made," Washington University sociologist Stuart Queen and geographer Lewis Thomas summarized scholarly sentiment in their 1939 book *The City*, it is possible to identify, tentatively at least, a typical life history or expected sequence of events."[102]

By the 1930s, the investigations that urban researchers undertook were street level, neighborhood based, citywide, and even regional, mirroring colleagues in plant and animal science who followed both the "monographic method" of studying a single species' distribution in various places and what Arthur Tansley called "intensive work," detailing a single area as it reflected phenomena such as competition or dispersal among several species.[103] Certainly, given the changing historical context, some exceptions to expected patterns could be found. Yet ecology's power as a mode of interpretation suggested the need to refine, rather than to discard, ecological perspectives and techniques. Thus, when Northwestern University economist Frederick Babcock questioned, in light of the prior axial-radial growth theories of economist Richard Hurd (called an "ecologist" by sociologists), whether Burgess's model could describe all cities, he proposed a new combination. "The two theories, when superimposed, do describe the grand pattern upon which cities are built," he explained, offering an alternative star-shaped concentric model equally consistent with visual representations in plant and animal studies in his text.[104] In an article based on student and faculty studies from his department, the work of economists at the Regional Survey of New York, and his own earlier field notes, University of Chicago geographer Charles Colby disputed the existence of five concentric zones, finding three instead. The American city remained "a dynamic organism constantly in the

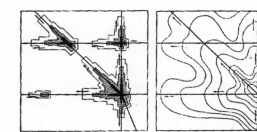 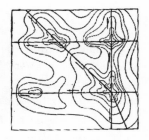

Left, Radial Growth. *Center*, Concentric Growth. *Right*, Theories Superimposed. From Frederick Babcock, *The Valuation of Real Estate* (New York: McGraw-Hill, 1932).

process of evolution," Colby explained, following plant and animal ecologists such as Frederic Clements and William McDougall who characterized ecological communities in organismic terms.[105] In a study of the Japanese in New York City, Columbia University political science graduate student Eleanor Gluck found no distinct Japanese neighborhoods — a notable "divergence from the usual procedure in immigrant groups." She reported that these conclusions confirmed several scholars' hypotheses about the behavior of Japanese versus other immigrant and racial groups, citing Stanford geographer Elliott Mears's observation that "the Japanese are eager to enter the best neighborhoods, which is not true of most of the Chinese."[106]

Such findings shared the spirit of efforts to develop scientific standards in urban studies. "Like a motion picture of a plant growing," Everett Hughes reflected, these general theories possessed "the scientific merit of illuminating the processes by which other things of the same general sort have arisen and may be expected to grow."[107] Standards offered averages or ideals that would not fit every context perfectly but could describe many quite well. As Harvard geographer Derwent Whittlesey had explained of sequent occupance, "normal sequences are rare, perhaps only ideal, because extraneous forces are likely to interfere with the normal course, altering either its direction or rate, or both."[108] So too, these were minor disputes within a broader context of agreement about naturalistic processes in the urban community, providing new opportunities to frame the interpretation of "exceptional" urban data in ecological terms.

As the new decade wore on, then, the power of ecological explanations for that ostensibly most "unnatural" environment, the American city, remained strong. Equally strong were scholars' convictions that limits to analogical thinking remained. Louis Wirth's 1930 review of William Morton Wheeler's *The Social Insects: Their Origin and Evolution* in the *American Journal of Sociology* did not flatter the entomologist's work.[109] So too Clark University geographer W. Elmer Ekblaw's 1936 assessment of *Geography: An Introduction to Human Ecology* from George Renner and C. Langdon White emphasized how the "application of the nomenclature and terminology of one science to another is always a hazardous, and rarely warrantable practice," given how "it extends the meaning and connotation of terms to qualities of objects not originally included, sometimes to distort them, sometimes to render them so ambiguous as to destroy their value."[110] Even as, or perhaps because of, this continued preoccupation with the limits to analogical reasoning, the few critics who suggested that the flexibility that permitted easy appropriation of ecological terminology had led to borrowings that were superficial, contradictory, or based on science that was out of date could be largely ignored.[111]

The lack of engagement with scientists that disturbed social scientific critics of human ecology appeared much less troublesome to scientists themselves. Paying at least some attention to the growing vogue for ecology within the social sciences, as they witnessed these developments, the plant and animal scientists whose casual reflections on human communities had laid the foundations for building the human ecology tradition seemed pleased.[112] In a 1935 article in *Ecology*, for example, Charles Adams — who had called for a closer association between sociology and plant and animal ecology two decades earlier — offered his evaluation of recent work in urban studies.[113] "It is no longer a rare event to find the word [ecology] in geographic, economic, and sociological literature," Adams observed, concluding that social scientists had learned a great deal, methodologically, from their colleagues in the natural sciences.[114] "It is significant to observe that as the application of the scientific methods to human social problems progresses, the methods of the natural history sciences . . . are increasingly being applied to them." Citing Helen and Robert Lynd's *Middletown* (1929) and Robert Marshall's *Arctic Village* (1933), he singled out the human ecologists' field study and natural history survey methods for special praise: "For this purpose the methods of the regional and ecological surveys are particularly appropriate and illuminating."[115] A 1937 review of British human ecologist John William Bews's *Human Ecology* in the journal *Ecology* by Alfred Emerson student Edward Haskell lauded the author's sound "initiation to the ecological attitude of mind."[116] In 1939, Adams together with botanist and conservation educator Paul Sears invited Ernest Burgess and Roderick McKenzie to participate in a symposium at the American Association for the Advancement of Science "On the Relation of Ecology to Human Welfare— The Human Situation."[117]

At universities around the United Kingdom as well as in the United States distinguished ecological scientists took an interest in the human ecology field. "More and more frequently of late," wrote Arthur Tansley in 1939, "we have been hearing talk about 'human ecology' . . . evidence of the increasing attention that is being paid to the ecological point of view." When it came to the question of ecology's validity for studying human society, the answer for Tansley was clear. "Some human communities are in relatively stable equilibrium, others are in process of rapid development; others again are in disruption or decay, sometimes from internal, sometimes from external causes." In short, "the principles of ecology are unquestioningly applicable to mankind . . . the basic principles are the same."[118] The term "ecology" is "itself not infrequently replaced by biology, sociology, geography or geobotany," agreed Frederic Clements and Victor Shelford in a book published that same year, predicting that "students of ecology will continue to be trained primarily as botanists, zoologists, sociologists, or economists for some time to come."[119] In the 1930s as in the 1920s, it was rare to find plant and animal ecologists publishing in the professional social science literature or social scientists publishing in plant and animal ecology.[120] Nonetheless, ecologists clearly indicated they saw new life for their ideas in urban scholars' work.

Thanks to these new sources of data and intellectual support, during the 1930s momentum for the further formalization and refinement of ecological approaches to urban study continued to build. Courses in human ecology became standard fare at many colleges and universities, most often in departments of sociology such as University of Michigan (with Roderick McKenzie), University of Washington (with Jesse Steiner) and University of North Carolina (with Rupert Vance). New textbooks soon followed reflecting the robust interdisciplinarity of this scientific vision for understanding cities. Sociologists Roderick McKenzie and Jesse Steiner, for example, compiled readings from plant and animal studies, sociology, geography, and economics in sourcebooks for their students.[121] Among these selections were many strategic juxtapositions, for example, the chapter on "Dominance" in the McKenzie text began with readings on plants and animals from William McDougall and Charles Adams; the section on "Succession" opened with an essay by Frederic Clements. Indeed, throughout the reader, articles from scientists and social scientists proposed the extension of plant and animal ecology to human affairs. Thus, while geographers such as Robert Dickinson complained that sociologists received disproportionate credit for developments in human ecology, ample evidence documents how, in their research and in teaching, sociologists relied on geographers' as well as economists' work.[122] As Roderick McKenzie had explained to his sociologist colleagues in 1925, "research in economics and commercial geography on land value, marketing, transportation, commerce, factory and business location frequently has ecological significance."[123] University of California sociologist E. Gordon Ericksen, reflecting on the discipline's evolution, concluded it "found the writings of economists, geographers, historians, and philosophers an added impetus to the concepts developed by plant and animal ecologists."[124] A look at textbooks on cities from this period finds, then, that while the predilections for reading plant and animal studies in the original was strongest among sociologists, colleagues in geography and economics shared a common commitment to understanding the "scientific" principles of urban structure and evolution in ecological terms. With cities now recognized as the nation's leading frontier, scholars across disciplines found that ecology introduced "some of the spirit and much of the substance and methods appropriate to the natural sciences into the study of social phenomena."[125]

During the 1920s and 1930s, a proliferation of ecological studies provided data to scholars seeking to build a respectable tradition for urban research. From these efforts came a firm belief in the cause-and-effect dynamics in cities and the possibilities for urban prediction. Yet, for two professional communities whose interest in the future of U.S. cities was more than academic, human ecologists' recognition of "the undertow" that "pulls outward from the beach as the waves dash in," as Massachusetts Agricultural College economist Robert McFall put it in 1935 — that an inevitable urban life cycle pushed populations away from city centers as physical, economic, and social problems accumulated there—became a matter of grave concern.[126] "The evil results of allowing a city to develop 'naturally,' suggested University of Wisconsin economists Richard Ely and Edward Morehouse, even before the stock market collapse of 1929, "is the best evidence of the need of social control of urban land utilization."[127] Cities might have "grown more or less along natural lines without any concerted direction or interference on the part of the community," as Charles Male put it in *Real Estates Fundamentals*; yet, the depression highlighted how natural was not necessarily ideal.[128] Witnessing the federal government's uneven investments in action on urban versus rural problem solving, city planners and real estate appraisers were anxious for a national city improvement strategy to appear. How these urban professionals drew inspiration from the federal response to America's natural resources crisis to transform aspects of human ecology into an applied science of urban conservation is the subject of chapter 2.

TWO

The City Is a National Resource

"The ultimate social, economic, and structural decay, requiring a major surgical operation in the form of complete demolition, need not necessarily be the last chapter in the history of every residential community," wrote U.S. Home Owners Loan Corporation Deputy Manager Donald McNeal in 1940.[1] Defiant in the face of claims that hope for cities' future lay only in total reconstruction, McNeal insisted that, thanks to experts in real estate science, an alternative solution had been found. A " 'Department of Conservation' should be established in every large city, whose sole function it will be, by precept, example, and inspirational activity, to promote community stabilization projects in potentially and partially depreciated sections throughout the city."[2] If government could educate citizens about the "predictable structural, economic and social life cycle through which the average American residential community passes," and in turn prod them to act "before neighborhood corrosion first begins actually to show itself," together they could "reverse community disintegration before it attains a definitively destructive momentum."[3]

How, between 1920 and 1940, did leading figures in the American real estate community and their colleagues in city planning come to see cities as resources to be conserved? As a research tradition for the scientific study of cities was unfolding, two other communities of urban professionals fell under the spell of naturalistic thinking that academic colleagues had cast. For city planners seeking to maintain their field's scientific standing while attending to social and economic concerns, and for real estate appraisers seeking to deduce scientific principles of property valuation, the predictability of the urban life cycle central to ecological interpretations of cities proved a promising addition to their work. This interest in human ecology, first piqued in the search for scientific standards during the 1920s building boom, deepened with the economic depression of the 1930s. The urgent need for a response to the nation's growing urban problems suggested to both professions new reasons for importing these social scientists' ideas. Eager to motivate federal and local investments in city improvement, they pushed the analogy between cities and nature a step beyond their social science colleagues to articulate how solutions to urban challenges ahead would succeed if they kept the "nature" of cities in mind.

Historians already have observed the lure of ecological thinking in the early twentieth-century real estate and city planning communities. Scholars have described how social scientific concepts borrowed from plant and animal studies, including "natural area," "invasion-succession," and the "life cycle" became part of the common wisdom in these fields.[4] Such accounts of the diffusion of an "image of the city" that highlighted close correspondences between patterns of life in the urban environment and patterns of life in the natural world emphasize how, at a time when the applied urban professions were seeking greater authority, the ecological approach to understanding cities that preoccupied scholars in academic urban studies provided new scientific rationales for policies such as racial segregation to which these fields' practitioners already subscribed.[5] Yet the support that academic urban studies offered for these professionals' claims about the "nature" of cities was not the only motivation for their choice of rhetorical turn. In a cultural context where specific theories from plant and animal ecology gained prominence with the nation's burgeoning interest in the scientific management of natural resources, these urban professionals saw in the increasingly accepted analogies between plant, animal, and human communities an opportunity at hand. The Clementsian school of plant ecology, on which notions of the urban life cycle were based, became one of the intellectual underpinnings of this push for resource conservation, suggesting to city planners and real estate appraisers concerned about the future of declining urban residential areas how they might achieve for cities what plant and animal ecologists were closer to accomplishing for America's forests and farms. From the mid-1930s, the Research Committee on Urbanism of the National Resources Planning Board, which insisted that the pioneer mentality that had caused the Dust

Bowl equally plagued cities, pressed for a national program of urban resource conservation. A few years later, the National Association of Real Estate Boards set out to demonstrate how technical experts and grassroots volunteers might collaborate toward the conservation of city neighborhoods at the local level. In an era in which science played a growing role in public policymaking, these figures went beyond the recognition that natural resources were essential factors in the rise and decline of cities to portray cities themselves as "resources," deserving equal attention and an equivalent program of scientific conservation.

Ecology and Conservation in the New Deal

This book has documented how the sociologists, geographers, and economists who set out to build a science of cities from the 1920s were attracted to ecology for its resonances with extant social science concepts and methods — and its possibilities for urban prediction. What they could not predict was how, in the wake of the depression, the plant and animal ecologists they admired would gain stature on the national policy stage. Historians of environmental science have characterized the two decades before World War II as a boom period for the interdisciplinary study of plant and animal communities, in large part because of the belief that this knowledge applied to the nation's increasingly urgent resource planning questions.[6] In the 1920s, following the close of the frontier and the recognition that cities were expanding into rural areas, land use issues became matters of growing concern.[7] During the depression of the 1930s, environmental problems from soil erosion to agricultural surplus spawned visible rural poverty, tax delinquency, and the abandonment of farms, bringing a new level of attention to the scientists seeking to understand nature's complex laws.[8] Investments in ecological research at land-grant institutions, agricultural experiment stations, and federal agencies initiated on President Herbert Hoover's watch blossomed under President Franklin Roosevelt — who made conservation his signature issue.[9]

At its origins in the late nineteenth century, America's conservation movement emphasized the inventory and set-aside of natural resources for the future, most notably creating national forests and parks. Such preservation-focused activities continued through the early twentieth century as states established departments of conservation.[10] In the 1930s, with many professional ecologists participating in the movement — among them Charles Adams, Frederic Clements, Victor Shelford, John Weaver, and Herbert Hanson — conservation took on new meaning, expanding beyond inventories and set-asides to attend to the scientific management of land with its long-term productivity in mind. "The war cry of the earlier conservation movement was 'save our natural resources,'" agriculture department economist L. C. Gray explained, "The slogan of the new land policy is 'use the land for its most beneficial purpose.'"[11]

In the new generation of forest, range, and farm management at the U.S. Department of Agriculture and U.S. Department of the Interior, scientific research backed policy planning, and plant ecologist Frederic Clements's model of the universal processes of succession to climax lay one important foundation for action on resource problems.[12] The appeal of his theory that "communities arise, grow, mature, attain old age and die" in a predictable set of stages was its capacity not only for description and diagnosis but also prescription and remedy.[13] The nation must "meet change with change" and "apply it to the problems of readjustment with scientific foresight," Clements elaborated with Ralph Chaney. "Man will never approach mastery of his environment . . . until he understands the universal ebb and flow of processes and uses them to his own advantage."[14] In a climate of fear that America's wasteful practices had consumed too many resources, Clements's assessment was that humans had disturbed the balance of nature — but equally could restore nature's equilibrium state if, in this stabilization program, ecologists played starring roles. "The special contribution of ecology," explained Herbert Hanson in his 1938 presidential address to the Ecological Society of America, "is to ferret out relationships with the environment so that man, using this knowledge in conjunction with that obtained from other fields, can strive intelligently to secure balance and stabilization, a goal essential for the attainment of the 'abundant life' and the building of a culture far beyond our present dreams."[15]

Cartography was an essential technique in the quest to secure this balance, as classification schemes evolved from simply identifying the locations of different types of crops and terrain to maps that were plans by another name. The Soil Conservation Service's "land use capability" approach, for example, which settled on an eight-class color-coded scheme, deemed some areas "suitable for cultivation," others for "occasional or limited cultivation," and others "not suitable for cultivation."[16] Attending to natural rather than political boundaries to map areas in terms of their capacities for present and future development, these efforts sought the scientific understanding of land's "best *possible* use, as contrasted with its *present* use."[17]

At the same time the Roosevelt administration broadened conservation's meaning to encompass scientific management, it called attention to the need for conservation on the privately held lands alongside the public lands that were the movement's traditional focus. As the Civilian Conservation Corps planted trees and constructed recreation areas on public lands, then, the Soil Conservation Service and Resettlement Administration rallied farmers to establish conservation districts and to develop short- and long-term plans for the "wise use" of their holdings. Recognizing that conservation's efficacy in these contexts would depend on more than the participation of government officials or scientific experts, and the resistance of many property owners to changing their ways, the administration geared up to convince the public to organize their lives with conservation in mind. Informational brochures, radio broadcasts, postage stamps, and school curricula were among the components of its educational campaign to explain how land wore out, how scientific study could predict the rate and location of future decline, and how citizen-government collaboration in a comprehensive and continuous program organized around nature's predictable patterns would enable resource conservation to succeed.[18] Historians disagree about the effects of ecological knowledge on the outcomes of resource planning projects during this period, but it is clear that the popular frame for conservation efforts drew heavily from ecologists' scientific work. In the calls for a "permanent agriculture" and "permanent conservation" from agriculture department staff such as Hugh Bennett and Rexford Tugwell, and in the multimedia materials—plays, films, and photographs—their agencies used to enlist the public's participation in land use strategies from replanning to retirement, ecological themes featured prominently.[19]

Although the effort to transform ingrained land use practices was a challenge in areas such as forests with few human occupants, settled farmlands presented greater obstacles to correcting what the government called "maladjustments."[20] Under Rexford Tugwell at the U.S.

Land Use Capability Mapping for Farm Conservation. From U.S. Soil Conservation Service, *What Is a Conservation Farm Plan?* (Washington, DC: U.S. Soil Conservation Service, 1948).

Federal Emergency Relief Administration's Rural Rehabilitation Division and Harold Ickes at the U.S. Public Works Administration's Subsistence Homestead Division and, later, Tugwell's guidance at the U.S. Department of Agriculture's Resettlement Administration, the initial plan was to resettle those farmers volunteering to move away from the most unproductive plots, while helping those struggling to rehabilitate their holdings.[21] Finding it unexpectedly difficult to assemble adequate land on which to relocate families, the administration refocused its program on rehabilitation in place.[22] Social scientists participated alongside ecologists in these efforts to address rural poverty, providing government agencies and landowners with expert advice on community development and household management.[23] With their participation, the meaning of conservation expanded still further, joining the rehabilitation of humans to the rehabilitation of nature, the "conservation of human resources" of which Roosevelt so frequently spoke.[24]

Among these social scientists were a few human ecologists whose chief interest was rural areas. From the field's origins, scholars, including A. B. Hollingshead, Paul Landis, and Carl Taylor, had expressed interest in studying the ecology of nonurban populations, research that was ongoing in the 1930s.[25] To the plant and animal ecologists participating in the nation's conservation efforts, these scholars' orientation to relations between humans and the land was an ideal complement to their work. "A paper entitled 'The People of the Drought States,'" wrote Herbert Hanson of the work of two agriculture department–sponsored sociologists, showed "the similarity of sociological thinking with the ecological . . . Successful 'invasion' has, of course, been accomplished in some localities and stabilization is beginning. Attainment by the people of equilibrium with the environment, or the development of a 'climax' community will be a slow process, but the essential steps are being taken."[26]

Supported by a well-regarded president, backed by enabling legislation in many states, and showing promising early results, the massive mobilization to scientifically manage America's natural resources made conservation "a household word."[27] "The rural population is more conservation-minded than previously," explained Paul Landis as early as 1936.[28] Recognizing that progress would be slow, yet confident that many of the nation's resources would be renewable through a program of continuous and comprehensive care, scientists, public officials, and citizens came to share the belief that contemporary environmental problems could be reversed, and future problems prevented.[29] "So favorable to the ideas of conservation" was popular opinion, geographers Almon Ernest Parkins and Joe Russell Whitaker observed that year, that diverse policy proposals increasingly were "represented by their promoters to be in the field of conservation in order to secure popular support."[30]

Among these promoters would be an apparently unlikely group: city planners and real estate appraisers whose focus was that most "unnatural" territory, the American city. Despite the public's recognition that America was an urban nation, and that the economic depression affecting farms and forests was causing harm to cities as well, a federal government whose chief priority was natural resources had offered comparatively few investments in urban affairs.[31] Like plant and animal ecologists who were concerned about the forces decentralizing U.S. cities, yet instead focused on developments in urban centers rather than at the fringe, the urban professionals advocating a federal commitment to city improvement for several years with no avail soon saw in the nation's enthusiasm about conservation inspiration for a new rhetorical campaign.

The Nature of City Planning in the Early Twentieth Century

This book has described how, for the social scientists captivated by ecological models of urban systems, the conviction they had identified the "natural laws" of city growth and decay — and, in so doing, they had turned urban studies into a predictive science — was a thrilling intellectual achievement, signaling the successful fashioning of the social sciences after the natural sciences. For American city planners, human ecology's predictive power held even greater appeal. Confronting a growing array of social and economic problems in an emerging profession with roots in engineering and public health, these practitioners sought a social science remedy that could map onto the "scientific management" they already espoused.[32] From the late 1920s, as human ecologists made increasingly frequent use of planners' maps, many began to speculate about how their theories might meet this need. "City planning and zoning, which attempt to control the growth of the city," explained Harvey Zorbaugh, "can only be economical and successful where they recognize the natural organization of the city, the natural groupings of the city's population, the natural processes of the city's growth."[33] Proposing that any effort to exert control over the future would require organizing remedies with cities' natures in mind, some suggested social scientists would have to play a role in planning interventions. "The time may come," wrote Eduard Lindeman and Nels Anderson, "when each growing, progressive municipality will employ a sociologist as well as a city planner; working together these two might be able to deduce certain formulae" to provide for scientific urban improvement.[34] Already committed to concepts including environmental determinism and city as organism and laboratory, and using methods from field studies to statistical maps, planners soon suggested their growing agreement in texts such as Stanley McMichael and Robert Bingham, *City Growth Essentials: A City Planning Text* (1928), which explicitly or implicitly acknowledged the findings of human ecologists across social science fields.[35]

The two professional communities grew closer in the wake of the depression as the city surveys and federal committees that brought human ecologists to work on policy matters also brought them into contact with planners. As social scientists, including Nels Anderson, Charles Colby, Louis Wirth, Ernest Burgess, and Ernest Fisher, offered advice on planning matters in academic journals and at professional meetings in both fields, planning primers, including Karl Lohmann's *Principles of City Planning* (1931), Harland Bartholemew's *Urban Land Uses* (1932), and Mabel Walker's *Urban Blight and Slums* (1938), drew out the implications of ecological knowledge for practitioners' concerns.[36] Human ecologists' interest in natural as opposed to administrative boundaries as analytic categories and their attentions to interactions between land use and social and economic concerns offered planners new scientific explanations for the failures of past problem-solving efforts and in turn new ways of thinking about programmatic interventions. Although they recognized limits to ecology's explanatory power, for example, that Burgess's idealized model did not fit all cities, these professionals were uninterested in the assessments of analogical reasoning that preoccupied their social science colleagues.[37] Instead, they focused on how ecological knowledge could help them continue to place science above politics as their quest to improve the nation's cities expanded beyond physical planning to include social and economic issues.

Witnesses to the advance of urban blight and the movement of middle-

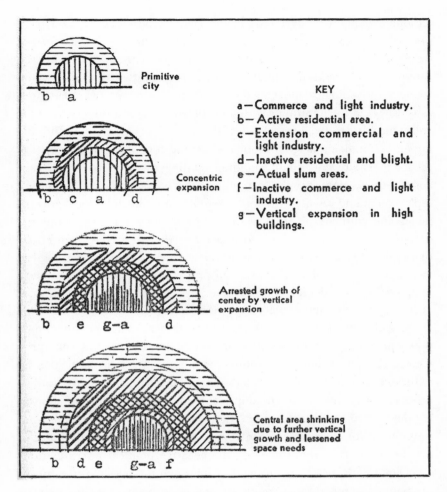

Henry Wright, Typical Spread of Modern American Cities and the Accumulation of Increasing Areas of Blight and Slums. From Mabel Walker, *Urban Blight and Slums: Economic and Legal Factors in Their Origin, Reclamation, and Prevention* (Cambridge, MA: Harvard University Press, 1938), 9.

class residents toward suburban areas that academic human ecologists described in their studies, many planners looked to the post–depression era with a specific plan in mind. Expanding the application of eminent domain toward comprehensive planning for cities — in short, rallying new support for the scientific management of land, was high on the profession's national agenda. Architect and planner Henry Holsman's 1932 observations of Chicago, where the "blighted areas and slums" in the city's center were "due to an ecological movement of people without destroying the rubbish they leave, a movement with as many subtle causes as may be found in the ecological movement of plants along streams, over plains and into valleys in the natural growth and spread of vegetation," typified his colleagues' frustrated assessments of the nation's urban areas.[38] With the hope of revitalizing neighborhoods where the cost of public services exceeded the tax revenues paid, and of forestalling further decay in adjacent communities, many sought to mobilize a massive program of housing replacement and repair, with land assembly undertaken by local governments and with city planners as their expert guides.[39] Holsman pointed to the "sufficient surveys and reports available to warrant the organization of a group of responsible business, professional, and civic minded citizens to begin the assembling of funds and properties" to improve Chicago's "blighted areas under a comprehensive physical and financial plan," and yet the legal impossibility of its implementation.[40]

In a climate of concern about socialistic dimensions to planning, such policy proposals had little traction at national or local levels. Geographer George Renner explained, "national planning in America can neither go to the extremes nor employ the methods used in Russia, Germany, Italy, or other countries under dictatorships."[41] Despite some history of support for the scientific management of cities, it was difficult to articulate a national need for comprehensive planning through government takings of private property without evoking the negative political interpretations associated with these foreign regimes. Through the mid-1930s, eminent domain laws limited the capacity for land assembly in cities by public or private entities; in 1937, federal legislation made the taking of lands acceptable only for constructing public housing.[42]

For the city planners who occasionally joined their natural resources planning colleagues at conferences and other events, it would be hard to escape noticing how not all calls for national planning were receiving a similarly cold reception.[43] Programs for the scientific management of natural resources on public and private lands, characterized as quintessentially democratic in their focus on efficient economic development and grassroots political action, offered evidence that national planning and even applications of eminent domain for comprehensive planning could thrive in the United States.⁴⁴ In the wake of this recognition, several prominent advocates for city improvement began a concerted effort to define cities as "resources" on par with the forests and farms that had become a national preoccupation. Characterizing the movement of residential populations toward suburbia and the discarding of urban lands as evidence of the American pioneer mentality alive and well in an agricultural-turned-urban nation, Harold Ickes offered the model argument in 1935: "As it was with the land itself, so it was with our cities. Expanding in a time of swift growth, they spread—altogether aimlessly at first, then with only a modicum of direction." Ickes had worked on urban reform in Chicago before assuming dual federal posts as director of the Federal Emergency Administration of Public Works and secretary of the Interior during the Roosevelt administration. From his unique position at the helm of two agencies, Ickes moved beyond common claims that urban and rural lands had little in common to insist that their problems of overdevelopment were one and the same: "Sensible plans for city development were neglected as unheedingly as were plans for the intelligent development of our natural resources."⁴⁵ Overused and subsequently abandoned land, haphazard developments mixing residences and businesses, and the continued expansion of city boundaries as speculators sought new profits were among the problems these observers identified. The wasteful use of resources that had spawned the nation's environmental problems equally was the cause of urban decay, he argued — "slums and blighted living areas were becoming an integral part of the municipal scheme, with little done to prevent their growth. The fungus took root and its inevitable economic and moral drain began." Ickes urged that in discussions of planning for national recovery the scientific management of cities should assume the same importance as the scientific management of nature.⁴⁶ His call, in other words, was to reframe planners' extant hopes for the urban future in the more compelling language of city as "resource."

Although Ickes's emphasis on the similarities between urban and rural areas was at odds with critics' popular portrayals of the artifice of urbanization, his was not the first call for the scientific planning of cities on the model of the scientific planning of nature. During the 1920s, as organizer of the Joint Committee on Bases of Sound Land Policy (a private organization), the president of the American Civic and Planning Association, Frederic Delano, had advanced a similar proposal in the context of discussions about the future of the nation's agricultural lands.⁴⁷ By the 1930s, the increasingly common view that continuities existed between urban and rural areas — from the popularity of ecological models in urban planning theory and practice to a growing belief in the "unity of land" that motivated regional planners — would provide ample support for this argument's growing use.⁴⁸ With both Delano and Ickes named to prominent positions on the National Resources Planning Board, the organization became headquarters to the view that cities were national resources to be conserved. Its chief proponent was the board's Research Committee on Urbanism, where social scientists such as Louis Wirth joined with city planners such as Arthur Comey and Ladislas Segoe to transform the comparisons between forests, farms, and cities earlier drawn by scientists and social scientists into an argument for "restating conservation in urban terms."⁴⁹

The Research Committee on Urbanism and the Conservation of Urban Resources

Franklin D. Roosevelt created the National Resources Planning Board in 1933 as part of the mobilization for post–depression era resource planning. The board was a study group rather than a unit with authority to approve or implement policy. Its charge was a scientific study of America's resources to support agencies, including the agriculture and interior departments, and to provide the president with recommendations for coordinating resource planning across the United States. "National" resource was interchangeable with "natural" resource in the administration's mind, but this terminology left a window of opportunity open for advocates of city improvement to have their say.⁵⁰

In the board's first report in 1934, urban land received only the barest mention, as part of a discussion on "miscellaneous uses of land."⁵¹ Pressure from the U.S. Conference of Mayors, frustrated by the emphasis on the "rural or agricultural standpoint" in New Deal programs, prompted the organization of a Research Committee on Urbanism the following year to devote its full attention to city concerns.⁵² Assembling social scientists and city planners to prepare an overview of the state of the nation's cities, the committee assessed past and present trends with an agenda for city improvement in mind: that expert intervention guide the stabilization of urban areas. Characterizing itself as a parallel to Teddy Roosevelt's earlier Country Life Commission on the problems of rural living, the committee followed Delano's and Ickes' rhetorical precedent, pairing charges about the pioneer mentality that spawned the wasteful planning of cities with the continued insistence that urban affairs did not differ substantially from the matters of rural concern that already had garnered national attention. Both rural and urban areas "have the problem of order, health, welfare, education, and the maintenance of democratic participation in the communal life."[53] Yet "it may be questioned," whether the federal government "has given sufficient attention to some of the specific and common problems of urban dwellers as it has for farmers through the Department of Agriculture."[54] Observing "restrictive provisions of State laws . . . inadequate grants of power to local governments" and "adverse court decisions," committee members concluded "attempts at improving the physical environment in the city have been blocked at almost every step."[55] They made these accusations accentuating the similarities and continuities — "the interests of farmers as farmers are not always the same any more than the interests of city dwellers as city dwellers are always the same" — rather than the differences and disjunctures between urban and nonurban lands.[56] "City planning, country planning, rural planning, State planning, regional planning, must be linked together in the higher strategy of American national planning and policy," they urged, "to the end that our national and local resources may be best conserved and developed for our human use."[57]

This was not the first call for conservation in urban areas. Beginning a few years earlier, the Civilian Conservation Corps had set out to create recreation areas in many cities as part of its paired program of natural and human resource conservation.[58] Yet, the focus of these actions was on saving nature in cities. By contrast, with participants chiefly concerned about the future of cities, the urbanism committee was proposing to conserve cities by emphasizing their commonalities with the natural world. Complementing the calls from Harold Ickes and the Brownlow Committee to unify the nation's broad program of resource management in a single cabinet-level department (University of Chicago public administration professor Louis Brownlow was a member of the urbanism committee), they proposed expanding the category of conservation to include cities as well.[59] "From the point of view of the highest and best use of our national resources our urban communities are potential assets of great value," the collaborators insisted in their landmark report *Our Cities*, "and we must consider from the point of view of national welfare how they may be most effectively aided in their development."[60] In agreement with human ecologists' assessment that "the new American frontier is in the city," these figures accentuated the characteristics and values that urban and rural areas shared to articulate how the nation's earlier and ongoing experience with the management of natural resources provided both rhetorical support for intervention and a general template of how to proceed.[61] Their broad proposal sought a research base for action on property improvement and poverty reduction on the model of "the conservation of physical and human resources" under way in rural areas.[62] "A section for urban research" should be "set up in some suitable Federal agency which should perform for urban communities functions comparable to those now performed for rural communities by the Department of Agriculture."[63]

Under Secretary of Agriculture M. L. Wilson, part of the urbanism group, did not dispute the legitimacy of these proposals. The report was "a very commendable piece of work," he commented on its penultimate draft, suggesting the authors' general argument would be strengthened by making several small adjustments in the text.

> The Bureau of Agricultural Economics is properly referred to, but there are many other Bureaus in the Department, for example the Land Use Planning Section of the Land Utilization Division of the Resettlement Administration, which carry on programs of far more immediate concern to rural communities than the Bureau of Agricultural Engineering. Possibly the difficulty could be overcome by merely saying ". . . to those now performed for rural communities by the Bureau of Agricultural Economics and other Bureaus of the Department of Agriculture."[64]

Thus, as the National Resources Planning Board's land use planning group cited ecologists, including Charles Adams and Victor Shelford, to support its call for comprehensive resource management in rural areas, urbanism committee participants made an analogous argument about cities.[65] Richard Ratcliff noted in a review of *Our Cities* how, from its depictions of the zonal forces of outward growth and inward decay to its observations of the relation between natural areas and crime, "the ecological and physical structure of a typical community is described... Scholars in the field of land economics and human ecology will discover few unfamiliar facts in the descriptive parts of this report."[66] With consulting planners, including Mabel Walker, characterizing neighborhood transformations in animalistic terms — "like a migratory flock of black birds resting and feeding temporarily, so groups of immigrants as well as individual families and isolated individuals stop in this transitional area on their way up or down the social scale," later committee publications seeking to document connections between impoverished land and impoverished people liberally cited the work of human ecologists on urban living conditions and quoted from Ernest Burgess's analysis of cities' zonal structure at length.[67] Building on the work of the social scientists for whom Frederic Clements's model of succession to climax provided a predictive template for explaining urban patterns, then, these planners saw an opportunity in ecological perspectives on cities — rhetorical links to natural resources planning and with these links a new rationale for insisting that the problems of cities and their inhabitants receive more substantial public consideration. "It is seldom that we become so enthusiastic over a report — in these days of many reports — as to say that you SHOULD read a particular one, but in this case we say as an official, in any level of government, you MUST study this report," wrote American Society of Planning Officials Executive Director Walter Blucher to the organization's membership in 1937 following the report's release. "We hope to obtain enough copies so that every one of our members who wishes to read the report may have a personal copy."[68]

Although planning professionals launched the campaign to align the restoration of cities and their populations with the national program of natural resources management already under way, their real estate colleagues reaped the rewards of these efforts. As sociologists Nels Anderson and Eduard Lindeman noted from the vantage point of the late 1920s, "real estate men" rather than planners or social scientists were the guiding force in American urban development, a trend that, despite planners' best efforts, showed no evidence of abating.[69] Borrowing from planners the idea that cities were resources to be conserved, the real estate community pushed the analogy a step further. If urban resources, as natural resources, possessed predictable life cycles, then the successes of the conservation movement to date suggested that individual homeowners persuaded to act collectively on matters of common concern might interrupt the "inevitable cycle of urban birth, life and death."[70] Looking at the work of the National Association of Real Estate Boards to build support for urban conservation reveals how the ecological thinking that scholars have described as providing the conceptual basis for appraisals, redlining, and planned abandonment, simultaneously offered an organizational model for community-based efforts at property repair.

The Science of Real Estate

Like their colleagues in the planning community, members of the American real estate industry were seeking new professional status during the 1920s building boom.[71] Lacking equivalent intellectual foundations—their poor public image most notably was satirized in Sinclair Lewis's *Babbitt* (1922)—a major aspiration was to be scientific. "We cannot claim that our business is a profession unless we constantly endeavor to improve," argued Kansas City real estate developer Jesse Clyde "J.C." Nichols at a meeting of the National Association of Real Estate Boards in 1924. "Laboratories of research are needed in the real estate profession. 'Realology' should be as much an established science as geology and zoology."[72] Headquartered in downtown Chicago at Northwestern University's Institute for Research in Land Economics and Public Utilities, where director Richard Ely's claims about the "unity of land economics" emphasized continuities between urban and rural areas, the effort to build a rigorous research basis for real estate practice saw an important role for ecological ideas.[73] Evidence suggests there were some occasional interactions between Northwestern economists and University of Chicago sociologists and geographers; Ely and Ernest Burgess, for example, participated together on the Chicago Housing Commission from 1926, and researchers at each university frequently cited colleagues on the other side of town.[74] At the same time that human ecologists such as Roderick McKenzie increasingly relied on economic data for their arguments, then, real estate researchers, including Herbert Nelson, found human ecologists' studies of "where the various nationalities are congregated" possessing "more than sociological interest."[75] Less interested in the limits to analogical reasoning than in how the new "science" of cities could identify patterns and make predictions to serve their needs, for these professionals, concepts, including "natural area," "competition," and "symbiosis," and the possibilities for "visualizing such facts" using maps proved "helpful in a practical way."[76] In a city where the vocabulary of real estate professionals already included "invasion" and "competition," for example, ecologists' findings offered new scientific rationales to support Realtors' longstanding preference for maintaining homogenous neighborhoods.[77] When the Chicago-based real estate association asked the institute to develop a Standard Course in Real Estate, researchers such as Arthur Weimer, Frederick Babcock, Ernest Fisher, Herbert Dorau, Albert Hinman, and Herbert Nelson crafted a twelve-book series to introduce scientific concepts and methods to future professionals.[78]

The depression-era call for research on policy action that brought social scientists and city planners to Washington, DC, for temporary participation on committees and in conferences provided opportunities for interactions with the growing community of real estate professionals. More significant for the long-term influences of ecological ideas in real estate circles were the federal agencies that the Roosevelt administration established to secure the financial base of America's housing market, the U.S. Federal Housing Administration and U.S. Home Owners Loan Corporation.[79] Despite its successes with developing the intellectual core of real estate economics, Northwestern University's Institute for Research in Land Economics and Public Utilities closed in the early 1930s, sending its staff in search of employment. Frederick Babcock, Ernest Fisher, and Richard Ratcliff were among those who moved to Washington, DC, to offer their expertise as the two agencies set out to standardize techniques of predictive risk rating for mortgage insurance and mortgages on residential property in the United States.[80] In contributions to the Federal Housing Administration's *Underwriting Manual* and the Home Owners Loan Corporation security maps, they and other researchers translated ecological theories already appearing in the national real estate curriculum into an applied science for insurers' and lenders' use, with Burgess's idealized model of invasion-succession and the city's predictable life cycle the focus of attention.[81] With real estate professionals essential partners in the risk rating work, and with both agencies taking action to publicize the new federal standards, this community, like its planning colleagues, would be hard-pressed to escape learning about ecological ideas.[82] University of Chicago sociologists may be better known than their Northwestern colleagues for their innovations in ecological theory; yet, Northwestern economists took the lead when it came to translating these theories into policy practice.[83]

Although the real estate community was successful in its bid to build an applied "science" for the profession and was comparatively more influential than planners on the national stage (as evidenced by the agency status of the Home Owners Loan Corporation and Federal Housing Administration versus the committee organization of the urbanism group at the National Resources Planning Board), in the mid-1930s, it still had other outstanding goals. With many in real estate concerned like their planning colleagues about the future of urban residential areas, the profession came to agitate for the expansion of eminent domain for city repair.[84] Yet if in planners' estimation, eminent domain powers should be turned over to planners (or at least be subject to their review) so that technical experts might "guide effectively the physical, social, and economic structure of a comprehensive plan broadly construed," real estate professionals focused on property values more than coordinated citywide redevelopment advocated that such authority be granted to private interests — homeowners associations — to address local neighborhood needs.[85]

In 1935, as the National Resources Planning Board's urbanism committee was gearing up for its research overview of the state of U.S. cities, the National Association of Real Estate Boards organized a conference to advance its agenda, inviting representatives from federal agencies, including the Home Owners Loan Corporation and Federal Housing Administration together with the Federal Home Loan Bank Board and Public Works Administration.[86] Follow-up discussions from the meeting culminated in a proposed model statute for states to adopt, "Neighborhood Protective and Improvement Districts," and the association's creation of a Committee on Housing and Blighted Areas to promote it. Drafted by association general counsel Nathan William MacChesney, with the assistance of Ernest Fisher and Miles Colean of the Federal Housing Administration and Donald McNeal of the Home Owners Loan Corporation, this "Suggested State Statute Authorizing the Creation of Such Districts by Property Owners in Cities, Towns, and Villages" expressed the real estate community's desire that individual neighborhoods become the basis for self-reliant planning. "The neighborhood, an entity hard to define but which everyone understands more or less clearly," association secretary Herbert Nelson observed in the foreword, "is the logical unit with which to begin planning. The people with whom to begin such planning are those who own the property. They have the power and incentive to make sound plans effective."[87] The act proposed to permit groups of property owners to organize neighborhood improvement districts. With 75 percent enrollment, homeowners associations would gain eminent domain rights to condemn other properties in the area and assemble these lands to initiate demolition and repairs.

In crafting this statute, the National Association of Real Estate Boards advanced its belief that the urban life cycle so often characterized as "inevitable in all residential districts," in fact could be slowed if not completely halted or reversed. Although analyses of the influences of ecological thinking on real estate valuation to date have focused on the deterministic aspects of the urban life cycle as it moved from growth to decay, and how such ideas became the basis for risk evaluations at the Federal Housing Administration and Home Owners Loan Corporation, among a few prominent thinkers in the period this cycle's inevitability was not guaranteed.[88] For example, in the same volume for the Standard Course in Real Estate in which Frederick Babcock laid out five variations on the patterns whose "end result is a blighted or slum community . . . occupied by the poorest, most incompetent, and least desirable groups in the city," the author speculated about the circumstances that might interrupt "the incorrigible tendencies here described." Noting a few "rare but encouraging exceptions" in "deliberate programs of slum rehabilitation," Babcock, whose influences on real estate practices were particularly significant because of his work as Federal Housing Administration underwriting division head, concluded a broader lesson might be drawn. If "the gradual decline of residential districts" could be "arrested in many instances in future years," then a "marked reversal in the fundamental tendencies of city growth" might "provide a welcome exception to the tendencies" previously "described as 'inevitable.'"[89] The National Association of Real Estate Boards hoped that, by formalizing voluntary practices for property protection already found in many middle-class residential communities, homeowners could make more common the rare episodes of stabilization Babcock identified, reversing the cycle of decline. Like their planning colleagues in the 1930s, then, many real estate professionals saw potential solutions to urban problems in interventions that kept the nature of cities in mind.

Tensions between Urban Professionals on the Matter of Eminent Domain

Despite the proposal's grounding in "neighborhood unit" theory, a scheme also being advocated by the Research Committee on Urbanism and American City Planning Institute, and one defended on ecological grounds, reaction from planners was swift and unflattering.[90] It was "legislatively undesirable," wrote American Society of Planning Officials Executive Director Walter Blucher in a private evaluation, because "it does not follow what we conceive to be fundamental planning principles."[91] The city planning and real estate communities, while both using ecological analysis in their work, did not agree on what constituted "natural" urban development. Real estate professionals believed, as Frederick Babcock put it, "in the competition of uses for sites, the highest and best uses tend to win . . . This competition results in a distribution of uses throughout the city which has been called 'natural zoning.'"[92] Planners, however, frustrated by the dominance of business interests in urban development, saw the activities of their colleagues interfering with cities' natural patterns. "If land were allowed to ripen rationally and were taxed equitably, there would be a steady or orderly accretion to the area of a city," *American City* editor and urbanism committee contributor Harold Buttenheim observed, "It would grow, insofar as barriers did not prevent, with few gaps from the center to the rim . . . in successive rings or belts on the periphery of the city and bearing a logical relationship to the city's core," but this "natural process" had been "upset by wildcat speculation."[93] Superficially similar to planners' calls to make land assembly for large-scale redesign easier — "it is becoming more generally recognized that the owners of small parcels of property are helpless before the forces of neighborhood disintegration," Buttenheim explained — the real estate community's proposal for neighborhood units perverted the concept by leaving planning experts and in turn comprehensive planning for cities out of the equation.[94] Equally troubling was how the National Association of Real Estate Boards threatened to hijack their work pushing the natural resources conservation model as a guide for city improvement. "The procedure provided in the Statute is modeled largely on the procedure which has been used for the creation of special purpose districts, such as flood conservancy districts, drainage districts, reclamation districts, and the like," Blucher grumbled, and suggested that the recommendation to involve 75 percent of the owners was "apparently based on the procedure in irrigation and reclamation districts . . . The arguments for allowing the taking of property for slum clearance rested very heavily upon that sort of statute."[95]

In this context, planners' reactions reveal more about their professional anxieties than about the specifics of the real estate association's legislative aims. Like the human ecologists whom critics suggested saw concentric zones in data where none were present, planners saw a proposal for urban conservation where no references to cities as "resources" or urban "conservation" appeared. The call for 75 percent enrollment and for the collective control of territory by neighborhood property owners instead extended practices already widely supported by the National Association of Real Estate Boards. MacChesney was the author of the infamous Article 34 of the Realtors' Code of Ethics, which instructed its members to avoid introducing "into a neighborhood a character of property or occupancy, members of any race and nationality, or any individual whose presence will clearly be detrimental to property values in the neighborhood."[96] This ongoing practice, which depended on the use of restrictive covenants, operated with similar district creation provisions: Property owners would divide their neighborhoods "into sections of six to eight blocks with separate covenants for each section; set the agreements to go into effect once 75 percent of the property was signed for, and *then* drive for total coverage throughout the entire district."[97] By calling to expand the powers of restrictive covenants the association sought to grant urban property owners the same appealing powers that were drawing residents to suburban areas.[98] It "merely proposes to give old neighborhoods powers similar to those which new subdivisions or small suburban village governments now have," Herbert Nelson explained, confirming the comparison was to suburbs rather than farms or forests.[99]

Happily for America's planning professionals, this policy proposal, renamed the Neighborhood Improvement Act before being introduced into several state legislatures, was largely unsuccessful.[100] Even though the real estate community's vision of eminent domain based in collective action toward private property protection escaped the charges of socialism that dogged planners' proposals for government-organized land assembly, the nation remained unsympathetic to its ideas. With many powerful figures in the industry still committed to this legislative agenda, the Realtors' association searched for alternative approaches to demonstrate to public officials and private property owners that they had found a compelling solution to some of America's mounting urban woes. Convincing the Federal Housing Administration and Home Owners Loan Corporation — already promoting repair among individual beneficiaries of their housing programs — to assist, the association initiated a test program in two cities whose results could be widely copied if the model statute became law.[101] Recognizing, however, that even federal support would not necessarily guarantee their experiments' success, project leaders sought a new frame for their arguments about the urgency of land assembly and collective repair. Now, planners' work to redefine cities as natural resources to be conserved — increasingly well-known in real estate circles through the much-discussed *Our Cities* as well as a later National Resources Planning Board report on land policies to which staff from the Federal Housing Administration and Home Owners Loan Corporation, including Miles Colean and Arthur Goodwillie, contributed as consultants, suggested a new direction for their rhetorical campaign.[102] When the demonstrations opened in Baltimore and Chicago, they would chiefly be characterized not as property "protection" or "improvement" but as neighborhood and community "conservation."

From Property Protection to Community Conservation

"Because we believed that America was ever 'marching onward,' the conservation of physical resources, already created, was frequently ignored or wholly forgotten in the haste to develop new wealth and novel amenities of life," wrote Home Owners Loan Corporation staff in an untitled report as planning for the pilot program was getting under way. The depression-inspired "opportunity for sober reflection" that had made clear how "in more than a geographical sense are our old frontiers gone," also had brought with it a new attitude toward resource use. Yet, America's "increasing effort" to apply science to derive "maximum benefts from what we already posess" was better developed in applications to natural resources concerns.[103] The corporation, following in the footsteps of the National Resources Planning Board Research Committee on Urbanism, with its call for "the substitution in place of the philosophy and aspiration of bigness the philosophy and aspiration of quality," was now aware of how "while America was growing, it also was wasting away" and insisted on a national program of urban conservation. "Near the top of the list of existing values" to be conserved "stand our older established residential districts . . . beginning to show evidence of incipient blight and . . . the poisonous seed of decay."[104] With appropriate treatment, "the constant cycle of birth, life, and death" in urban neighborhoods, previously considered "an inevitable process — a natural heritage" might no longer be guaranteed.[105]

The belief that "conservation was needed" in cities did not represent a complete departure from the real estate research tradition.[106] Some of the earliest public urgings for a science of real estate, for example, from University of Wisconsin President Glenn Frank, had compared Realtors unfavorably with pioneers, pointing to their predilection for abandoning areas of declining profit instead of seeking knowledge about their repair. "I suggest that the Realtor rises to the dignity of a professional man only as his methods cease to be dominated by the mind of the pioneer," Frank had insisted at the annual Realtors' convention in 1928, "that is, until he becomes more than an exploiter of site values, leaving behind him, as did his pioneer predecessor, a trail of ugliness and impermanence."[107] So too among real estate economists, as among regional planners, were some efforts to promote the alignment between rural and urban land use concerns. "The mistakes of the past . . . are largely traceable to the failure to recognize the essential unity of land economics" Richard Ely had explained in the inaugural issue of the *Journal of Land and Public Utility Economics*, criticizing policies that ignored how "the utilization of agricultural, forest, mineral, and urban land is characterized by a network of interrelationships."[108]

In the New Deal context of national support for conservation, however, such arguments took on new urgency and new appeal as city planners' work to redefine urban areas as "resources" suggested to the real estate community how conservation as the preservation of individual properties and conservation as the scientific management of neighborhood values might be combined.[109] For the Realtors' association, eager to advance its policy agenda, the hope was that a push to conserve urban resources might inspire states to take legislative action as many already had in support of natural resource conservation.[110] For the Home Owners Loan Corporation and Federal Housing Administration, seeking to protect their financial interests and increasingly aware of the limits to home repair at the individual level, the hope was that a call to conserve urban resources might inspire "vigilant groups of homeowners" toward the kinds of voluntary collective action their rural counterparts already were taking as federal policies shifted from an emphasis on resettlement to rehabilitation.[111] Thus, when the experiments in support of the model legislation made their public debut, they suggested to observers how the real estate community's knowledge of the ecology of cities could help to stabilize decaying urban areas by putting the concepts and methods previously employed for risk-rating studies to "halt and reverse the process of community decline."[112] "Through the HOLC Division of Research," McNeal explained, "we have developed maps of hundred of cities which show areas that are doomed unnecessarily — which can be saved."[113] The shared language of ecology and conservation belied the fact that the real estate community's agenda differed substantially from their planning colleagues'.

Experiments in Waverly and Woodlawn

The first experiment got under way in Baltimore's Waverly neighborhood in 1939.[114] According to Marc Weiss, it was Baltimore Realtor and developer and Roland Park Company head John McC. Mowbray, National Association of Real Estate Boards Housing Committee chair, who first proposed this community as a demonstration site. A subdivision three miles north of the central business district, Waverly was a neighborhood in decline, but not yet a slum — in the words of officials at the Federal Home Loan Bank Board, "the incoming tide of neighborhood disintegration in Waverly had only just begun its advance," such that its "useful life" now was "threatened."[115] With the local housing stock deteriorating, residents had encountered difficulties selling their properties and some had begun to default on mortgage payments. "The district is occupied by white owners and tenants of moderate means and substantial character, predominantly American born, who, in general, give definite evidence of social and civic pride," described one Home Owners Loan Corporation report. Although there were a few new houses, "the majority of them grade B&C," it was activities in adjoining neighborhoods that gave the community pause. To the west lay "one of the finest residential districts in the city." To the north and east were "comparatively new Grade A neighborhoods." Yet to the south "and definitely threatening it with infection through infection and by actual infiltration—is a fully developed, though not congested, slum area, containing a predominantly Negro population."[116] With both the Home Owners Loan Corporation and Federal Housing Administration in agreement that the "infiltration of inharmonious racial or nationality groups" human ecologists had identified as signaling neighborhood transition marked imminent community instability and property value decline, the corporation, which first "became interested" in Waverly, according to Federal Home Loan Bank Board head (and former U.S. Chamber of Commerce director) John Fahey, "because of properties it had been forced to acquire there," set out to guard against "the infiltration of a progressively undesirable type of occupant" on account of the implications of such trends for the area's financial base.[117] The project's goal was to "restore the district to a general parity of condition with its principal surroundings to the West, North and East," and simultaneously create "a barrier, through stabilization, against the encroachment of substandard conditions from the South."[118]

In this "test stabilization project" the Baltimore Housing Authority "acted as sponsor, the Works Progress Administration conducted a survey of conditions and needs, and the U.S. Housing Authority and Home Owners Loan Corporation contributed technical service."[119] Corporation "technicians," working under the supervision of former Chicago Regional Reconditioning Supervisor Arthur Goodwillie, designed a three-part attack on area problems.[120] First, they resurveyed the area; the effort to update residential security maps got under way in March 1939, "under the direction of a supervisor provided by the HOLC with the assistance of an HOLC appraiser, [and] WPA enumerators."[121] Second, they analyzed this survey information to designate areas for strategic improvement as part of a community-wide master plan, a plan for public as well as private areas, which offered proposals for revisions to local zoning. Third, following an educational campaign to call attention to the program, the

Home Owners Loan Corporation organized area homeowners into the Waverly Conservation League to ensure community buy-in; this twenty-five-person group organized further neighborhood action by selecting block captains within the area to canvass property owners and encourage their participation.[122]

A few months later, a similar "experimental approach to the problem of saving 'borderline' neighborhoods from deterioration," one "to be known as the Neighborhood Conservation project," got under way in Chicago's Woodlawn neighborhood.[123] "Borderline" areas had two meanings in real estate appraisal—communities on the precipice of decline or improvement, and communities that divided "inharmonious" land use categories. Woodlawn, a community near the University of Chicago whose residents were keen on taking group action to maintain property values, fit both definitions. Despite a long history of efforts to maintain neighborhood homogeneity—"property owners in Woodlawn were pioneers of community containment" from the late nineteenth century as historian Thomas Philpott has described—Home Owners Loan Corporation surveys had documented in more recent years how Woodlawn faced decline due to trends in the migration of African Americans.[124] All-white areas such as C-216 ("quite a desirable place to live") and D-78 (where restrictive covenants were enforced) nevertheless were graded C and D on account of the projected directions of neighborhood change.[125] Noting an infiltration of "Negroes, Jews, and Swedes," for example, an anonymous appraiser described the situation in the Washington Park subdivision that comprised area D-78: While this "semi-blighted area" remained "restricted to Whites" for another decade, there was "a constantly increasing encroachment of Negroes from both the west and south." For this reason, it was "expected ultimately that this entire area will revert to the Colored race."[126] Elsewhere in Woodlawn, for example, in D-79, the appraiser categorized an all-white neighborhood (albeit largely comprising immigrant groups) as fourth grade, reporting that property values already had been declining "due to Negro encroachment on the west."[127]

Thus, the Woodlawn experiment, like that in Waverly, was "organized against a definitely anticipated intrusion," with consequences for property valuation—an effort to halt the erosion of property values as much as the erosion of property itself.[128] Its attack on faltering neighborhood conditions similarly began with a survey, followed by a plan, and

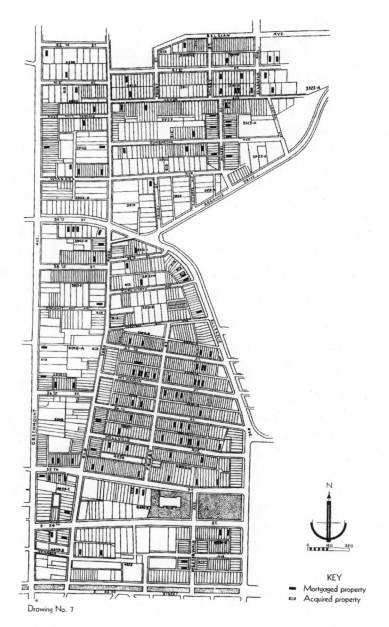

HOLC Mortgaged and Acquired Properties in the Waverly Area of Baltimore. From Federal Home Loan Bank Board, *Waverly: A Study in Neighborhood Conservation* (Washington, DC: Federal Home Loan Bank Board, 1940).

subsequently by community action — in this case from the Woodlawn Association (also referred to as the Associated Clubs of Woodlawn, Inc.). Yet, with the majority of the nation's real estate and planning organizations, including the National Association of Real Estate Boards, headquartered in Chicago, the participating cast of characters in this endeavor was even broader.[129] As the Works Projects Administration ran the survey, "neighborhood organizations, large and small Chicago business interests, the Chicago Plan Commission, and other city departments, and federal government agencies," including the Home Owners Loan Corporation, U.S. Housing Authority, and the Reconstruction Finance Corporation, offered technical advice and manpower.[130] Representatives of the appraisal community, for example, corporation district manager, Society of Residential Appraisers president, and Northwestern University economist Herman Walther, and the University of Chicago, also were involved.[131] The project was under the direction of corporation staff member and former University of Chicago Business Office Manager Robert Mitchell. "Forest conservation is the process of improvement, cultivation, intelligent use and replacement of forest resources in the interest of sustained yield . . . Sometimes it involves elimination of unfit or unimproved elements," Mitchell explained in a draft of the first Woodlawn plan. The conservation of urban areas could be conceived "likewise as the process of producing and maintaining neighborhood resources — social, economic, and physical."[132] Previously at work for the Housing Division of the Federal Emergency Administration of Public Works as well as an advocate for the Realtors' model neighborhood improvement law, Mitchell was one of the first members of the real estate community to draw links between neighborhood improvement and the notion of "conservation" previously associated with natural resources planning, offering more public reflections on such connections in 1938 comments to a conference at the National Association of Housing Officials.[133]

Defining and Promoting Neighborhood Conservation

Environmental historians have described conservation as a shape-shifting concept. Associated with political positions on the left and right; with technocracy, democracy and patriotism; with policies from set-asides to scientific management, this flexibility of meaning was central to its enduring appeal.[134] During the New Deal, the expansive movement for resource planning held contradictory ideas in tension in the lead up to program implementation. Although agency heads such as Hugh Bennett and Rexford Tugwell hoped to achieve a program of long-term resource management affecting rural residents across the economic spectrum, farm groups sought relief rather than reform in economic, technical, and legal assistance for property protection.[135]

For two communities of urban professionals who shared the belief with agricultural researchers that "sub marginal men are often closely associated with sub marginal land," yet disagreed on the appropriate remedy for city problems, the protean meanings of conservation that cemented its influences in natural resource planning contexts equally were apparent in early discussions of its applications to urban settings.[136] City planners at the Research Committee on Urbanism espoused a vision of conservation as a centralized, scientific endeavor, one with comprehensive aims, including human rehabilitation toward poverty reduction. The real estate community, more concerned about financial security and suspicious of large-scale government intervention, preferred conservation as a decentralized operation for property protection whose purpose was to maximize the economic productivity of land. Although the experiments in Waverly and Woodlawn drew from planning the concept of "neighborhood units," then, they did so absent an appeal to citywide master planning.[137] Combining the home improvement focus of the do-it-yourself movement and the collective orientation of homeowners associations organized to maintain racial homogeneity in city neighborhoods, urban conservation aligned these endeavors with the campaign for wise resource use toward efficient production whose myriad benefits had been well articulated.[138] Conservation strategies in Waverly and Woodlawn included establishing and policing ordinances on housing standards and sanitation; spot demolition; public improvements to schools, streets, parks, and municipal services such as trash collection; rezoning and replatting areas; closing streets; attracting new construction; and restrictive covenants—all at the neighborhood level. Thus, while the real estate community's two test programs shared with planners an ecological understanding of cities (indeed using the model of succession to climax not only as a theoretical paradigm for explaining the urban life cycle but also as a guide to returning urban communities to more stable states), and even sought approval from local planning commissions, organizers never made problem solving beyond neighborhood borders a priority.

Early reviews of both experiments were enthusiastic. Although organizers encountered initial resistance from homeowners "unwilling to believe" their property might "already be involved in an obscure process of community decline" — and in fact some initially refused surveyors' entry — publicity soon rallied community members to act.[139] In Waverly, after just a year in operation, participants expressed satisfaction about how the conservation project had improved the housing market. The Home Owners Loan Corporation, for example, "pointed to the fact that the number of paid-in-full loans on its books had nearly doubled during the year, while the ratio of borrowers in default had dropped from 37 per cent to 14 percent."[140] Work in Woodlawn was slower to get off the ground, but National Association of Real Estate Boards president Newton Farr praised his city's efforts as he urged the extension of such measures to other urban areas.[141] The low cost of these undertakings held special appeal: activities in Waverly had cost only $50,000 and in Woodlawn only $70,000, expenditures that could be partially recouped through Title I of the Housing Act of 1934.[142]

With ecological theory finding every city operated according to similar laws, observers were confident in the generalizability of the pilot program. As the *Federal Home Loan Bank Review* reported in 1940, participants in the Waverly experiment had "traced the life cycle of a residential neighborhood — the growth and decay of the hundreds of 'Waverlys' in the cities of this country," meaning that "the results of the Waverly program have demonstrated that neighborhood disintegration in hundreds of cities can be halted to the benefit of the blighted areas themselves and to the city taxing authorities which would be affected by a loss of property values."[143] Assessments of the Woodlawn experiment pointed to similarly sweeping lessons: "It is hoped that this study will clarify the problems and point a way toward their solution not only in Woodlawn, but also in other neighborhoods and cities," Robert Mitchell explained.[144] For any city, if subjected to "a coordinated and sustained neighborhood conservation program" might "measurably extend its period of usefulness and long postpone its disintegration."[145] Providing scientific instruction so that the public would "question the thesis that complete decay and eventual structural demolition are necessarily the final chapter in every residential neighborhood's life cycle" would be critical to these efforts' success.[146] For "too often," McNeal lamented, the attitude of mortgage lenders, city governments, and citizens alike had been "one of patient acquiescence in a natural phenomenon which cannot be controlled and even has not been particularly feared."[147]

To fulfill this educational mission, Arthur Goodwillie, who had directed the Waverly work, created a Conservation Service within the Home Owners Loan Corporation.[148] The agency prepared a *Field Report and Operating Manual* for "the information of students of municipal problems and for the possible guidance of urban communities which, aware that prevention is the best cure for neighborhood decay, are seeking means for the effective application of that remedy."[149] And while confidentiality remained a priority — the security maps, "intended as an aid in servicing the Corporation's loans, have not been and, for obvious reasons, cannot be made public" — staff offered to make these cartographic surveys available "to any city or organized community agency by which the subject of local urban blight is being studied as the precedent to curative action."[150] Finally, the conservation experiments offered another opportunity to promote the Realtors' model statute, which appeared at the conclusion of the 1940 Waverly report.[151]

McNeal took his message on conservation back to the planning community that had inspired it, speaking to the National Conference on Planning in 1941 where he called for "a municipal department of conservation — established to promote, direct and coordinate neighborhood conservation groups."[152] Hoping to capitalize on the growing interest in urban conservation these experiments had spawned, the National Resources Planning Board created an Urban Conservation and Development Unit in its director's office, headed by former National Association of Housing Officials executive director Charles Ascher. If "Urban conservation and development is a 'fourth dimension' in the work of the National Resources Planning Board," observed Ascher soon after the unit's creation, then "a restatement" would be needed "of the application of the sovereign powers of the state — eminent domain, taxation, and the police power — to the control of the use of urban land," noting that "a comparable — and admirable — statement is that in the rural field by the Department of Agriculture."[153] Ascher drafted an organization chart with his vision for a unified Federal Urban Agency comprising a Bureau of Urban Economics, Office of Plans, Urban Land Administration, Urban Works Administration, Urban Housing Administration, and Urban Dwelling Insurance Administration to mirror the organization of the agriculture department.[154]

By the early 1940s, then, members of the American city planning and real estate communities agreed that federal investments in urban concerns on par with natural resources concerns were urgently needed.[155] Although the exact nature of the "public purpose" for which to permit eminent domain for the scientific management of urban lands remained a point of dispute, both professions rallied around the idea of conservation for cities. Had the federal government immediately thrown its weight behind their proposals, "urban conservation might have taken its place of honor with soil conservation," Charles Abrams opined. "Yet that did not happen," as the bombing of Pearl Harbor and America's subsequent entry into World War II diverted its attention to national security concerns.[156] With the push for urban conservation having stalled in federal circles, enthusiasts hoped to jump-start it from below. Among them was economist Homer Hoyt, watching the Waverly and Woodlawn experiments unfold from his vantage point on staff at the Federal Housing Administration and later the Chicago Plan Commission. How Hoyt embraced the war as an opportunity to devise a life cycle–based plan to repair the city of Chicago, and how he rallied diverse constituencies to unite in support of his vision of urban resource management, are the focus of chapter 3.

THREE

A Life Cycle Plan for Chicago

"An analogy might be made between the city and a summer garden," wrote Chicago real estate appraiser Richard Nelson in a report on the "conservation of middle-aged neighborhoods and properties" for the U.S. National Housing Agency in 1944. "If properly planned," a garden would be "always beautiful and always balanced, though at any given moment it will consist in . . . budding plants, plants in full bloom, others which are past their blooming time and are fading, and others which must be cut to the ground or uprooted." Highlighting cities' similarly orderly march from growth to decay—"each of our cities goes through the same sort of kaleidoscopic interior movement," with "development areas (buds)," "stable areas (blooms)," "middle-aged or conservation areas (wilting flowers)," and "clearance areas (dead plants)" —Nelson insisted the management of urban areas should follow their predictable cycle of birth and death and that federal authorities show their commitment to reversing the cycle of decline.[1]

How did it come to pass that Nelson, like so many contemporaries, would press for a national program of city improvement organized with the life cycle of urban neighborhoods in mind? Faced with wartime delay on action during the early 1940s, a loose association of Chicago-based social scientists, city planners, and real estate professionals combined their visions of cities as "ecological" communities and cities as national "resources" to lay the groundwork for a comprehensive postwar urban revitalization scheme. Pivotal in bringing together these actors was Chicago Plan Commission research director Homer Hoyt, who left the Federal Housing Administration with the hope of adapting his contributions to grading neighborhoods and delimiting urban districts to a plan for Chicago in which "each part" of the city stood "in a real functional relationship to each other part, the whole forming an organic entity."[2]

To date, accounts of city improvement efforts in the United States have separated the campaigns for demolition versus rehabilitation. According to this view, supporters of clearance and advocates for modernization—like the planning and real estate professions more generally—were two factions in conflict about the one best way to manage the future of U.S. cities.[3] Yet, from the early 1940s, a growing Chicago community built on the work of the National Resources Planning Board and the National Association of Real Estate Boards to articulate the need for both strategies if cities were to be saved for the longer term. Under Hoyt's direction, their shared conviction that cultivating the city in a garden (Chicago's motto: *Urbs in Horto*) required attending to its resemblances with an actual garden transformed ecological theories of urban structure, behavior, and evolution into an applied science of urban resource management, specifying the timing and sequencing for future "clearance" and "conservation" projects throughout the city.

Homer Hoyt from Chicago to Washington and Back

Homer Hoyt exemplifies why, as Charles Abrams has described, real estate appraisers were "the intellectual hierarchy of the real estate group."[4] Hoyt, who worked as a real estate investor during Chicago's building boom, returned to the University of Chicago, his alma mater, to study for a PhD in economics after the market crash.[5] In keeping with institutional tradition, he simultaneously took a job at the nearby Cook County Assessor's Office to gain access to essential data for his dissertation research. Published as *One Hundred Years of Land Values in Chicago* (1933) even before its formal acceptance for the PhD degree, the dissertation took a historical overview of trends in Chicago real estate from 1830 to 1933, synthesizing massive quantities of information from published and unpublished records. Presenting his findings in charts and maps, as well as in text, Hoyt's overarching argument was that Chicago had a "real estate cycle," and he traced five repetitions of this pattern from the city's origins to the present day.[6]

Hoyt's work did not mark the first time an economist had observed cyclical trends in cities. By the early 1930s, institutional economists such as Wesley Clair Mitchell had identified the existence of business cycles in numerous contexts, a finding that sociologists, including Roderick McKenzie, had cited in connection with their work to develop models of urban life cycles.[7] Real estate economists, for example, Ernest Fisher and Frederick Babcock, had adapted actuarial notions of the anticipated life cycle of individual buildings ("useful life") to appraisal.[8] Hoyt's contribution was to merge these macro-scale and micro-scale theories. Taking advantage of the intellectual resources around the city to draw on the findings of Chicago-affiliated sociologists, including Ernest Burgess, Robert Park, Roderick McKenzie, Louis Wirth, and Earl Johnson, and Northwestern University economists, including Herbert Simpson and John Burton, Hoyt characterized his investigation as "the first scientific account of the interaction of all the different elements in the so-called real estate cycle," one with the potential to be used as a forecasting tool.[9]

One Hundred Years of Land Values in Chicago set the tone for Hoyt's many later explorations of how naturalistic perspectives on cities from across the social sciences might productively be combined into an applied science for the real estate and city planning communities' present and future work. Indeed, it was the enthusiastic reaction from Ernest Fisher — then head of research and statistics at the Federal Housing Administration — that led to Hoyt's invitation to join the agency's staff. He became chief of the section on city growth in Fisher's division and later, principal housing economist in that division.[10]

Hoyt's first job at the new housing agency involved refining aspects of the risk rating methods that underlay the system of mutual mortgage insurance established with the Housing Act of 1934. Chiefly designed by Frederick Babcock, now Federal Housing Administration underwriting director, these methods had established national standards for grading mortgage applications by profiling the risk of each borrower, property, and property location — and for those applicants deemed adequately safe in these three categories, the risk of mortgage pattern. In light of his pioneering work on urban economic trends, Hoyt was charged with improving the location rating process to speed insuring officers' risk assessments around the United States and to increase the uniformity of their results. He proposed that risk raters create for every city a map of "mortgage risk districts." Such maps, prepared at the local level and then sent to Washington headquarters for review, would complement the location rating techniques already laid out in the agency's *Underwriting Manual*.[11] Essential data for these cartographic assessments was widely available, he observed, in the Real Property Inventories under way in cities across the nation.[12]

In the wake of the nation's economic depression, the U.S. Department of Commerce and the U.S. Civil Works Administration had initiated Real Property Inventories, surveys of urban areas to parallel the federally organized surveys of rural areas prepared by government agencies and commissions, including the U.S. Department of Agriculture and the U.S. National Resources Planning Board. The promise of these large-scale investigations lay in their future applications to housing planning. At the end of 1934, sixty-four cities had conducted house-to-house field studies to gather information on area physical conditions (for example, age of building and number of rooms) as well as local social conditions (for example, number of occupants and race of household) — ample evidence on housing trends for Federal Housing Administration insuring officers to use. By 1936, assessments of 203 American cities, including New York City, Washington, Philadelphia and Boston (but not Chicago, Los Angeles, or San Francisco) were complete.[13]

As of 1937, Federal Housing Administration headquarters possessed mortgage risk district maps for the nation's seventeen largest urban areas.[14] Alongside these cartographic assessments, local insuring offices sent on to Washington, DC, the range of source materials each had used to profile neighborhoods — sources that included Real Property Inventories, U.S. Census data, and the time-interval studies that Hoyt had demanded as part of the new risk rating scheme.[15] Upon "seeing giant stacks of maps" from these efforts, Hoyt grew increasingly interested in identifying the overarching urban housing patterns such a diversity of records, when juxtaposed, might reveal.[16] Cartography had played an essential role in his research to date: *One Hundred Years of Land Values in Chicago*, like the dissertations of his colleagues in sociology, had depended on maps as sources of data and as analytic tools. Hoyt had shown interest in map work from natural resources planners as well as the appraisal community, reviewing cartographic studies in forest and farm management and studying the work of field agents at the Home Owners Loan Corporation.[17] In his continuing work for the federal government, as in his dissertation, Hoyt set out to synthesize the massive quantities of data that insuring offices had gathered, eventually in hundreds of cities. He published his initial findings in a series of articles for the housing agency's journal *Insured Mortgage Portfolio* and later a more comprehensive report, *The Structure and Growth of Residential Neighborhoods in American Cities* (1939), which offered both a theoretically oriented documentation of techniques to guide future "scientific analysis of city structure" and a practically oriented handbook that extracted from the agency's distributed data gathering effort general principles to direct building development and mortgage lending.[18] Echoing observers such as Columbia University economist Robert Murray Haig, who earlier noted that each city "operates like a Bunsen flame under a test tube to produce" urban patterns, Hoyt argued that, in the depression's aftermath, scientific study revealed natural laws continued to guide cities' structure and evolution.[19] "There is not an indiscriminate mixture of homes with varying characteristics in every part of the city," he explained, but rather "a definite series of patterns according to which dwelling units that are similar with respect to a given factor tend to be concentrated in certain areas."[20] By taking a bird's-eye view of U.S. cities with the benefit of access to new evidence, Hoyt aimed "to bring order out of chaos" through statistical techniques and reach levels of generalizability greater than any predecessors had achieved.[21]

Beginning with a discussion citing sociological theorists of urban land use, relying on terminology such as "invasion" and "transition," now common in real estate analyses, and including maps to illustrate the predictable cycle of residential growth, decay, and change, *The Structure and Growth of Residential Neighborhoods in American Cities* validated the hopes of ecologically oriented urban analysts that findings from specific cities would generalize to other settings. That transportation routes and land values were closely correlated, that heterogeneous communities were often a sign of decay, and that cities' internal repatternings gathered blight in central areas—the maps of mortgage risk districts, Real Property Inventory data, and a range of other cartographic sources supported social scientists' earlier observations. For example, it was in this report that Hoyt looked more closely at the details of invasion-succession. Observing the frequency with which "the lower and intermediate rental groups filter into the homes given up by the higher income groups," he developed a "filtering model" of community change, which held that building change and population change simultaneously occur, lowering the value of a neighborhood until it inevitably became a slum.[22] Here, Hoyt declined to define higher- and lower-quality occupants, a question he explicitly tackled in his earlier work. (Chicago Realtors, he reported in *One Hundred Years of Land Values in Chicago*, generally worked with the following ranking of racial and nationality groups in mind—from best to worst: English, Germans, Scotch, Irish, Scandinavians; North Italians; Bohemians or Czechs; Poles; Lithuanians; Greeks, Russian Jews [lower class]; South Italians; Negroes; Mexicans.)[23] Other interpreters would similarly demur, joining Hoyt's new theory with the kinds of numerical assessments already found in the Federal Housing Administration's underwriting grid. As Richard Nelson summarized in a report to federal housing officials, "neighborhood desirability" could be "represented by a series of index numbers from '0' to '10,'" rankings defined by the character of area inhabitants: "A '6' neighborhood becomes a '5' neighborhood at the moment when a substantial number of '6' people leave and their places are taken by '5' people," he explained, "The way in which a neighborhood deteriorates is through the withdrawal and replacement of its population."[24]

Hoyt's analysis confirmed the possibility of developing an urban model that was generalizable. As to what specific aspects of urban growth, decay, and change generalized, however, Hoyt and his academic predecessors did not completely agree. The Real Property Inventories offered land use data on a scale across cities and at a level of detail that academic researchers had not previously seen. Together with the supporting records from insuring offices' risk assessments, this new information suggested the need to revise some established beliefs, namely, that cities grew not in complete concentric rings with successively nicer districts as one moved outward from the center but rather in sectors, where high-grade or low-grade residences could be found clumped together. It was here (as well as in the earlier articles in *Insured Mortgage Portfolio*) that Hoyt first

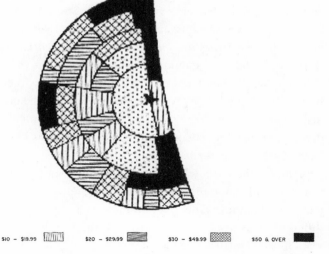

LESS THAN $10 $10 – $19.99 $20 – $29.99 $30 – $49.99 $50 & OVER

Sector Model of Chicago, Illinois, Based on Average Rent. From U.S. Federal Housing Administration, *The Structure and Growth of Residential Neighborhoods in American Cities* (Washington, DC: Government Printing Office, 1939).

presented his famous sector model as an alternative to Burgess's concentric vision.[25] In a wide range of urban areas from Miami to Detroit to Washington, DC, Hoyt explained, this natural sector pattern was repeated, but the geographic arrangement of these sectors varied widely across cities.

    Hoyt would not be the first scholar to propose a model of "the city" to compete with the zonal scheme. Noting in the same real estate text where he praised the sociologist's work how Burgess's zones were neither "rigidly determined" nor "of uniform width . . . they interpenetrate each other," Hoyt's boss, economist Ernest Fisher, had been among those who wondered whether "the city" could be modeled with greater precision.[26] Researchers earlier had offered up variants that were star shaped or resembled moons circling a planet—but none of these alternatives had stuck.[27] Because of its grounding in an avalanche of data, Hoyt's sector model would be widely used. The same year that he reprinted Burgess's concentric model in his textbook *Principles of Urban Real Estate*, then, Hoyt offered an alternative interpretation of the ecology of urban processes, patterns, and trends.[28]

Social scientists, real estate appraisers, and city planners of the 1920s and 1930s, while sharing a view of cities as ecological communities, differed in their assessments about the uses for such naturalistic interpretations. In *The Structure and Growth of Residential Neighborhoods in American Cities*, the culmination of his work for the Federal Housing Administration, Hoyt paired innovation in urban theory development with some discussion of the local problems these predictive theories might help to solve. From "the selection of areas for slum clearance," to "the determination of mortgage lending policy by areas," to "decisions in regard to zoning or rezoning of sections for given types of areas," Hoyt explained, "there is a whole series of vital urban problems," whose solution "depends upon proper analysis of the apparent riddle of the internal nature of American cities . . . and the past and prospective movements of different types of neighborhoods."[29] Understanding the ecology of cities was "more than an academic matter."[30]

During his time at the housing agency, Hoyt was offered a public platform to shape urban redevelopment efforts across the nation by educating real estate professionals, city planners, and other concerned actors in the public and private sectors about how recognizing patterns and trends in the urban context could guide housing and planning decision making in the coming years. Although Frederick Babcock was chief author of the government's first *Underwriting Manual*, Hoyt's conclusions appeared in subsequent editions. The seven key principles of city growth and mortgage risk that he outlined in *Insured Mortgage Portfolio*, for example, featured prominently in the discussion of location rating in the 1938 manual. Later versions reprinted parts of Hoyt's 1939 report nearly verbatim.[31] Yet, as a federal administrator, even from his vantage point overseeing the production of risk rating maps, Hoyt's ability to transform local conditions in U.S. cities were limited. This would change in 1941 when Hoyt returned to Chicago. Hired to direct research at the city's plan commission, his first task there was to write up the results of Chicago's Real Property Inventory — the *Chicago Land Use Survey*, one of the last in the nation. A large-scale collaboration with the Works Projects Administration that employed nearly three thousand, this study and the follow-up plans for residential redevelopment prepared under Hoyt's direction extended his work reading urban information through the lens of ecology into planning with the natural resources model in mind.

The *Chicago Land Use Survey* as Applied Ecology

Early discussions to organize Chicago's survey date to 1935, around the same time that other U.S. cities initiated their Real Property Inventories; a preliminary technical advisory board was assembled that included University of Chicago sociologist Louis Wirth and Northwestern University economist Coleman Woodbury, scholars with experience in qualitative and quantitative field studies and survey work.[32] The actual survey did not get under way in earnest, however, until 1939, during a recession throughout the Midwest.[33] This delay frustrated city officials in several agencies, such as the Chicago Housing Authority, who bemoaned the "lack of Real Property Inventory" in discussions about Chicago's future—comments suggesting that the device, which created a case record of current land uses and sought to discern general patterns of stability and change, was by that time seen as a necessity for effective planning.[34]

In fact, this delay was pivotal in distinguishing Chicago's Real Property Inventory from its predecessors in other metropolitan areas. For once the survey moved past the planning stage, two important events occurred. First, Homer Hoyt was lured from Washington, DC, to direct the Research Division of the Chicago Plan Commission, the unit charged with organizing the survey and writing up its results. Hoyt came to this Real Property Inventory late, after survey director Hugh Young and survey technical director Robert Filley got the project off the ground.[35] Yet, Hoyt wrote much of the multivolume report of the survey's findings, and the ways in which his colleagues came to understand the information they had gathered would reflect his intellectual influences.[36] Second, with Hoyt's arrival, the plan commission hired the first of many alumni and continuing students from the University of Chicago's social science programs to staff its research division. No longer the academic headquarters for real estate economics, the city remained a center of urban sociological and geographic research, and university documents suggest faculty viewed such extracurricular collaborations as essential to students' training.[37] Hoyt created an atmosphere wherein the traffic in ideas flowed in both directions: graduate students such as sociologist Donald Foley and geographer Robert Klove used the data they gathered through plan commission field studies and surveys as a research base for theses and dissertations, as academic theories simultaneously shaped the development of city plans.[38] Given the persistent popularity of naturalistic interpretations of cities, including studies based on Real Property Inventories in Cleveland and New York, and the deepening relations between social scientists and planners more generally, without these developments Chicago's survey might have taken a similar turn.[39] Together, these developments guaranteed that naturalistic interpretations underlay the model of Chicago the survey and subsequent plan commission reports eventually produced.

Methodologically, the *Chicago Land Use Survey* shared much in common with academic work in urban sociology, urban geography, and real estate economics as well as prior Real Property Inventories' appraisal techniques. Its information-gathering efforts combined historical research with field surveys, and in the analysis of this quantitative information, statistical maps were the central analytic tool. Improving on earlier tabulation procedures by using eighty-column punch cards rather than index cards to help automate their work, surveyors created 378,305 cards for residential properties, 53,210 for nonresidential properties, 608,261 cards for each unit of multifamily dwellings, and 20,200 cards representing city blocks. Although it was one of the nation's last Real Property Inventories, Chicago's land use survey became "the first, in this country, to tabulate and map land use by individual parcels."[40]

There were other notable differences from predecessor surveys. In his work to refine location rating practices at the Federal Housing Administration, Hoyt had articulated the importance of obtaining detailed demographic information on city residents, insisting that risk raters take note of the locations of "desirable" and "undesirable" racial and nationality groups.[41] Real Property Inventories, standardized at the U.S. Department of Commerce and Works Progress Administration before Hoyt's map-based system was complete, showed only the location of white or nonwhite areas, however, sending insuring officers to find such population data in other sources (most often, it was supplied by the local real estate industry). In a nod to the expectation that its data would be applied to update the city's mortgage risk district map, Chicago's land use survey was revised to inquire as to the race or nationality of each head of household, aggregating information on predominant racial and nationality group by block as well.[42]

With so much information gathered, this large-scale study lacked the sense of intimacy of the ecological analyses most familiar to student workers. Like the many plant, animal, and human ecologists who prioritized direct experience, Hoyt acknowledged that "no amount of statistics can ever take the place of sound experience and common sense." He took his research staff into the field so that future users of the data could become more confident that "facts such as those compiled by the Land Use Survey are the tools by which judgment can be applied."[43] Donald Foley, who worked under Hoyt's supervision while studying for a master's in sociology with Louis Wirth, reminisced:

> As junior staff, we would have been working on the statistics describing an area within Chicago. We'd not be sure what the statistics meant. He'd say, "Let's go look." And he'd call for a city car and a chauffeur. Two or three of us would pile into the car with him and go out and look at or, perhaps, walk the area. For a graduate student it was a remarkable experience of seeking to match what the statistics reported and what we saw with our own eyes. Our group discussions of what we saw were like mini-seminars. I suspect he was equally engaging in promoting ideas to other senior staff members and to members of the Plan Commission.[44]

Drawing analogies to scientific "inventories" and "reference atlases," whose value would only increase over time, and to "life histories" allowing city planners to diagnose neighborhood problems, the *Chicago Land Use Survey* repeatedly emphasized that its work was scientific.[45] The study of urban areas might be "a relatively new science," but it was one that could identify the presence of "natural laws" in cities.[46] The hope was to make the *Chicago Land Use Survey* a source of continuously updated information: "In the belief that a permanent source of data on land use will be of great value to the city of Chicago, civic and business organizations, and others, recommendations for keeping the land use survey up to date" were "being prepared" alongside the final report.[47]

Explicitly referring to the ecological theories advanced by University of Chicago scholars, and citing their local work as a benchmark against which to analyze contemporary change, the final report mapped block-by-block information on the age, condition, type, and value of residential structures throughout the city, with specific reference to sanitary facilities, duration of tenant occupancy, and race of household.[48] (This latter category included white, nonwhite, and "Negro" despite the more nuanced demographic data gathered by surveyors.)⁴⁹ A second volume provided more detailed neighborhood maps, including data on types of structure at the parcel level.⁵⁰ And closely resembling the "community fact books" earlier published by the University of Chicago's social scientists, seventy-five supplementary volumes offered additional text, statistics, and maps charting the "life history" of each community area as well as a comparison on measures such as rental price and quality of housing to the Chicago average. As with *The Structure and Growth of Residential Neighborhoods in American Cities*, some findings contradicted researchers' expectations about the coincident factors of urban decay. For example, in Community Area 29, North Lawndale, where "the proportion of units in single-family structures was less than one-third of that in the city as a whole," simultaneously "over 91 per cent of the dwellings" in the area, "as compared with approximately 80 per cent of all dwelling units in the city, were in structures in good condition or needing only minor repairs, had installed heating equipment, gas or electric lighting, and a private inside toilet and bath."⁵¹

In light of Hoyt's earlier revisions to Burgess's work in his recent research for the Federal Housing Administration, among the most striking features of the *Chicago Land Use Survey* was the finding that, when it came to residential development, Burgess's 1925 model of the city's concentric zones retained its relevance. "Viewing the city from the center outward, the relief-map effect of the distribution of types of housing is strongly apparent," the survey explained. The Chicago Plan Commission used the terms "belts" and "arcs" rather than "zones," but the text accompanying the maps makes clear they were essentially the same thing: "Radially, the types are distributed in a pattern of communities which rise more above average in condition, median rental, and proportion of single-family structures as distance from the center of the city increases."⁵² Hoyt and his colleagues had risen to the challenge economist Robert McFall had posed in 1935, to gather information to show that while studies suggested U.S. cities were growing, on closer inspection, the picture was more complex. As McFall had hypothesized — and as the National Resources Planning Board had confirmed in 1937 was the case for many U.S. cities — Chicago's boundaries had expanded as businesses and residents fled the urban center for outlying areas.⁵³ The concentric template thus might not fit all cities, but it continued to describe Chicago rather well.

As a comprehensive inventory of land use information, the *Chicago Land Use Survey* created a multivolume model of the city that repeatedly would be tapped for later reference. Preliminary reports enabled many public agencies and private organizations to use community area data even before the survey's official publication.[54] The survey helped to put a new public face on the city's plan commission as an information-gathering and information-processing agency. Yet, what this repository of urban information did not offer was a plan for intervention.

To determine appropriate corrective measures would require reducing the survey's massive amounts of data into a more manageable form. Plan commission research staff accomplished this task under Hoyt's direction in the *Master Plan of Residential Land Use of Chicago* (1943), which reframed the encyclopedic inventory to counsel local leaders on the short-term and long-term strategies they needed to undertake to address the problems of decentralization and deterioration. Breaking with local appraisers whose 1940 assessments for the Home Owners Loan Corporation suggested the city's future did not look bright — "To liken Chicago to a human being, the city is suffering from all the diseases characteristic of old age ... The future of Chicago is definitely not promising" — their work to prepare this document during World War II conveyed great optimism that postwar Chicago could be a model of scientific organization.[55]

American Cities at War and the Evolution of Conservation Policy

Despite some attention to the problems of U.S. cities, Franklin D. Roosevelt's New Deal never placed comprehensive urban improvement near the top of its agenda. Although the real estate community wielded some political power, city planners were frozen out of federal circles of influence in a climate of associations between national planning and fascist and socialist political regimes. With America's 1941 entry into military conflict came the opportunity for many professions to reconsider their place in the hierarchy of federal priorities. "The war has given a new intensity to thinking about the future of cities," noted National Resources Planning Board conservation unit head Charles Ascher, suggesting to planners that the ice was cracking as public concerns about the survivability of U.S. cities in an atomic age came to the forefront.[56] Although Burgess's idealized model now served as an eerie reminder that urban areas were enemy targets, interest in the problems that planners had identified as critical to the future of U.S. cities had little place in the national discussion about matters of urban security. In this context, urban professionals adopted two distinct approaches to addressing the implications of war for their work.

Some urban analysts and reformers found in the conflict immediate opportunities to redefine longstanding interests in new national security terms — from the academics called to Washington to advise the government on planning for suburban dispersal to the city officials who used the Lanham Act to garner funds for pet projects.[57] Arthur Goodwillie is one example. A leader of the Waverly experiment and later head of the Conservation Service at the Home Owners Loan Corporation, Goodwillie quickly recognized the potential power of aligning the real estate community's vision of neighborhood-based city improvement with the nation's most pressing concerns. In 1942, seeking to expand conservation to the southwestern portion of Washington, DC, to stop the continued decline of property values associated with an influx of racial minorities, he suggested to colleagues that defense workers (implicitly white) would not perform their jobs effectively if they did not have adequate places to live.[58] "Maximum war production — which includes the adequate supply of housing for war workers — is now one of the two paramount national objectives to which all other considerations must give way," he explained, and argued for the value of "the conservation of neighborhood assets for the production of wartime housing."[59] Only a few years earlier, this former "Regional Reconditioning Supervisor" had adjusted his rhetoric from "reconditioning" and "modernization" to argue for urban community "conservation" in the Waverly neighborhood. Now, conservation — redefined as a wartime housing issue — became a means to satisfying the nation's security needs. The recognition that "America has grown beyond the pioneer stage when we can build, destroy, and rebuild in a more or less mad scramble to produce new wealth," no longer a sufficient rallying cry on its own, was married to analyses of national defense: "Particularly in view of the economic costs of the war, we are going to have to make increasing use of existing community wealth."[60]

Such a shift mirrored tactics in the natural resources conservation movement as environmental historians, including Hal Rothman, have described.[61] When America entered World War II, President Roosevelt adapted a program earlier focused around rescuing the nation from its economic depression to providing for the conflict. Although comparatively few ecologists were drafted for service on studies related to wartime needs, human conservation initiatives such as the Civilian Conservation Corps were dismantled to recruit citizens for military participation and efforts to conserve other resources were reoriented around the "maximum war production" to which Goodwillie referred.[62] (Some cities had odder requests: in 1942, proclaiming Halloween "Conservation day this year and each year until after the war is over," Chicago mayor Edward Kelly placed an official moratorium on holiday "mischief," asking residents "to refrain from soaping windows and destroying property and to engage, instead, in conservation and salvage efforts to speed the winning of the war.")[63]

With security having outstripped scientific management as a national priority, conservation's promoters altered the rhetoric of their appeals. For those citizens not directly involved in the military campaign, action on conservation — once an emblem of America's grassroots democracy — became a symbol of patriotic contributions to national needs. As farmers heard messages such as "defense of the soil is an inseparable part of national defense," "conservation farming is a necessity in time of war," and "efficient use of land is of first importance under wartime demands for food, oil, and fiber," urban residents were asked to save scrap metals and seed victory gardens.[64] In this context, Goodwillie's updating of neighborhood conservation seemed like a compelling choice. The Washington Housing Authority threw its weight behind the Goodwillie Plan to conserve housing for war workers, a plan whose emphasis on "community wealth" took a different tack from the earlier Waverly experiment by combining the expert upgrade of physical property with the relocation of minority families so that whites could take their place.[65] Convincing a broad audience that his proposal for Washington, like the earlier undertakings in Waverly and Woodlawn, would be generalizable — "Preliminary surveys made in Boston, Bridgeport, Philadelphia, Baltimore and Norfolk disclose the existence of many substandard residential areas . . . that can be re-used for the production of modern, low cost housing, in such a way as to advance the war effort and, at the same time, assure the permanent improvement of the communities involved," he wrote — Goodwillie was awarded $500,000 from the Lanham Act to move ahead.[66]

Another stream of thinking about cities in the wake of the military conflict similarly recognized that national security was now the nation's top priority. Yet these urban analysts and reformers awaited the conflict's official end — from the many professional associations of architects and planners that spoke passionately of "postwar planning" to the home improvement industry that helped consumers dream of a better future.[67] Among them was Homer Hoyt. At work on the *Master Plan of Residential Land Use of Chicago* (1943), Hoyt and his staff at the Chicago Plan Commission's Research Department used the war to ruminate about "the development of Chicago during the first generation after the war — the period up to 1965."[68] Residential land use was at 60 percent, the most substantial allocation of territory within the city; therefore, the future of housing would be pivotal to the city's long-term fate. "In the course of the next generation or two," the authors anticipated, "a large amount of construction will take place." The "people of Chicago" had an important decision to make. Would "this building . . . be fragmentary and sporadic, repeating all of the errors of the past?" Or would it be "in accordance with a well-conceived, comprehensive plan for a better city"?[69] The choice was rather obvious it appeared. Framing the continuing departure of middle-class residents and in turn businesses from the city center as an urgent problem with no simple solution, the plan proposed a remedy with housing improvement at its core: A sequence of actions from government and citizens, including demolition and modernization, would be coordinated with the life cycle of neighborhood growth and decay. Like the Home Owners Loan Corporation staff who came to question whether urban decline was "as inevitable as is the rising of tomorrow's sun" and in turn put the agency's residential security surveys to new uses in Waverly and Woodlawn, Hoyt adapted his contributions to urban analysis toward a citywide program of urban reform.[70]

Taking Life Cycle Planning Citywide: The *Master Plan of Residential Land Use of Chicago*

The *Master Plan of Residential Land Use of Chicago* marked Hoyt's own intellectual growth and shift away from the view that urban life cycles moved inexorably in the direction of decline toward a broad vision of scientific management for cities. Setting aside one of the foundational assumptions guiding his prior risk rating work at the Federal Housing Administration, Hoyt went several steps beyond what his colleagues had undertaken in Waverly and Woodlawn and planned for Washington, DC, where the focus was alleviating "neighborhood deterioration... which invariably precedes and eventually produces the slum."[71] Hoyt's rhetoric of resource planning would be muted in this time of war. Yet his insistence on a multistage citywide program rather than a single-stage neighborhood-based attack sought closer resemblance to the nation's earlier land use planning efforts in the natural resources field. Taking a cue from Woodlawn project leader Robert Mitchell's observations (citing "a number of studies in urban sociology and land economics") that "you can't plan for the future of a neighborhood unless you know its probable place" in the future city pattern, "natural areas" became "planning areas" as the *Master Plan of Residential Land Use of Chicago* transformed ecological models of Chicago neighborhoods into a quantitative grading system to locate and order interventions throughout the city.[72] In an era when confidence ran high, as H. G. Wells put it on a visit to the United States, that "one of the most significant contributions of the land-management biologist is the prediction of consequences resulting from a knowledge of natural principles," its effort to work with the "laws" of the urban system makes clear how naturalistic approaches to urban analysis and reform continued to work their way into the ground-level planning activities of the city plan commission and allied agencies.[73]

Following some discussion of the city's natural and man-made resources was the plan's central argument: depending on the condition of a neighborhood and the type of treatment adopted, the life cycle might be slowed, halted, or even reversed. In this blueprint for two decades of residential development in the city, the staff synthesized *Chicago Land Use Survey* field observations and statistical data from planning commission research with the now-common notion of a predictable urban life cycle to delineate eight "types of planning areas." In keeping with the social scientists who stressed the importance of natural areas—and the resource planners who called to "lay out controls along natural rather than artificial lines," these areas bore no relation to official administrative boundaries.[74] Instead, they integrated "three qualitative elements capable of quantitative measurement"—age, condition, and rent—to project the life span of different neighborhoods and, by extension, the sequential order in which urban problems should be attacked.[75] In four, Blighted Areas, Near-Blighted Areas, Conservation Areas, and Stable Areas, 50 percent or more of land already was being used for residential building. Planners also named four categories where less than 50 percent of land use was residential but where development was expected: Arrested Development Areas, Progressive Development Areas, New Growth Areas, and Vacant Areas. These categories moved from worst to best in quality, and from shortest to longest in what Hoyt called "life expectancy."[76]

Preparing the *Master Plan of Residential Land Use of Chicago* was a team effort, which drew on the expertise of many of the University of Chicago affiliates who had worked with Hoyt on the *Chicago Land Use Survey*. Echoing the observations of natural resources planners such as Clarence Wiley, who characterized "good land" as "only a relative rather than an absolute concept," they argued that it was "necessary to measure and to map all the grades of housing from the best to the worst" before delimiting any land use categories.[77] Hoyt, who wrote most of the text, devised the statistical technique for determining the planning areas. Sociologist Donald Foley and geographer Robert Klove assisted with the definition and demarcation of blighted and near-blighted areas.[78] Geographer Elizabeth Broadbent drafted several chapters and tables and helped to measure the planning and type-of-structure areas. Geographer Harold Mayer worked with Broadbent and took field guide–style photographs for the report. Sociologist Gerald Breese mapped the vacant land areas, and sociologist Mae Schiffman Maizlish — who had prepared maps and charts for Hoyt's dissertation work — prepared the plan's maps and charts.[79]

The *Chicago Land Use Survey* had identified in the portions of the city devoted to residential usage a pattern of concentric zones; in their subsequent analysis, these collaborators discovered how the eight types of "planning areas" also appeared to move from the center toward the periphery of the city. The "blighted and near-blighted areas" were found "mostly in a wide belt encircling the central part of the city" to the north, south and west (east of the city lay Lake Michigan). The "conservation areas" formed "a belt outside of the blighted and near-blighted areas." Beyond these belts were the "stable areas." Finally, at the "fringe" of Chicago lay the several categories with limited residential development to date.[80] Of course, there were exceptions scattered throughout the city, for example, blighted areas overlapping the South Chicago steel mills and African American neighborhoods on the South Side; pockets of near-blight in parts of West Englewood, Hegewisch, and Gresham; and conservation areas in Norwood Park, Beverly Hills, and Pullman. Yet, Hoyt explained that despite some distinctions between his sector model and Burgess's concentric model of "the city," his own thinking about a plan for Chicago had been significantly influenced by Burgess's work.[81] What Burgess had described as the Zone in Transition, Zone of Independent Workingmen's Homes, and Zone of Better Residences, said Hoyt, became the "blighted and near-blighted," "conservation," and "stable and other" areas in the *Master Plan of Residential Land Use of Chicago*, reformulating predictive academic theories into prescriptive action plans and, in so doing, invoking the land use capability maps of the nation's ongoing efforts at natural resources conservation.[82] Moving beyond the earlier ecological mapping efforts of staff at the Home Owners Loan Corporation, as well as his own work on predictive risk mapping while he was at the Federal Housing Administration, now his diagnosis of neighborhoods' life stage specified treatment as well. "By classifying each portion of the city in accordance with the nature and urgency of the problem that it suggests," plan commission executive director H. Evert Kincaid explained, "a time sequence for planning in relation to the needs of each area is set forth."[83]

Urban Resource Management for the Postwar Period

What precisely was the course of action that Hoyt proposed? If, as he had suggested in *The Structure and Growth of Residential Neighborhoods in American Cities*, the "apparent riddle of the internal nature of American cities" had now been solved, and if, as many urban professionals currently believed, the course of urban blight might be interrupted, an orderly sequence of clearance and rebuilding (slum removal) together with conservation and improvement (slum prevention) would be the obvious line of attack.[84] "By a selective process," he explained, "the sub-standard and economically unfit structures should be weeded out."[85] Clearance, administered by government authorities, would demolish the city's most blighted neighborhoods and relocate these populations to better living conditions. Conservation, organized by homeowners associations, would invest communities in the repair process to slow neighboring areas' decay. With the plan commission fearing the outcomes should "nature take its course," both strategies, applied sequentially in neighborhoods throughout Chicago, would improve the city by planning with its "natural" tendencies in mind.[86]

The 1943 document did not mark the first call to combine the removal of obsolete structures with the repair of structures with a remaining "useful life." The Illinois Emergency Relief Commission, for example, had pursued both strategies in the early 1930s.[87] Nor was it the first mention of clearance or conservation in urban programs in the United States. In Chicago and other cities, slum clearance had been a precursor to the construction of New Deal public housing—the only legally acceptable application of eminent domain for federally organized housing planning to date.[88] Already, conservation was under way in the city; even before Hoyt's arrival in Chicago, the preliminary release of data in connection with the *Chicago Land Use Survey* had referenced its potential applications to "programs of conservation and rehabilitation," likely a nod to Woodlawn's continuing work.[89]

Hoyt's original contribution was to link these strategies to a long-term citywide improvement plan rooted in the era's belief that there were "natural principles of land use," to use the title of one popular book.[90] His vision of the city as an "organic entity" and calls for a multistage intervention focused on the life cycle of urban lands followed Frederic Clements's work. For ecology to be central "to the processes of recovery," Clements had written together with Ralph Chaney, "the organization of farm and ranch must be reshaped into an organic entity with all the parts present and coordinated to bring about optimum results."[91] The resource planners who agreed with Clements that the nation's agricultural depression resulted from unscientific methods of land use had set out to encourage farmers to classify their properties toward retiring some lands, letting others lie fallow, and selecting new crops for still other areas. Restoring nature's balance in cities demanded a similar set of classifications. With its classification system alternatively characterized as having eight categories or four, and with the types of planning areas requiring actions from citizens and government, Hoyt's vision for Chicago approached a compromise between the limited property improvement techniques espoused in the Waverly and Woodlawn demonstrations and the broader hopes for comprehensive planning and human rehabilitation laid out by planners on the National Resources Planning Board.[92] Thus, as the plan's terminology built on that of the real estate community, with conservation following the narrow definition of physical improvements established in Waverly and Woodlawn, it did so as part of an ambitious program of scientific management for urban physical resources in which local authorities had some oversight — closer to the goal for which the planning community had campaigned. The lack of attention to poverty reduction through human conservation was consistent with its declining priority status at the national level.

In addition to its innovation in quantitative analysis for housing and neighborhood rating, three aspects of the *Master Plan of Residential Land Use of Chicago*'s "scientific approach" commanded significant attention. First was its insistence on citywide planning.[93] Recognizing that "a new residence placed in the midst of a slum takes on, chameleon-like, the character of its environment and loses upon completion half or more of its cost because it is surrounded by slums" and that the "very popular practice now to single out the blighted areas for special treatment and to speak of them as if they could be isolated from the rest of the city and the entire study of urban problems be concentrated upon them alone" would simply displace rather than remedy problems, Hoyt diverged from colleagues in the real estate community who saw the value of ecological models applied at the building or neighborhood level.[94] Although the *Master Plan of Residential Land Use of Chicago*, like other plan commission visions for Chicago's future saw value in the neighborhood unit organization, the authors made clear that the challenge of urban decay was not one that any individual homeowner or even community association could fight alone.[95] "Fragmentary and sporadic" planning would repeat "all of the errors of the past."[96] Instead, these problems required a large-scale approach attuned to the dynamics of individual buildings within the context of neighborhoods and each neighborhood within the context of the entire city. Regardless of whether public agencies or private developers were changing the face of the city they needed to keep the logical planning sequence in mind. "Planning for this time sequence of development is one of the surest ways of preventing the city from being engulfed in the blight which creeps forward with haphazard growth," Hoyt explained. "All neighborhoods of the city must move forward together, growing and developing in harmony to provide the maximum benefit for its citizens."[97]

A second point was the need for continuous planning. As human ecologists such as Nels Anderson observed of the slum, this "natural resultant of the forces of rapid city growth" would "continue despite the mitigating effects of housing reform, zoning laws, and social welfare agencies."[98] Some architects and planners had communicated a similar message in their calls for "time zoning" — the demolition of buildings based on structures' anticipated life.[99] The *Master Plan of Residential Land Use of Chicago* acknowledged the city's work to stem the tide of blight would need to be ongoing, like the nation's programmatic natural resources planning efforts designed around dynamic ecosystems. Following the observation, in the land use survey, that "neighborhoods are never static; they move from one stage to another in one of several possible *life cycles*," that the stable areas of today would be the conservation areas of tomorrow, and that even the city's healthiest areas, its "reservoirs" of good housing, eventually would require treatment, the plan insisted that long-term success required city improvement efforts without end.[100] In proposing these interventions, Hoyt did not promise to stop nature completely from taking its course. Rather, by insisting on a continuous program that anticipated the direction of neighborhood change, he offered instead a plan that would create a scientifically organized Chicago that, in his words, "will be constantly renewing itself, weeding out the unfit and obsolete structures," and in turn the economic and social problems that accompanied them.[101] Building on earlier work with Arthur Weimer in which he had modified Burgess's zonal map to depict a "City of the Future" where the "slum area," having been improved, was now the "former slum area," Hoyt outlined Future Planning Areas for Chicago. This vision of the city as a renewable resource anticipated a "second cycle of rebuilding in Chicago" that likely would "start around 1965."[102]

A third point of attention was the importance of community participation. As essential to the success of the master plan as the contributions from technical experts such as city planners, building and health inspectors, and police, Hoyt outlined, would be the work of city residents — especially in conservation areas. Whereas urban professionals might be drafted to tackle the city's most significant problems in its blighted areas,

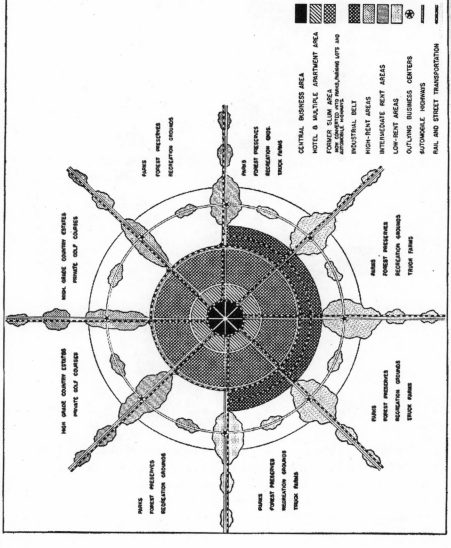

A City of the Future. From Arthur Weimer and Homer Hoyt, *Principles of Urban Real Estate* (New York: Ronald Press, 1939). Courtesy John Wiley & Sons Inc.

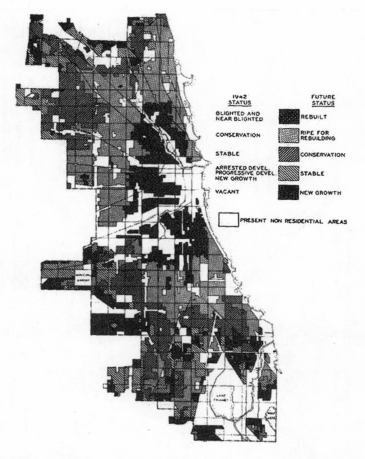

Future Residential Planning Areas of Chicago, *Master Plan of Residential Land Use of Chicago*. From John B. Blandford Jr., *Housing Charts Accompanying Testimony of John B. Blandford, Jr., Administrator of the National Housing Agency before the Senate Banking and Currency Committtee*, November 27, 1945 (Washington, DC: National Housing Agency, 1945).

"urgent measures" would be "necessary to prevent further deterioration" in the near-blighted and conservation areas, measures that included building repairs, trash removal, alley clearing, painting, masonry repair, landscaping, and sidewalk improvement. The Federal Housing Administration and Home Owners Loan Corporation had confirmed in their small-scale experiments how collective action maintained neighborhood values far better than individual efforts alone—that, at least in the short term, "vigilant groups of home owners can themselves halt and reverse the process of community decline."[103] Bearing in mind the design of natural resources conservation efforts and the Waverly and Woodlawn projects, developed by experts but implemented through the work of educated lay volunteers, Hoyt's vision of community-based collaborations extended to near-blighted and conservation areas across an entire city. "Neighborhood conservation can, indeed, be accomplished by democratic processes," he insisted, calling for "a wide-awake and enlightened citizenry" and encouraging churches, schools, and other local institutions to serve "as clearinghouses for information on improved home operation and maintenance, as media for cooperative action, and as the rallying points for greater social consciousness in each neighborhood."[104] The block groups now mobilized for action on civil defense, Hoyt proposed, might take on responsibility for conservation in the postwar period.[105]

Hoyt's influences are particularly apparent in the new direction that the Woodlawn conservation project began to take as the *Master Plan of Residential Land Use of Chicago* was prepared. Although Chicago Plan Commission chief engineer Hugh Young had participated on the original Woodlawn study's advisory board, that experiment largely had been directed by the real estate community.[106] With "the real estate people and urban planners" traditionally "at odds with each other" in Donald Foley's eyes, the plan commission had little say in how this experiment unfolded.[107] "Hoyt managed to wield disproportionate influence" among both professional communities, however, such that by 1943 the Chicago Plan Commission had assumed control of this "test case of rejuvenation by modern cooperative city planning methods."[108] Research staff set out to develop a plan for the neighborhood that embraced the values of citywide planning, continuous planning, and community participation central to Hoyt's vision for the future Chicago. Reexamining the prospects for Woodlawn in this new context, the plan commission laid out a program of property improvements that integrated some public amenities with the private investments of homeowners that had been a focus of the original demonstration. Costs were substantially higher, more than $4 million compared with $70,000, but planners argued, "To Woodlawn and the city, conservation becomes not a luxury but an investment in survival."[109] What the revised approach to conservation in this neighborhood shared with the earlier experiment was the sense that its results would be generalizable. The Woodlawn Plan would not only "rehabilitate Woodlawn," but it would also "encourage other communities" to undertake "similar efforts to save their areas from depreciation."[110] "Many of the suggestions," the *Chicago Daily Tribune* would report after the neighborhood conservation plan's completion, "are expected to be adopted by other civic groups" — but only "within the framework of the Comprehensive City Plan."[111]

Delay on Implementation

Hoyt and his colleagues provided an explicit and implicit naturalistic lens to frame the redevelopment task ahead. Physical and social surveys, analytic reports, and plans produced at the Chicago Plan Commission reveal the central influence of ecological and natural resources planning concepts in framing the understanding of interactions among forces in the urban community. Later reports and plans from the Chicago Plan Commission and allied public and private agencies (for example, the Chicago Housing Authority and the Planning Committee of Chicago's Metropolitan Housing Council) followed this lead in their vision for the city, despite Hoyt's 1943 departure to New York to work as director of economic studies on the area's regional plan.[112] (He was replaced by Harold Mayer.)

Yet as historians have documented, even plans with broad support are not always implemented. Publishing in the midst of a national security crisis and requiring new legal powers for their plan's execution, planning commission research staff anticipated the *Master Plan of Residential Land Use of Chicago* would not be relevant until the postwar period. By 1945, the City Council had approved the 1943 document, prioritizing the need to redevelop the city's blighted areas, but federal and local policies would impede the realization of this vision for longer than expected.[113] With wartime concerns taking priority, the first half of the 1940s saw little action on eminent domain for land clearance — even as the city planning and real estate communities set aside their differences in a push to expand the accepted definition of "public purpose" to include takings of urban as well as rural land.[114] On Chicago's local stage there was some progress; the Illinois Neighborhood Redevelopment Corporation Law, passed in 1941, helped to create private redevelopment organizations tied to specific neighborhoods. According to this legislation, any organization that could purchase 60 percent of the land in a proposed redevelopment area would be granted powers of eminent domain to acquire the rest — provided that "the predominant racial group in any development area must not be displaced."[115] Concerns about the constitutionality of such actions, however, meant further delay on redevelopment such that by the end of World War II (indeed, as late as the early 1950s), the law had not yet been used.[116]

Support for urban conservation fared no better. At first, the recruitment of former Home Owners Loan Corporation staff member Robert Mitchell to head the National Resources Planning Board urbanism group during World War II signaled a growing détente between the planning and real estate professions. "Desiring the greatest possible public benefit to accrue from its Chicago and Baltimore studies," Mitchell wrote to Charles Ascher in 1941, "the Federal Home Loan Bank Board is prepared to transfer the important products of its neighborhood conservation program, techniques and organization," to the National Resources Planning Board, because "no institution, local or national is as well equipped" to lead future activities in this area.[117] As urbanism staff discussions continued about the value of an "Analogy for Urban Development Based on Dept. of Ag. Org.," Mitchell supervised the start-up of several "Urban Progressive Planning Projects," intended to showcase possibilities for redevelopment in the postwar period.[118] Yet, the board concluded that other government agencies had captured the market for defense and postwar planning, and disbanded in 1943.[119] At the Home Owners Loan Corporation, little action followed the opening of the new conservation service and federal officials soon shut it down.[120] With government agencies focused on military priorities, even the Waverly and Woodlawn experiments flagged. In Waverly, for example, where early proclamations by Federal Home Loan Bank Board head John Fahey had reported on the "energetic and sympathetic local leadership and unified neighborhood support," only four years later, "it was found that the Conservation League never had become an organized and operating body and that it had apparently languished after federal sponsorship and support was withdrawn" — results exacerbated because "the improvements requiring municipal attention were not accomplished."[121] Some evidence suggests other cities, including Cleveland, flirted with programs of conservation ("Neighborhood conservation is your problem as well as the municipal government's —for we live in a democracy, which is based on citizen cooperation" explained that city's conservation handbook). Yet, wartime concerns, together with the decline of support for the national conservation effort, stopped them soon after they began.[122] "Many such attempts have been made," wrote National Housing Agency staff member Ruth Berman in the *Federal Home Loan Bank Review*, "not one of which ... has succeeded in achieving more than a limited portion of its goal."[123]

Obstacles to legislative change were not obstacles to neighborhood change, and as the decade went on, the processes of decentralization, central city decay, and racial migration that observers from social science, city planning, and real estate had identified in the 1920s and 1930s continued in full force.[124] In Chicago, for example, between 1930 and 1940, a "net exodus of 143,458 white persons" alongside a gain of 43,828 "in the Negro population" had begun to alter the face of the city; from 1940 to 1950, the African American population grew from 277,731 to approximately 400,000.[125] With the search for war jobs accelerating African American migration to Chicago in the first half of the decade, the plan commission estimated in 1945 that 300,000 people were living in an area that should hold only 225,000—a state of affairs resembling that in many other cities.[126] In light of these trends, together with slowed immigration, earlier concerns about the dynamics of racial and nationality groups ceded to a focus on the impact of nonwhite populations on the future of U.S. cities.[127]

As observers watched these developments, and became increasingly aware of the implications of the growing housing shortage for keeping the existing supply in adequate condition, many grew anxious that a national program of urban resource management had not yet become a legal reality. "We have come to the end of the type of physical development that faced the founders of our country ... the conquest of the frontier," MIT architecture school dean Walter MacCornack explained. Although "our natural resources have been skillfully developed ... We still face a frontier, one that will call upon the very qualities of pioneering and individual effort which were essential in that earlier frontier struggle."[128] This "new frontier in America" was cities, and it "must be mastered if our way of life is to be assured."[129] Planning must "become a part of ecology—a sort of artificial ecology," wrote North Carolina planner Miriam Kligman, "rather than an action program that is fairly independent of a descriptive field of study."[130]

In the postwar period, a community of urban professionals would seek to reopen the national discussion about what the National Resources Planning Board urban conservation unit had termed "social controls of the use of urban land."[131] Their base of operations would be Chicago, home to America's planning, housing, and real estate organizations, and a continued center for urban social science research. How this community successfully fashioned state laws in the image of the natural laws of urban growth, decay, and change to move ahead on a program of scientific management for Illinois cities is the subject of chapter 4.

FOUR

From Natural Law to State Law

"The Illinois Act (Ill. Rev. Stat. Chapt. 5, 106–138) was approved in 1937," wrote Harvard planning professor Reginald Isaacs and Chicago lawyer Jack Siegel of the state's Soil Conservation Act in 1953, "when emphasis by the federal government on soil conservation was at its peak."[1] In an era in which "the state legislature was convinced" that "the evil" of soil erosion "justified the granting of drastic powers," Illinois lawmakers — like their colleagues in other states — insisted on securing new protections for natural resources.[2] A close study of Chicago suggested the time had come to mobilize for urban resource management on a similar scale. "The problem facing our cities today is no less drastic and requires just as drastic an action program," Isaacs and Siegel insisted.[3] State legislators soon expressed their agreement in an Urban Community Conservation Act, signaling the growing importance of the city problem-solving cause.

Why, in 1953, did more than a decade of urgings to provide legislative backing for the scientific management of urban resources finally bear fruit in Illinois? A community of persistent idea brokers who moved between the ivory tower and local government in the immediate postwar period, and in turn mobilized interest from the private sector and citizen organizations, played a pivotal role in the implementation of two state laws that aligned the possibilities for city redevelopment with ecological understandings of patterns of urban change. The 1947 Blighted Areas Redevelopment Act, which certified blighted areas for demolition, and the 1953 Urban Community Conservation Act, which certified conservation areas for rehabilitation, paved the way for a program of citywide planning, continuous planning, and citizen participation in Chicago bearing close resemblance to Hoyt's 1943 plan.

Most accounts of the history of human ecology as a dominant mode of analysis in urban studies end in the 1930s, the height of criticisms of this tradition, with Robert Park's departure from the University of Chicago and Ernest Burgess's turn to other research topics.[4] A few scholars have noted the continued vitality of the life cycle perspective in urban policies of later periods.[5] In fact, new opportunities for academics to blend scientific urban theories and experimental city applications helped to sustain the view of cities as "ecological" communities and national "resources" through the 1940s and into the 1950s at the University of Chicago and beyond. Bolstered by increasingly broad support, Chicago's postwar experiments to slow, halt, and reverse the urban life cycle lay the groundwork for federal urban renewal policy.

Truman Steps In

This book has described President Franklin Roosevelt's role in changing the tenor of conversations about conservation during World War II. The "wise use" of public and private resources viewed during economic depression as essential to America's long-term prosperity was refocused to address concerns about the national defense.[6] Sarah Phillips has documented the competing visions of conservation during this period — for example, among scientifically oriented land use planners versus business-minded farm owners — and how in its focus on maximum production the wartime period gave priority to the latter point of view.[7] Following Roosevelt's death in office, Vice President Harry Truman sustained his predecessor's stance on national resources planning. As Truman served out the remainder of Roosevelt's term and a subsequent four years, with the Korean conflict and the Cold War front and center, his conservation policy emphasized economic development with a national security cast. "No part of our conservation program can be slighted if we want to make full use of our resources and have full protection against future emergencies," the president explained.[8] Even as plant and animal scientists expressed growing concerns about human impact on the natural world, arguments for the immediate or delayed use of specific resources became framed as critical military issues—a subject of some anxiety late in Truman's second term when the Paley report suggested that America might suffer from future shortages of key materials.[9]

Despite such resonances with Roosevelt's rhetorical and policy precedents on matters of natural resource management, however, Truman differed with his well-respected predecessor in a few areas. Among them was urban housing and city improvement. Truman had a keen personal appreciation of the need to repair race relations.[10] Early in his presidency, Truman agitated for a federal housing act as part of the package of programs termed the Fair Deal. Yet, if the new president was ready to make postwar planning for cities a national priority, the larger body politic, still more attuned to questions of planning for natural resources and national security, was not yet committed to such an approach.[11] Truman's housing policies, introduced into Congress several times, also were stymied by a real estate lobby whose support for expanding eminent domain within cities and directing new funds to urban revitalization did not encompass the public housing provisions of the proposed legislation, charging that they would bring socialistic policies to the United States.[12] In light of the continued power of conservation interests, as well as a desire to steel his legislation against such unflattering charges, Truman and his colleagues soon revised their proposal to encompass farm housing and to emphasize the role of private industry in the U.S. housing market. Simultaneously, they trumpeted the links between these proposed domestic policies and the nation's security aims.[13]

Compared with the prewar period, by the late 1940s, when it came to the national conversation about the possibilities for a program of city improvement, Americans witnessed a deliberate shift at the highest levels. It was in this context that academic social scientists, city planners, and the real estate industry, waiting patiently to initiate postwar planning, restarted their discussions about eminent domain and urban redevelopment. Many of the individuals who had found in the American city of the 1920s a predictable life cycle and who had set out to demonstrate in the 1930s the possibility of its interruption enjoyed the opportunity to bring the nation's housing laws up to speed with the tradition of scientific research on cities that recognized similarities between patterns of life in urban environments and patterns of life in the plant and animal world.

The Illinois Blighted Areas Redevelopment Act

This book has described Homer Hoyt's plan for postwar Chicago and the delay on its implementation. While a diversity of advocates backed the idea of an urban program that paired clearance and conservation, with some supporters objecting to specific proposals—not to mention an array of other opponents—legal obstacles persisted. The concerns about the changing face of Chicago that led Mayor Martin Kennelly, as one of his "first official acts," to appoint a Housing Action Committee "to study the problems and make recommendations," together with a recognition that similar problems were becoming apparent in other Illinois cities, finally spurred the state legislature to take action in 1947.[14] Anticipating the passage of the federal housing policy that so far had languished, the Blighted Areas Redevelopment Act gave to Chicago and other cities new powers of eminent domain in the face of areas deemed "blighted" or "slum."[15] Following from the view that the best way to improve the state's worst slums would be to demolish them and start over, and with the American real estate industry—headquartered in Chicago—seeking to deemphasize public housing, the act provided for the creation of new public bodies to supervise the acquisition of land and its redevelopment by private developers in Illinois cities: land clearance commissions. Chicago's Land Clearance Commission began its operations that year.[16]

Historians have emphasized the role of Chicago business and real estate interests concerned about deterioration around the city's central business district in the law's passage and subsequent selection of areas for improvement.[17] The Chicago Land Clearance Commission, for example, largely comprised a rotating collection of prominent members, including Herman Walther, a former Home Owners Loan Corporation Chicago district manager, Society of Residential Appraisers president, and Woodlawn experiment participant, as well as a real estate instructor at Northwestern University. Many members of this commission and of smaller neighborhood planning boards represented local business and real estate interests. Yet, it is essential to recognize how, in their discussions about the next steps in Chicago development, these figures' concerns about interrupting the urban life cycle to maintain property values, city planners' eagerness to organize a scientific program for comprehensive urban improvement, and more academic visions of the future ecology of the city continued to be aligned. This was owed in part to the lasting legacy of Homer Hoyt and to new opportunities to link the work of university social scientists to Chicago's redevelopment work.

Despite some longstanding tensions among varied communities of urban professionals, leadership from business in passing the 1947 law does not mean that other parties' visions of the value of a naturalistic framework for redevelopment suddenly disappeared. Although the new Illinois law addressed only a single category in the urban life cycle, its implementation enabled the Chicago Plan Commission research staff and these researchers' academic collaborators to have their opinions heard about next steps. With the 1947 act defining blighted and slum areas as urban territories sized between 2 and 160 acres where conditions were "detrimental to the public safety, health, morals, or welfare"—less technically precise terms than the 1943 plan—a key opportunity was adjudicating how the new legislation applied to the local situation.[18] Nearly a decade had elapsed since the land use survey on which the *Master Plan of Residential Land Use of Chicago* was based had been completed, and in this time the real estate community had become increasingly cognizant of the value of citywide planning. With this backdrop, one of the first acts of the Chicago Land Clearance Commission was to invite plan commission staff to offer their scientific reassessment of the most blighted areas in the city.

Ten Square Miles of Chicago (1948), the staff's report of its work, combined qualitative and quantitative analysis to revisit the findings of the *Chicago Land Use Survey* and the *Master Plan of Residential Land Use of Chicago*.[19] Like the neighborhood profiles appearing in Chicago's sociologists' *Local Community Fact Books*, the area descriptions and numerical appraisals of the HOLC's city surveys, and the neighborhood photographs and community life histories of plan commission reports such as *Forty Four Cities in the City of Chicago* (1942), in this document, each square mile became the subject of a "square mile narrative," and a score to determine its respective ripeness for redevelopment.[20] The analytic tools used—field studies, surveys, and statistical maps—were not much different from those of five years before, and the conclusions were essentially the same: merely a single square mile area on the city's South Side, bordered by Fifty-fifth Street, Sixty-third Street, Ashland Avenue, and Halsted Street, was added to the extreme blight category.

Site designation reports from the land clearance commission's research staff followed. Explicitly characterized as follow-ons to the *Master Plan of Residential Land Use of Chicago*, these reports used the maps and blight designations made by the Chicago Plan Commission under Hoyt as the basis for their work—even though his findings were ten years old.[21] Hoyt underscored the enduring relevance of his earlier analyses when the land clearance commission invited him to consult on several site designations.[22] Working on the Regional Survey of New York, and teaching at Columbia and MIT since his departure from the plan commission, Hoyt had maintained close ties to Chicago's planning and real estate communities. From 1948, for example, he began to coauthor issues of the *Savings and Homeownership* newsletter with Morton Bodfish, who was the Chicago Plan Commission vice chair, Chicago First Federal Savings and Loan Association president, and U.S. Savings and Loan League chief executive. Together, they lobbied for continued relevance of the principles laid out in his master plan for the city.[23]

As these projects moved ahead, Congress finally approved Truman's housing proposal. Cities gained federal support to condemn property and to relocate populations for the newly recognized "public purpose" of slum removal for private housing construction. Chiefly administered by the U.S. Housing and Home Finance Agency (the successor to the National Housing Agency), the Housing Act of 1949 pledged federal contributions to two-thirds of redevelopment costs if local governments would pay one-third.[24]

The first large area in Chicago to begin transformation under land clearance commission authority was on the "Near South Side" of the city, where major institutions included the Michael Reese Hospital (Chicago's largest private hospital) and the Illinois Institute of Technology (whose president, Henry Heald, was the first chair of the Chicago Land Clearance Commission). The New York Life Insurance Company signed on as developer. As the redevelopment process went on, area institutions came to express the conviction that the project was achieving scientific planning for the city. "A pattern for the automatic redevelopment of the South Side Blighted Area into a stable community" had been identified, declared the South Side Planning Board, invoking Hoyt's earlier vision of a city constantly renewing itself, "a pattern which can be used anywhere."[25]

The Second Chicago School

Homer Hoyt's work did not provide the only intellectual support as the land clearance commission, plan commission, and collaborating public and private institutions set out to consider next steps for Chicago. Social scientists and city planners at his alma mater also played a leading role. In fact, the 1947 passage of the Blighted Areas Redevelopment Act coincided with two institutional developments at the University of Chicago that would help to keep the naturalistic vision of cities alive: the founding of the Chicago Community Inventory and the Program for Education and Research in Planning.[26]

Gary Fine has written of the "second Chicago school" of postwar sociology when luminaries such as Howard Becker, Joseph Gusfield, Herbert Blumer, David Riesman, and Erving Goffman joined the department faculty.[27] The 1947 openings of these two units marked the arrival of another important "Chicago School." Scholars, including Edward Banfield, Donald Bogue, Ernest Burgess, Otis Dudley Duncan, Philip Hauser, Harold Mayer, Martin Meyerson, Harvey Perloff, Rexford Tugwell, Louis Wirth, and their students, all committed to the belief that there were "mutual advantages" to the "close cooperation between university research workers and planning and administrative personnel," worked to transform academic understandings of the nature of urban structure, behavior, and evolution into an applied science for city improvement.[28] As the city of Chicago geared up to implement its new legal powers, these homes for scientific urban analysis provided research support and ultimately pressured the Illinois legislature to make conservation as well as clearance part of the arsenal of standard urban improvement techniques.

Human Ecology in the Postwar Era: The Chicago Community Inventory

The Chicago Community Inventory was the follow-on to the University of Chicago's Local Community Research Committee and Social Science Research Committee, which historians have cited as pioneering units for multidisciplinary research.[29] The Inventory was created by Ernest Burgess and Louis Wirth with a grant from Chicago's Wieboldt Foundation and inherited their orientation to the local environment.[30] Wirth signed on as the new unit's first director, followed by Philip Hauser, a former student whose own work on the ecology of urban communities had brought him into contact with city planners in the 1940s.[31] The center's research staff comprised mostly graduate students in sociology, one of whom — Gerald Breese — already had worked for the plan commission under Homer Hoyt.[32]

Competing models of "the city" from Homer Hoyt in the 1930s and Chauncy Harris and Edward Ullman in the 1940s had not displaced the primacy of Burgess's work. As E. Gordon Ericksen explained in 1949, the concentric ideal continued to function "as a frame of reference to ecologists much like the perfect vacuum and perfect circle function in the fields of physics and mathematics."[33] Burgess had little interest in a static urban research tradition, however. Investigations organized by the Chicago Community Inventory such as *Residential Rental Value as a Factor in the Ecological Organization of the City* and *Ecological Aspects of the Labor Force in the Chicago Metropolitan Area* were not merely efforts to replicate prior studies for the postwar era.[34] Rather, while they acknowledged their debt to an earlier generation of scholars, and continued to focus on topics such as racial interactions, population distributions, neighborhood change, and the urban life cycle, faculty and students aimed to retain the university's cutting-edge reputation by extending previous researchers' qualitative and quantitative work. Their innovations were both conceptual and methodological. In *Chicago's Negro Population: Characteristics and Trends*, for example, a report prepared by Otis Dudley Duncan and Beverly Davis Duncan in connection with a graduate sociology seminar on "Research Methods in Human Ecology," the authors refined Burgess's four-stage model of succession by laying out three different categories for the process: invasion, consolidation (early and late), and piling up.[35] Reflecting the growing use of statistics at mid-century in plant, animal, and human ecology, multiple Chicago scholars and their students — Otis Dudley Duncan, Philip Hauser, Donald Bogue, Beverly Davis Duncan, and Richard Redick, among them — extended the social indices whose significance Ernest Burgess long ago had established to improve quantitative analyses of segregation and population distribution.[36] Most notable were their contributions to indices of dissimilarity and segregation, which measured the separation of populations in specific geographic areas.

With the power of ecology remaining strong, such refinements were widely found as scholars at the University of Chicago and beyond, discovering exceptions to standard patterns in urban theory, offered new ecological interpretations.[37] The flexibility of ecological approaches to urban study that accommodated these intellectual developments embraced political changes as well. With the burgeoning support for the rights of racial minorities evident in the Supreme Court's 1948 *Shelley v. Kramer* ruling that restrictive covenants were illegal came revised interpretations of what exactly constituted "natural" urban processes, even when ecological ideas remained at the foundation of the argument. In contrast to Robert Park, who earlier had cited Eugenius Warming on plant communities' separation by species to emphasize the naturalness of the "social selection and segregation" of populations by nationality and race, scholars now used ecological rationales to argue that segregated neighborhoods — restrictive covenants allegedly covered 80 percent of Chicago as of 1947 — had disrupted "the city's 'normal' pattern of growth."[38] Such intellectual continuities were most apparent in sociology. Yet reflecting its status as a subject for cross-disciplinary inquiry — "Human ecology is not an exclusive branch of sociology, but rather a perspective, a method, and a body of knowledge essential for the scientific study of social life . . . a general discipline basic to all the social sciences," Ericksen observed — the field's influences persisted in geography and in economics as well.[39]

Ecologically oriented urban studies after 1940 had few of the reflections on analogical thinking that had preoccupied an earlier generation.[40] Even as a few prominent plant and animal scientists renewed their calls for a human ecology to attend to "man's role in changing the face of the earth," social scientists' direct encounters with plant and animal specialists were rare.[41] Among Chicago faculty and students such intellectual traffic, while not common, did occasionally occur. For example, when zoology professor Warder Allee organized a 1941 conference on "Levels of Integration in Biological and Social Systems," he invited sociologist Robert Park to speak.[42] Following the paper presentations, the sociology department's Society for Social Research hosted a roundtable about the conference papers, with the discussion jointly led by zoologist Alfred Emerson and sociologist Everett Hughes.[43] In the early 1950s, Burgess convened a meeting of natural and social scientists at the American Association for the Advancement of Science to discuss the challenge of urban research and its applications—a gathering that included his geography department colleague Harold Mayer, together with ecologist Charles Adams, zoologist Frances Evans, and several others.[44] Otis Dudley Duncan sought legitimation from the scientific community by participating in the Cold Spring Harbor Symposium on Quantitative Biology.[45] And the recommendation that Chicago students in the social sciences interested in urban and population studies enroll in Zoology 304, Thomas Park's course on Animal Ecology, alongside classes in Human Ecology with Otis Dudley Duncan, Urban Geography with Harold Mayer, and Urbanism with Donald Bogue, persisted through at least 1958.[46]

More appealing than building bridges to plant and animal science, though, was engaging with urban professionals. As their careers went on, Burgess and Wirth expressed growing interest in how urban studies on the model of the natural sciences might inspire a new kind of scientific planning for U.S. cities. Burgess had participated on the Chicago Housing Commission in 1926; at the 1934 meeting of the American Sociological Society, his presidential address had examined the possibilities for planning during the New Deal.[47] Wirth, invited to consult for city, state, and federal planning commissions, including the National Resources Planning Board's Research Committee on Urbanism, the *Chicago Land Use Survey*, the Illinois Postwar Planning Commission, and the Planning Committee of the Chicago Metropolitan Housing Council, had even closer ties to communities of practitioners.[48] If the urbanism committee's original vision of a federal department for cities on the model of the agriculture department could not be achieved, Wirth had told the National Resources Planning Board in 1941, perhaps a new agency could be created in the U.S. Department of the Interior: "The Department of the Interior comes perhaps closest of any of the government departments to having the greatest experience with the problem of reclamation," Wirth explained, "which is essentially the problem of our cities."[49] The work at the Chicago Community Inventory reflects its founders' escalating confidence in the possibilities of human ecology as an explicitly applied urban science, an important shift from the field's earliest years. Faculty and student affiliates developed close relationships between their academic work and city agencies' perceived needs, and the Chicago Plan Commission was their chief customer.[50] Whether it was local officials who sought the counsel of these researchers or efforts by university researchers to locate funding and data sources for their ongoing work, from the late 1940s and continuing through the 1950s, the unit pursued close studies of Chicago expressly for application—a "cooperative arrangement" that was "designed, on the one hand, to stimulate research on important local problems and, on the other, to provide city agencies with a sound factual basis for policy formulation and administration."[51] That the Chicago Community Inventory's studies of population and housing distribution became essential reading at redevelopment units, including Chicago Plan Commission, Chicago Housing Authority, and Office of the Housing and Redevelopment Coordinator is clear. In the language used to explain concentric urban development and waves of out-migration and in the sources cited for further reading, these agencies and their staffs came to follow—sometimes verbatim—these scholars' intellectual lead.[52]

National Resources Planning in the Postwar Era: The Program of Education and Research in Planning

The Chicago Community Inventory was not the only campus-based unit whose work would encourage local officials to plan in line with the natural laws of the urban system "in lieu of letting nature take its course."[53] The university's short-lived Program of Education and Research in Planning (1947–1956) played a complementary role. With faculty, including Charles Merriam, Charles Colby, Gilbert White, William Ogburn, and Louis Wirth, having participated in government-organized efforts such as the National Resources Planning Board, the university had offered a range of courses related to planning even before the creation of its planning degree.[54] Situated in the university's Social Science Division, the program assisted in the planning discipline's continued transformation from an offshoot of engineering and architecture to a branch of the social sciences and from a practically oriented trade to an integrated discipline combining theoretical knowledge with hands-on practical work.[55] The program was initially composed exclusively of social scientists, including Rexford Tugwell, Louis Wirth, Edward Banfield, and Harvey Perloff, with the intent to outsource technical instruction to the Illinois Institute of Technology. This proved difficult to arrange, and a few faculty with planning experience, including Walter Blucher and William Ludlow, were later hired for the task.[56]

In the early 1940s, the National Resources Planning Board had laid out a vision for training future city builders in which "architecture, law, engineering, economics, physiology, government, biology, sociology" were involved.[57] In the wake of the board's disbanding, the University of Chicago's program kept its multidisciplinary ideal alive. Other institutions would show greater commitment to a view of planning as science, for example, the University of Michigan, which taught city planning in its School of Natural Resources (the former forestry program). Yet the social science orientation of University of Chicago's program was not incompatible with scientific approaches to planning.[58] Course curricula underscored the persistent vitality of the belief that cities were ecological communities and national resources, which made plant and animal studies relevant to the analysis of human communities and natural resources planning techniques relevant to the improvement of urban areas. For example, alongside Sociology 244 (Introduction to the Study of Population and Human Ecology) and Sociology 364 (Human Ecology), a range of classes in botany, geography, and zoology (Botany 331, Plant Geography; Botany 335, Forest Ecology; Botany 336, Ecology of Grassland and Desert; Botany 337, Ecology of Arable Lands; Geography 220, World Resources; Geography 226, Conservation of Natural Resources; Geography 301 Geography of the Land; Zoology 304, Animal Ecology; Zoology 313, Evolution; Zoology 411, Biological Background of General Sociology; Zoology 419, Genetics of Human Populations) were recommended electives within students' courses of study.[59] Martin Meyerson's Planning 341 (Comparative Physical Plans) stressed the acquisition of techniques "for natural resources, rural land, transportation, industry, education, communities, and other types and levels of planning."[60]

Founding chair Rexford Tugwell and his colleague Edward Banfield had particularly close ties to the natural resources planning tradition. A member of Roosevelt's brains trust and former economics professor at Columbia University, Tugwell had been one of the New Deal's key spokespersons for the science of land use planning. As Roosevelt's assistant secretary of agriculture and head of the Resettlement Administration, he had embraced Taylorist management techniques.[61] Early in his career, Tugwell had "been able to excite myself more about the wrongs of farmers than those of urban workers"; by the time he arrived at the University of Chicago, following a stint as head of the New York City Planning Commission, his outlook had shifted.[62] Ecological knowledge had much to contribute to planning for human communities, he insisted, suggesting in a 1948 presentation to the Michigan Academy of Science, for example, that planners read the work of Charles Elton, Warder Allee, Alfred Emerson, and John Bews (who had popularized human ecology in Britain).[63] Edward Banfield's career trajectory matched Tugwell's. Banfield came to Chicago from the U.S. Department of Agriculture, where he developed experience in both forest and farm management.[64] As he initiated new research on urban housing policy with colleague Martin Meyerson (published as *Politics, Planning, and the Public Interest* [1955], which established both scholars as experts on urban governance), he simultaneously taught a course on governmental planning for agriculture (Planning 311) and continued writing about topics in the natural resources field.[65] Banfield had witnessed firsthand how, from the 1920s, "the conservation and city planning movements had been growing and establishing lines of communication with each other and with other antecedents of the planning movement."[66]

In planning, as in social science, naturalistic interpretations remained dominant but not static. From research studies that continued efforts to marry ecological research to urban planning; to coursework in human ecology for planners at University of North Carolina and University of Michigan; to the curricula at schools such as University of Oklahoma and University of North Carolina, where the natural sciences were taught alongside social science, with the growth of interest in planning education, the perspective on cities that had developed during the prewar period would continue to mature.[67] Like their social science colleagues, many researchers questioned their predecessors' assumptions about the "naturalness" of racial segregation, increasingly criticizing the once popular "neighborhood unit" ideal.[68] "Too often an area is referred to" as a "Negro, Polish, Jewish, Catholic, German, or Swedish 'neighborhood,' and the people themselves referred to as Polish-Americans, Japanese-Americans, or Swedish-Americans rather than *Americans*," wrote Reginald Isaacs from Chicago in 1947. "It is true that the 'social islands' such as 'Fifth

Avenue,' 'Little Italy,' the 'black belt,' and the 'ghetto' did not 'just happen to come into existence.' But the vague neighborhood hardly seems to be an alternative to these islands—but rather a perpetuation of them."[69] Like Homer Hoyt's revisions to Ernest Burgess's model of "the city," and the Chicago Community Inventory scholars who modified their views on racial segregation, planners' rethinking of the "neighborhood units" earlier rationalized on ecological grounds as now constituting an "impediment to the democracy of the cities" did not mean discarding the ecological tradition whose flexibility was such that it could accommodate a variety of perspectives on urban affairs.[70] "Planning must be brought into accord with the processes of growth of the city as a living organism," Isaacs explained in the *Journal of the American Institute of Planners* the following year.[71]

Given its academic orientation, the Program for Education and Research in Planning has been memorialized as an example of planning's traditional focus on land use taking a back seat to planning theory.[72] Yet a closer look reveals how Chicago planners extensively combined the two. Because of faculty members' ongoing local consulting work, and the program's location across the street from the headquarters to twenty-three national planning and housing organizations (Chicago's planners shared a library with these groups), coursework in planning, like research at the Chicago Community Inventory, became intimately intertwined with Chicago's needs.[73] While the plan commission collaborated with the Chicago Community Inventory as its main university partner, other local and citywide development agencies were eager for these planners' help. Thus Martin Meyerson, who came to Chicago from the Philadelphia City Planning Commission (where he had worked with Robert Mitchell and Harold Mayer), organized Planning 359 around the possibilities for industrial development in Lake Meadows. Graduate students, including John Dyckman and Ira Robinson, offered recommendations to the South Side Planning Board that were incorporated into the South Side Industrial Study Committee's final report.[74] In keeping with tradition in the university's social science departments, students participated in extracurricular opportunities as well.[75] The academic tone of Chicago Land Clearance Commission's site designation reports, for example, reflected the presence of students and recent graduates of the planning program on staff—among them Ira Robinson and Walter Schilling.

Combining strength in interdisciplinary scholarship with ties to ongoing projects at public and private city agencies, researchers at the Chicago Community Inventory and the Program for Education and Research in Planning shaped local perceptions of area problems and solutions. Their work, echoing an earlier generation of urban professionals to characterize Chicago as an ecological community and national resource whose predictable life cycle continued apace, underscored the importance of a program for the scientific management of urban resources and the inadequacies of legislation targeting slum removal alone. The two units made their greatest impact on the city together, in the early 1950s, when they collaboratively set out to make conservation — already gathering informal support in the late 1940s — a legally authorized component of Chicago's arsenal of area improvement techniques.

Toward Urban Conservation in Chicago: The Movement Gathers Momentum

In the wake of the withdrawal of federal oversight in Waverly and Woodlawn, local commitment to neighborhood conservation waned and area decay continued. Although these disappointments were widely acknowledged, they did not sour enthusiasm among urban professionals and city residents for "conservation." In postwar Chicago, where Mayor Martin Kennelly proposed the Woodlawn effort "should be emulated throughout Chicago," interest grew among city officials and residents for taking conservation measures elsewhere in the city.[76] With the expectation that individual homeowners would apply for federal assistance or use their own funds to meet these obligations, several neighborhoods adopted voluntary Community Conservation Agreements, pledging to care for properties and forbid conversions of single apartments to multiple apartments.[77] Compared with "blighted area" designations, controversial because they meant clearance and relocation for neighborhood populations, the idea of designating "conservation areas" became increasingly popular.

The growing interest in community property maintenance was not limited to the white city dwellers whose fears of neighborhood racial change had helped to motivate participation in the Waverly and Woodlawn experiments. Following the passage of the Blighted Areas Redevelopment

Act in 1947 and subsequent Federal Housing Act of 1949, when Illinois city residents came to recognize their homes might be demolished if they could not stem the tide of blight, conservation increasingly drew praise from institutions devoted to African American concerns. At the Chicago Urban League (where, notably, both Robert Park and Harold Ickes had served as past presidents), figures such as Community Organization Department director Alva Maxey, known for her vigorous protests against many of the city's redevelopment policies, explained with pride that it was local block clubs in Chicago's African American neighborhoods that first encouraged the city and later the state to pursue legal action on conservation.[78]

The appeal of conservation extended to some racially mixed neighborhoods as well. As Frank Horne of the U.S. Housing and Home Finance Agency's Racial Relations Service described to administrator Raymond Foley, even before the *Shelley v. Kramer* ruling, Chicago's Oakland-Kenwood Property Owners' Association had been seeking to promote both community stabilization and interracial harmony and had found in conservation agreements the ideal tool.[79] The group had "recently announced the adoption of a unique Community Conservation Agreement, designed to replace racial covenants." Signatories to the agreement, "developed over a year of study and conference between representatives of the Association and of leading Negro organizations, brought together under the auspices of the Mayor's Commission on Human Relations and the Metropolitan Housing Council," would pledge to be vigilant about home upkeep and to avoid actions that would spark neighborhood decline. "Should this idea 'catch on,'" Horne noted, "it might well prove to be one of the most significant contributions of recent years in the fields of neighborhood conservation, urban redevelopment and sound racial relations."[80]

Like their white neighbors, African Americans in these communities as in other U.S. cities would have become acquainted with conservation as a tool for natural resource planning long before discussions of its applications to urban areas. Many had firsthand experience with New Deal programs such as the Civilian Conservation Corps.[81] Others would have enjoyed the fruits of their neighbors' labor; the Chicago area had been the site of many Conservation Corps projects, from the creation of recreation areas to the beautification of roads. Later wartime efforts among rural and urban populations to conserve a range of materials were the frequent subject of news reports in the *Chicago Defender*.⁸² The mobilization for urban community conservation as housing maintenance and repair linked these national priorities to older property-improvement efforts among middle-class African Americans, activities that the National Urban League and *Chicago Defender* had long promoted.⁸³ That the racial implications of neighborhood conservation, like assumptions about the "natural" character of racial segregation, had evolved over the course of a decade was apparent in the public reaction to Richard Nelson's bid to head one of Chicago's land development agencies. As Horne recounted to Foley, Nelson "was strongly and effectively opposed by groups concerned with racial minorities because of his previous record of advocating racial restrictive covenants as a device for neighborhood conservation."⁸⁴

In light of the groundswell of interest in conservation, the Catholic Church, one of the city's most powerful nongovernmental institutions and an organization hostile to clearance, threw its weight behind conservation. Support was especially strong in West Kenwood where local pastor Rev. John Ireland Gallery of St. Cecilia's spearheaded the efforts at community property protection, and in Lincoln Park, home to DePaul University's campus.⁸⁵ Indeed, so committed were Chicago church officials to the belief that "timely steps, possibly involving some new construction, could be expected to halt or reverse the trend" toward blight, that they gathered 120 area pastors for a discussion of conservation as a neighborhood improvement tool.⁸⁶ Among the presenters was prominent community organizer Saul Alinsky of the Industrial Areas Foundation, who "described the type of social organization that is necessary as a foundation for a program of neighborhood conservation."⁸⁷ To spread the gospel to colleagues in other cities, Chicago's Catholic leadership hosted a workshop on conservation at the annual National Conference of Catholic Charities meeting in Detroit in 1951. A subsequent session at the National Conference of Catholic Charities meeting in Cleveland followed in 1952.⁸⁸

Thus, in the immediate postwar period, conservation served as an informal community improvement tool in a diverse range of Chicago neighborhoods. Lacking legal standing, however, participation was voluntary and penalties for noncompliance did not exist or were not easily enforced. Despite some involvement from the Chicago Plan Commission in these undertakings and general agreement from the local real estate community that a successful revitalization effort depended on more than voluntary citizen action, the city had far to go before conservation could be applied as part of the program of continuous citywide improvements outlined in Homer Hoyt's master plan.

To give conservation areas the formal status the plan demanded would require a new push to broker relationships among urban professionals and public policymakers — and recruit vigorous citizen participation to the cause. Help arrived from Reginald Isaacs, whose leadership role in mobilizing a diversity of individuals and institutions to support legislative action on conservation picked up where Homer Hoyt left off. With Isaacs's work, a new set of interest groups — community groups and block clubs concerned about local property values, together with a range of public and private organizations throughout the city concerned about the future fate of their neighborhoods — joined the campaign for slum prevention.

Enter Reginald Isaacs

Reginald Isaacs came to Chicago from the Urban Development Division of the National Housing Agency in 1945. His first two years were spent working as a planner for the Michael Reese Hospital and the South Side Planning Board, preparing the report that would certify the area as the city's first for redevelopment under the state's Blighted Areas Redevelopment Act.[89] After he accomplished that task, Isaacs enrolled for three years of continuing studies at the University of Chicago, where he worked with faculty from both the Chicago Community Inventory and the Program for Education and Research in Planning, especially Louis Wirth and Rexford Tugwell. (A former student of Walter Gropius at the Harvard design school, Isaacs considered registering for a PhD but decided not to.) In 1950, Gropius invited him to return to Harvard as a visiting lecturer in city and regional planning.

Isaacs's new position in Cambridge did not end his connections to Chicago, however, as he continued to serve as planning director at Michael Reese. From this vantage point, he became convinced that the city's ongoing redevelopment efforts were being stalled because legislation to address multiple stages of the urban life cycle as well as a master plan integrating Hoyt's vision for residential areas with plans for commercial and industrial development were still lacking. In his first year at Harvard, Isaacs proposed that the Graduate School of Design, the University of Chicago, and the Illinois Institute of Technology collaborate on an ambitious interdisciplinary research project in which students and faculty would study Chicago's South Side and come up with proposals for its further improvement.[90] The Community Appraisal Study (or South Side Community Appraisal Study), like the Research Committee on Urbanism of the National Resources Planning Board, would bring together social scientists and planners to advocate for progress on a comprehensive program for the scientific management of urban resources. It would provide opportunities to reexamine issues related to the blighted area close to the Michael Reese Hospital and Illinois Institute of Technology since demolition and new construction had begun, and to assess the Hyde Park–Kenwood neighborhood surrounding the University of Chicago, since, as Isaacs noted, "ecologically, the area is one of the most interesting for examination, survey and analysis," but it had not yet been officially certified for redevelopment.[91] Like the decade-old Woodlawn experiment, study goals would be far broader than merely advancing the understanding of these specific neighborhoods. Consistent with Homer Hoyt's earlier press for coordinating planning efforts across the city in a multistage sequenced attack on the urban life cycle, the aim was to demonstrate the necessity of integrating conservation into Chicago's ongoing redevelopment work.

Taking advantage of his comfort in both planning and social science circles, Isaacs recruited colleagues from the three institutions to participate — distinguished figures who included Mies van der Rohe, Walter Gropius, Louis Wirth, and Martin Meyerson.[92] Most among them took the view expressed by Gropius that the obsolescence of "industrial slums, business slums, and dwelling slums ... like dry rot on wooden buildings" might "often be quite 'natural,' " in its cause, and advocated for "an organic-evolutionary procedure" for the reshaping of postwar cities.[93] Together, their work would set the agenda for the city's conservation program, linking ongoing efforts to deal with blight to a broader mandate for the scientific management of urban resources.

The Community Appraisal Study

The Community Appraisal Study got under way in the fall of 1950, with simultaneous coursework at the three institutions.[94] Sociologists (mostly affiliates of the Chicago Community Inventory; some at the National Opinion Research Center) worked with Philip Hauser and graduate student Leopold Shapiro to prepare a sample survey of 1,600 families on Chicago's South Side.[95] Geographers, guided by Harold Mayer, drew upon existing repositories and gathered new information through field studies to prepare maps of the area. Planners at University of Chicago, directed by professors Martin Meyerson and Richard Meier, collaborated on both types of the surveys: as graduate student D. Reid Ross coordinated the interviewing and coding of the sample survey with statistical assistance from fellow graduate student Ira Robinson, classmates, including George Cooley and Loring Moore, conducted further field checks of local buildings.[96] Planners at Harvard working with Isaacs and planners at the Illinois Institute of Technology working with Ludwig Hilbersheimer focused on developing model proposals for the future of each neighborhood. The Harvard group created scale models and land use maps with alternative redevelopment scenarios, and the Illinois Institute of Technology built models and maps showing how the stages of redevelopment might proceed.[97]

In their work to comprehensively analyze an area that "evidences all of the stages of urban physical deterioration within its boundaries . . . here urban physical and social change can be studied with examples of each important stage," participants portrayed their work as an extension of the University of Chicago's ecological research tradition, an action-oriented addition to the legacy of Robert Park, Ernest Burgess, Harvey Zorbaugh, and Louis Wirth.[98] Focusing on "the related planning problems of a blighted area ripe for redevelopment and those of an adjacent 'middling aged' area" would revive the longstanding interest in conservation and its potential for aiding in a comprehensive strategy for urban revitalization by planning with the city's natural tendencies in mind.[99] Although their specific proposals for achieving community standards in an area conservation program varied — for example, with some participants emphasizing the importance of local governments' police power tactics and others promoting voluntary neighborhood enforcement, they agreed on the racial implications of the proposed strategies for property improvement. In contrast to conservation enthusiasts who saw in these techniques an opportunity to achieve the racial homogeneity that restrictive covenants no longer could legally pursue—most of them, like Richard Nelson, members of the real estate community—these social scientists and planners saw in conservation the preservation of property values in racially mixed or changing neighborhoods as an alternative priority.[100] The draft Community Conservation Agreement contained in one study report, for example, refers to the hope "to develop and improve the said area and the surrounding community for themselves and all persons irrespective of race, creed, or national origin."[101] Despite the continued exodus by whites to suburban areas in the wake of the Supreme Court's *Shelley v. Kramer* decision, and enduring practices of discriminatory valuation and lending at the Federal Housing Administration, members of the Community Appraisal Study explicitly rejected the associations with neighborhood unit planning of earlier conservation proposals, expressing optimism that the community action provision of conservation might help to repair race relations in this and other neighborhoods instead.[102] "As this process goes forward, there will be less pressure for restrictive real estate practices, less reason for mistrust and discrimination."[103] Diagnosing disappointments of earlier slum prevention efforts as the result of an inadequately trained citizenry, alongside their analyses study participants turned their attention to enlisting residents in the neighborhood improvement cause. This portion of the project, sponsored by the Wiebolt Foundation, was the job of Herbert Thelen, educational psychology professor and Human Dynamics Laboratory director, and focused on cultivating participation and improving human relations through the Hyde Park–Kenwood Community Conference.

Hyde Park–Kenwood Community Conference

Created in 1949, the Hyde Park–Kenwood Community Conference brought together neighbors on matters of community concern in an area where it was "difficult to distinguish between citizens and officials" as "many of the influential residents of Hyde Park–Kenwood wore more than one hat."[104] As of the early 1950s, several of the conference's working groups had academics as heads. Sociologists St. Clair Drake and Everett Hughes led the Community Survey Committee. Planner Martin Meyerson and lawyer Oscar Brown Sr. headed the Planning Committee. Other affiliates, most serving on the conference's Planning Committee during the period of the appraisal study, included geographers Harold Mayer and Gilbert White; planning faculty Martin Meyerson and Harvey Perloff; and sociologist Philip Hauser.[105] Herbert Thelen, an expert on small-group processes and how to teach scientific thinking, chaired the Block Organization Committee.

Like the Federal Home Loan Bank Board, Thelen maintained that "rarely does such a residential district develop into a slum because of factors beyond the control of those who live in it." His work to mobilize community conference members to join the conservation effort began from the assumption that "incipient blight need not run its course" if actions to interrupt the natural order of events could be promptly implemented.[106] Describing the decline of once pristine city neighborhoods into "unstable" middle-aged communities, and the fears that these events inspired, and recognizing "progress or deterioration in the area would depend upon the cumulative wisdom of the thousands of decisions made by citizens every day," the Human Dynamics Laboratory organized a training program composed of biweekly sessions at the lab, as well as a community clinic every three weeks designed to help form block groups for action on conservation.[107] Discussions at the National Resources Planning Board urbanism unit had praised the agriculture department's "tactical approach" to conservation, which required "the bringing along of the uninformed layman so that he can have his say in the making of plans and thereby understands them and is qualified to support them in an intelligent fashion."[108] The Waverly experiment had vested significant responsibility in individual block captains, each "furnished with a kit, made up of structural rehabilitation studies, cost estimates, sketches, street revision maps, interior play area plans, landscaping recommendations, etc. relating to his particular block" to "present visually, by means of an example which may be easily comprehended by, and is well known to, each property owner assigned to him, the project approach to community conservation."[109] In this "continuous program of education... to help the community see the blighting forces at work," the focus of attention was the collective dynamics of block groups.[110] As Thelen explained in *Neighbors in Action: A Manual for Community Leaders*, and offered to broader audiences in *Dynamics of Groups at Work*, citizen participation was a "technology," an opportunity to shape the action of a group's membership toward a purpose or goal, "a set of principles useful to bring about change toward desired ends."[111] Although Thelen recognized that expert advice would be needed to provide the information and training to help local groups recognize that deterioration was not inevitable and to organize themselves toward self-defined ends, he explicitly disavowed too activist a role for consultants.[112] "The 'expert' is not the leader or manipulator of the group," he insisted, urging instead that "the group itself has to be the engineer."[113] According to this view, like the scientific analyses of "types of planning areas" that would teach planners to think about the most rational sequence for intervention, knowledge about scientific approaches to urban theory and community participation would help neighborhood associations to achieve complementary goals.[114]

The political climate of recent years provided an additional rationale for citizens to embrace conservation. Truman's rhetorical stance on housing had created a link between urban affairs and the urgency of national security needs, but the reality was that "from the peak of redevelopment enthusiasm which followed passage of the National Housing Act of 1949," explained John Dyckman in one appraisal study report, "there has been a steady diminution during the past two years" as "acceleration of armament spending has claimed an increasing proportion of the resources which might have been invested in urban redevelopment."[115] In this context, the smaller civic investments that conservation demanded held the promise of improving cities at reduced costs and showcasing Americans' patriotic spirit. Echoing Thelen in his insistence that "it will not suffice for the leaders to do all the work or to make all the decisions. Responsibility must be broadly shared and cooperation well practiced," Dyckman suggested how action on conservation was the embodiment of democracy itself: "American democracy means we all should have a share in shaping the future of our community."[116]

Beyond the Ivory Tower

Early results from this work buoyed organizers' optimism about the future of conservation, as alongside physical property improvements, improvements in neighborhood human relations also could be seen.[117] Thelen's manual for community leaders included several descriptions of white residents' initial fear of their "different" neighbors and how such conflicts ultimately were resolved for the benefit of the entire community. Yet cognizant of the need to mobilize the support of a broader range of interest groups if legislative backing for conservation was to become a reality, Isaacs set out to popularize the group's work to public and private area organizations—and met with immediate success. University of Chicago faculty and students offered "a research resource for planning," observed South Side Planning Board executive director Morris Hirsch, pointing in particular to "more than a dozen thesis studies under Prof. Louis Wirth in the Department of Sociology" with relevance to local planners' concerns.[118] From 1951, responding to the growing interest in neighborhood organizing around the city, the university's Human Dynamics Laboratory initiated three-week community action workshops, "training in face-to-face group operation and urban sociology" that was said to have "markedly influenced programs of several agencies" in short order.[119] Program documents soon described the study as a joint undertaking of the Michael Reese Hospital planning staff, the Hyde Park–Kenwood Community Conference, and the three universities, together with the South Side Planning Board, the Mayor's Housing Coordinator, the Metropolitan Housing and Planning Council, the Chicago Dwellings Association, the Chicago Housing Authority, the Chicago Plan Commission, the Chicago Park District, the Chicago Land Clearance Commission, the Oakland-Kenwood Planning Association, Draper and Kramer Real Estate, PACE Associates Architects, and several community organizations.

In these efforts to build momentum for an integrated citywide program of conservation as well as clearance, Chicago's real estate community provided an important source of support. Now convinced that pockets of slum removal would be insufficient to stop the spread of blight, and eager to promote public and private action on city improvement, the professionals who once had appropriated and narrowed planners' ambitious vision for the scientific management of urban resources, including human populations, to conserve physical properties on a neighborhood scale took Hoyt's advice to push for research-based, coordinated citywide plans.[120] "It is excellent that there can be collaboration between the architect and the planner and the sociologist and those who are engaged in the social sciences," observed developer Otto Nelson at a meeting Isaacs organized at the Harvard Graduate School of Design, suggesting Nels Anderson's and Eduard Lindeman's hopes from an earlier decade finally had been achieved.[121] "REHABILITATION AND CONSERVATION OF URBAN PROPERTY open a vast new frontier for professional work in the real estate field," noted the president of the Chicago-based National Association of Real Estate Boards in 1952: "We found that neglect and abandonment of one site in order to move to another had come to a stop in agriculture, and so we have begun to apply deliberate measures to conserve our soil resources. It is clear now that the process of neglect and abandonment is equally wasteful of our urban resources. . . We have all the evidence we need that rehabilitation and conservation of urban resources can be done on an economic basis. The job is a challenge that offers unlimited opportunities to professional people in real estate."[122]

Thus, while Community Appraisal Study organizers did not work to return the definition of "conservation" to the associations with human rehabilitation and poverty reduction for which the National Resources Planning Board earlier had campaigned, this undertaking offered a new opportunity to promote an ambitious vision of the scientific management of cities — and hopes for improved urban race relations as well.

"I want to make a forecast," offered Walter Blucher, executive director of the American Society of Planning Officials and planning instructor at the University of Chicago, at the close of a meeting Isaacs organized in 1951. "I think time will prove this project to have been much more important to this school and to the students and to the City of Chicago than those of us here now realize."[123] He did not have to wait long for his prediction to come true. The following year, as the appraisal study was wrapping up, Mayor Kennelly created an Interim Commission on Conservation, appointing conservation enthusiast James Downs, principal at Downs-Mohl real estate and housing and planning coordinator for the city, as its chair.[124]

Simultaneously, the private Metropolitan Housing and Planning Council "set up a committee to study how declining neighborhoods could be shored up through government and private planning efforts."[125]

With networks between the two organizations already established, they quickly became collaborators on a conservation study for which the Metropolitan Housing and Planning Council took the lead.[126] Impressed with Isaacs's successes in calling attention to the importance of slum prevention, the council invited him to compile the official report. Input from members of both organizations and writing help from study assistant director Jack Siegel transformed the Community Appraisal Study, completed in late summer of that year, into the focus of the council's 1953 conservation study.[127]

The new study, which followed Homer Hoyt's decade-old work, characterized the city as an organic entity with a "life cycle" such that "the stable areas of today are the conservation areas of tomorrow."[128] The onward march of blight outlined was slow, with little change from Hoyt's earlier assessment.[129] What had changed in the intervening period, however, were the legal instruments permitting slum clearance to deal with blighted areas, and the authors expressed concern that instead of interrupting the process of community decline in fact applications of such ostensible remedies were making some neighborhoods worse. "The relocation necessitated by slum clearance and redevelopment," they explained, "is partially responsible for the population pressures exerted on the conservation areas."[130] To succeed in "preventing tomorrow's slums" (the interim commission report's official title), the city needed to take immediate action to certify conservation areas alongside blighted areas so as to arrest further neighborhood decay.[131] Executing this plan would require Illinois lawmakers to expand the possible uses for eminent domain. Noting in the case of soil erosion how state lawmakers had recognized the conservation of public and private lands to be an essential "public purpose" for natural resources planning, the authors called for new legislation to give equal status to urban conservation: "The acceptance of conservation or 'slum prevention' as a public purpose in the legal sense is vital," Isaacs and Siegel explained.[132] "Many of the provisions of the soil conservation program," from its "local initiative" to "the referendum device" easily "could be adapted to urban conditions."[133]

Illinois' Urban Community Conservation Act

Supported by "an impressive array of business and civic leaders" that included many of the public and private organizations involved in the Community Appraisal Study and in turn new advocates in the institutions and organizations that Isaacs, through a later Conservation Conference, rallied to the cause, the Illinois Urban Community Conservation Act, "drafted by a member" of the Metropolitan Housing and Planning Council staff "and pushed by that organization in the General Assembly," became law later that year.[134] Called the "Butler Bill" (it was sponsored by Illinois Senate Majority Leader Walker Butler of Chicago), the new legislation extended eminent domain to conservation areas.[135] The news was good for conservation enthusiasts: because of the difficulties of relocation, legislation had restricted the maximum size of blighted areas designated for slum clearance to 160 acres. Conservation areas faced no such limitation: the minimum requirements were forty acres, where 50 percent or more of the structures were residential, with an average age of thirty-five years or older.[136]

The 1947 Blighted Areas Redevelopment Act had called for the creation of "land clearance commissions" to oversee local efforts; Illinois' new 1953 law enabled similar oversight organizations: "community conservation boards." Like the Chicago Land Clearance Commission, the Chicago Community Conservation Board was chiefly composed of members of the business and civic communities, assisted by a staff of planners and draftspersons. The board carried the status of a city department, asked to coordinate its work with Chicago's other planning and housing agencies.[137] Calls for cities to create official departments of conservation had a long history. Even before the Woodlawn experiment had gotten under way, director Robert Mitchell had called for cities "to have some central agency or clearing house which could encourage the organization of groups in various neighborhoods, and more important, guide them in the establishment of their objectives and procedures."[138] During the 1940s, participants in the Waverly experiment had observed how, if adopted citywide, such tactics might "one day inspire the establishment, in every large city, of a 'Department of Conservation' . . . to promote community stabilization projects in potentially and partially depreciated sections throughout the city" through technical and financial assistance to individuals and groups."[139] In 1944, Richard Nelson of the Downs-Mohl real estate company, seeing citizen participation in Waverly's conservation effort falling off, suggested to the National Housing Agency that such programs would be more effective if guided by a central federal or local organization.[140] James Downs, following his appointment to the city's Interim Commission on Conservation, had issued a proposition "in November, 1952, calling for the creation of a permanent Community Conservation Board" for Chicago.[141] Harvard students working under Isaacs on the appraisal study had expressed enthusiasm about the possibility of creating a Conservation Authority.[142] Hyde Park Alderman Robert Merriam, in the spirit of taking conservation to the people, the ultimate arbiters of its success, had urged the city to organize "neighborhood conservation districts . . . with a local office under the jurisdiction of the Conservation Unit of the Building Department or the Mayor's Housing and Redevelopment Coordinator."[143]

The motivation for these proposals was not merely enthusiasm about conservation, but a recognition that, even in the short term, after initial bursts of energy, voluntarily organized citizen associations for the conservation of urban resources largely had disappeared. "When rents are high and dwellings scarce, tenants Protective Associations swarm up like Green Bay flies. Like the flies, they are gone when the sun rises." This course of events was to be expected in the natural life cycle of any institution, sociologist Everett Hughes explained. "If a group arises in crisis, and fails to define its role after its first failure or success, it will automatically die."[144] Thus, when Reginald Isaacs and Jack Siegel reported of the Hyde Park–Kenwood community three years on that "local organizations have not been able to halt the deterioration of the study area," these shortcomings did not lead enthusiasts to question the value of conservation but rather were the reasons why, when the tactic achieved legal status, Chicago officials gave it the official administrative oversight and technical assistance for which so many supporters had campaigned.[145] The nation's prior experience organizing citizens in natural resources planning projects had lessons for urban professionals, suggested William Slayton and Richard Dewey in a much-cited essay on alternatives to clearance. The disappointments of Waverly and Woodlawn had not occurred at the Tennessee Valley Authority where government technicians had helped private citizens to help themselves.¹⁴⁶ "Although test demonstrations on the use of fertilizers seem somewhat afield from redevelopment," they reflected on a paper by Saul Alinsky describing community organizing at the Tennessee Valley Authority, "the basic operation is the same, ie, presenting a proposal to local residents and securing their participation through local organizations."¹⁴⁷ Cities should deploy an analogous set of experts, they proposed, to ensure that citizens were adequately organized for the slum prevention task.

There would be two important differences between how the city dealt with its blighted areas and its conservation areas. A first difference was that while land clearance and redevelopment were the work of the Chicago Land Clearance Commission in collaboration with private developers, conservation was chiefly an activity of private citizens requiring substantial initiative from local community groups. Neighborhood outreach was formalized in the organization of local Conservation Committees (also called Neighborhood Committees) of thirty to forty people, five of whom (ideally a community council leader, design professional, community leader, real estate expert, and banker) were to establish a specialized local planning organization.¹⁴⁸ The Back of the Yards Council, which organized a Neighborhood Improvement and Conservation Committee, cast a wide net, inviting "Representatives of Church–Business–Labor–Industry–Savings and Loan Associations–Builders–Construction Men–Property Owners–Tenants" to a lunch meeting "to discuss a conservation program by the people to protect the area from blight and deterioration."¹⁴⁹ A second difference was that in contrast to the Land Clearance Commission's immediate review of the city's blighted areas in the *Ten Square Miles of Chicago* report, the Community Conservation Board did not pursue a similar citywide assessment.¹⁵⁰ Instead, officials left to individual communities the task of raising the money and hiring the experts for neighborhood surveys to certify their area as part of the conservation program. Like the earlier observations of extension agents participating in the organization of soil conservation districts that "the local people must operate the district if it is to be successful," this was a calculated move to encourage community investment in the city improvement effort — in James Downs' words to one local group, "that you mean business and that we can count on your on-going cooperation."¹⁵¹ As the Lincoln Park Conservation Association explained to its membership, "the legislative principle unique to this Act is that such planning should be more than a merely super-imposition of planning technique on a disinterested neighborhood."[152] Recognizing that without some government assistance, however, the program was likely to fail, city officials did not leave citizens entirely on their own. Following the community appraisal participants' insistence that "the nature of the decay and alternatives that lie ahead must become common knowledge," bulletins and brochures from city agencies publicized the program to the public to motivate them to participate.[153] The Community Conservation Board and the Chicago Department of Buildings together prepared an extensive survey schedule to provide technical guidance on this task.[154] And the Chicago Plan Commission expected ongoing interaction with Conservation Committees in its insistence that any neighborhood slum prevention effort conform to its master vision for the city.

By 1953, a variety of constituencies in Chicago agreed about the possibilities for and the value of interrupting the urban life cycle through a citywide program of neighborhood clearance and conservation—a continuous planning effort in which action from government and citizens would play a role.[155] Legal measures were in place to realize planners' longstanding ambitions for the scientific management of urban resources. Watching from Washington, DC, under a new Eisenhower presidential administration, federal policymakers, frustrated by the grand expenses and limited outcomes of slum clearance, were intrigued. How, even before the city's official conservation program got under way, the social science–based "experiment" in Chicago provided a model for the design of the federal urban renewal program—and in turn the design of cities across the nation—is the subject of chapter 5.

FIVE

A Nation of Renewable Cities

Speaking at a 1957 symposium on urban theory and practice at mid-century, Wayne State University sociologist H. Warren Dunham reflected on the limits of the Chicago PhD he had acquired in 1941. "The ecological image tended to put a damper on the zeal to act," he told the Detroit audience, arguing that action was needed on city problems and that the tradition in which he had been trained implied the futility of intervention.[1] Yet, even as Dunham spoke, his colleague Mel Ravitz was organizing area residents to participate in a program of urban renewal that other observers recognized to be "based on sociological concepts" — including an ecological image of the city.[2]

How could Dunham overlook the close connections between his city's redevelopment efforts and the ecological orientation of his graduate training? During the late 1950s, as they set out to address a nationwide audience, many of the figures who had worked for decades to implement a program of life cycle planning for cities would turn away from the naturalistic language their movement had long used. In the demand that cities classify neighborhoods according to present and future productivity, in the insistence on citizen education and organizing, and in the reliance on academic and other urban professionals to advise and assist local governments and community groups, renewal delivered a long-awaited slate of policies for the scientific management of urban resources to the national stage. Yet, after a final flowering of ecological and natural resources rhetoric just before the legislation's passage, arguments for joining citizens and government in citywide and continuous attacks on urban problems would be increasingly detached from their earlier rationales. In a period of declining cultural currency for the scientific management of nature, claims that the nation should invest in planning for its cities as it had invested in planning for its natural resources persisted. With the shift of focus toward city "renewal," however, came a revision to the explanations offered. Cities were sites for much-needed investment no longer because of their commonalities with rural areas but rather, as President John F. Kennedy suggested to an audience of mayors in 1960, because for too long the nation had prioritized agricultural over urban needs.

From individual city studies to overview analyses, urban historians already have devoted ample attention to this first federally sponsored program for city improvement in the United States. Although scholars have noted some connections between Chicago School sociology and the national renewal effort, their stories of relationships between ideas about cities and nature have focused on slum clearance aspects of the program. In these accounts, as successional processes prompted the middle classes to flee America's decaying urban centers toward the pastoral ideal that suburban areas represented, life cycle theories about the inevitability of neighborhood decline led to the "federal bulldozer" imposing order on the nation's cities.[3] In the postwar period, however, even as natural resources planners began to turn away from the popular Clementsian model of succession to climax, in the realm of urban policy this paradigm continued to provide inspiration for returning cities to equilibrium through a set of interventions of which clearance was only one.[4] Like the older tradition of scientific management for nature that conservation enthusiasts such as Rexford Tugwell espoused, renewal aimed to draw out order in cities by organizing a comprehensive and sequenced plan of action that kept the life cycle of land in mind. From "conservation areas" to "rehabilitation districts," then, the technical and legal categories that local public agencies employed to inform their analyses of urban data aligned the nation's cities' programs of acquiring land for demolition and modernization through eminent domain, relocating neighborhood populations, and enlisting citizens to repair their communities with the natural laws of urban growth and decay.[5] With a familiar cast of characters shaping efforts to take the renewal program nationwide, behind the new face for action on city improvement in the 1950s an older vision of urban resource planning from the 1930s remained.

Eisenhower Takes Over

The Truman era witnessed a shift in the national discussion about natural resources planning and conservation as calls to conserve for reasons of economic development took on new associations with national security concerns.[6] Environmental historians have documented the blossoming of disenchantment with conservation during this period as the opportunity to see the longer-term effects of Roosevelt's New Deal policies unfolded.[7] "What the nation did for the major resource industries during the depression was conservation only in the life-saving sense," noted one interior department staff member in 1952, rather than conservation "in the sense of assuring stabilization of the land or forest base, or less wasteful physical recovery of minerals."[8] Sarah Phillips has described New Deal debates about the aims of conservation — she stresses competing values of efficiency, equity, and sustainability — and the fallout from the growing recognition that conservation in practice addressed the interests of business and property owners instead of poverty reduction or scientific management of land.[9] In soil conservation, for example, despite Soil Conservation Service head Hugh Bennett's efforts to promote centrally managed land use planning, voluntary decentralized conservation measures became national policy so as not to restrict the authority of land owners on their holdings.[10] African American farmers, disproportionately bypassed by new production techniques, received an especially raw deal.[11] As a result of these and other policy choices, earlier confidence that "the waste of our natural resources . . . had been averted" gave way to the recognition that "many of the ghosts of social disaster, thought to have been laid forever by the New Deal, were still abroad in the land."[12] New environmental problems in the postwar period, including another dust bowl during the mid-1950s, seemed to confirm this interpretation.[13] Advances in ecological science increasingly suggested the theoretical underpinnings of a wide range of programs required reconsideration as well. With Frederic Clements unable to respond to a new generation of critics (he died at the close of World War II), the Clementsian theory of succession to climax, debated upon its debut, now was largely discredited.[14]

As Eisenhower took office in this context, the continued scientific management of nature took a back seat as a policy matter. Instead, the new president's environmental policies would echo the message he had trumpeted during his campaign — that federal efforts to control natural resources were tantamount to socialism and communism. "The federal government was so deeply involved in controlling water and power," he had told one audience, "that it did everything 'but come in and wash the dishes for the housewife.' "[15] While the administration's farm policies established a Conservation Reserve Program to remove land from production, Eisenhower saw little of the relationship between conservation and democracy or natural resources planning and national security planning that his presidential predecessors had espoused.[16] The former general, while interested in questions of human resources planning on account of his military experience, had limited patience for New Deal visions of centralized nature protection.[17]

The test of time that called into question the national record on natural resources planning policy also pointed to problems in the 1949 Housing Act. As implementation moved forward, several obstacles to slum removal became broadly apparent. One was the availability of financial support. Although the legislation offered cost-sharing, neither federal nor local governments possessed the means to redevelop more than a few blighted areas, making the idea of citywide slum eradication an impossibility. Another was the difficulty of relocation in those projects that were approved. With local governments required to rehouse affected residents, the law's stipulations about what constituted adequate living standards slowed progress on redevelopment, a challenge exacerbated by continuing segregation in many communities.[18] The policy also had revenge effects. Focused on the demolition and rebuilding of trouble spots alone, the repairs to individual neighborhoods that had been approved appeared to be displacing rather than removing problems.[19]

With the limitations of a national program of slum clearance alone now widely apparent, the Eisenhower administration took its first steps toward a reconsideration of this and other urban policies soon after coming to office in 1953 by creating a Committee on Government Housing Policies and Programs, headed by Housing and Home Finance Agency administrator Albert Cole.[20] Cole's earlier opposition to the 1949 legislation, especially its provisions for government-financed public housing, reflects the overarching orientation of this group. Together they worked to craft a new generation of policies consistent with Republican values —

more specifically in which action on property improvement from private industry and individual homeowners, directed by local rather than federal government, would play central roles.[21]

The Subcommittee on Urban Redevelopment, Rehabilitation, and Conservation

The Cole Committee, charged with offering recommendations on a broad spectrum of housing issues, organized study groups on topics that included veterans housing, low-income housing, housing credit, and the organization of housing activities inside the federal government. The task of assessing the 1949 Housing Act and developing alternative policy proposals fell to a Subcommittee on Urban Redevelopment, Rehabilitation, and Conservation, chaired by Baltimore mortgage banker James Rouse.[22] In the wake of the Waverly experiment's fade out, urban professionals in Baltimore, like their Chicago colleagues, had taken action to reinvigorate local slum prevention activities. Divorced from earlier associations with natural resources planning, however, their work had proceeded as a movement for neighborhood "rehabilitation"[23] The two cities' leadership work on slum prevention as "conservation" and as "rehabilitation" is thus reflected in the subcommittee name.[24]

Yet with contributors, including Jack Siegel, Homer Hoyt, Walter Blucher, Ferd Kramer, Reginald Isaacs, Miles Colean, Charles Ascher, and Robert Mitchell, having participated in ongoing national discussions about the scientific management of urban resources and having helped to turn Homer Hoyt's proposal for Chicago into Illinois law, when the group opened its conversation that city's experience loomed largest in their thinking, even though its paired program of slum removal and slum prevention thus far had yielded few results.[25] "The conservation of our natural resources is considered the proper function of government," observed Jack Siegel together with Baltimore lawyer C. William Brooks in their report on *Slum Prevention through Conservation and Rehabilitation*, which ignored Eisenhower's personal views on the subject. "As a nation, we have reached the point of maturity at which the conservation of our urban resources is equally warranted."[26] Echoing the urgings of the now-defunct National Resources Planning Board, they insisted, "Just as the nation's agricultural program is considered the proper concern of the central government because of its vital impact on the nation's economy, so must consideration be given to the future economic and social wellbeing of the cities . . . The federal government does have a continuing interest and responsibility in promoting economically sound and healthy urban communities."[27] The subcommittee concluded that, in cities across the United States, while actions on city improvement were being taken slums were growing faster than they were being removed—and that it was cheaper to prevent than to repair decay. Members called for slum prevention to join slum removal in the nation's arsenal of urban improvement tools, with new financial assistance extended to cities and homeowners to maintain the existing housing stock.[28] In these efforts at "conservation" and "rehabilitation," industry and citizens would play a central role, helping to reduce costs, limit relocations, and increase the effectiveness of urban interventions. Citing slum prevention efforts already ongoing in a number of U.S. cities, such as Baltimore, Philadelphia, and especially Chicago, the subcommittee observed how already at the local level leading figures in the urban professions and American citizens had thrown their weight behind such measures—but without support from federal authorities, these efforts would not likely succeed.

The major difference between these figures' contributions to the federal advisory panel and to earlier discussions of slum prevention was their tone. In his work for the subcommittee, for example, Siegel together with Brooks expressed a far greater sense of crisis than in his writings with Reginald Isaacs earlier that year. To Chicago's Metropolitan Housing and Planning Council, Siegel had reported little change in the status of the urban physical environment from Homer Hoyt's 1943 assessment. To the Presidential Advisory Committee, however, he and Brooks communicated a more urgent message. "Blight moves fast," they cautioned. "If a survey were made today," areas that had been classified as "stable" would now be classified as "conservation," and the "conservation" areas would in turn be "near blight."[29] Following observers who recognized the "accelerated deterioration" in Chicago's conservation areas, and the social and economic problems these changes had spawned, including "a growing crime rate, juvenile delinquency, and a serious population loss which has drained the city of needed tax revenues," Siegel and Brooks insisted the federal government act quickly to design a more comprehensive program for fixing cities across the United States.[30]

Mobilizing for Renewal: The 1954 Housing Act

The president adopted the subcommittee's recommendations nearly verbatim. Too many housing laws had been designed as temporary solutions to crisis problems, Eisenhower argued, and he proposed overhauling piecemeal policies in favor of a package that joined redevelopment, rehabilitation, and conservation to serve the Republican party's aims of enlisting private enterprise and average citizens in improving U.S. cities with local government authorities as their guides.[31] Although criticized by both Democrats and Republicans, this proposal found substantial support among parties hostile to the earlier 1949 Housing Act.[32]

The real estate industry, for example, which had vehemently opposed that legislation on account of its public housing provisions, saw new business opportunities for Realtors, real estate boards, and the building industry in conservation and rehabilitation. "In a frontier spirit, we mined our soil by farming one area to exhaustion, abandoning it, and moving to another, until finally we realized the need of soil conservation," wrote the National Association of Real Estate Boards in a 1953 brochure. "Just as the supply of arable agricultural land is limited, and so must be conserved, the supply of economically usable urban land is also limited, and is equally in need of being conserved." Organizing local and national clean-up campaigns and housing demonstration projects, gathering evidence on property protection activities of community groups across the country, and issuing a range of instructional materials, the association of real estate professionals — now like planners recognizing the inadequacy of depending on neighborhood associations alone — hoped to formalize "the process of bringing new vitality, livability, usefulness, and attractiveness into older areas that are adaptable to conservation measures, and a plan to arrest erosion in other areas so exposed."[33]

The religious leaders critical of the 1949 Housing Act on account of the demolition and relocation that followed neighborhoods' designations as slums saw new life for their congregations in the focus on improving communities in place through neighborhood services and neighborhood action that conservation and rehabilitation espoused.[34] Philadelphia Quakers organized an interracial "self-help" experiment in rehabilitation; in Detroit, slum prevention offered opportunities for interfaith collaboration.[35] With the National Council of Catholic Charities having adopted a resolution for "the establishment of an organization within the Church to work for the elimination of bad housing and slum conditions in much the same fashion as the Legion of Decency has attacked the conditions which blight the mind and soul," even the organization's Reverend Monsignor John O'Grady—who testified before Congress on the problems of the 1949 Housing Act—was sold.[36] Pointing to the "priests of Chicago" as "pioneers" in a campaign for urban revitalization "whose influence has already spread far and wide throughout the country," he later called conservation a "great movement . . . an opportunity of influencing the lives of people in the whole community."[37]

Backed by a diversity of interest groups, Eisenhower's proposal garnered the needed support to become law. A new housing act was passed in 1954, and a federal Urban Renewal Administration was created inside the Housing and Home Finance Agency to replace the Division of Slum Clearance and Urban Redevelopment established there to administer provisions of the Housing Act of 1949. Division head James Follin became the nation's urban renewal commissioner, the first of several who had pressed for a comprehensive national program for cities since the 1930s, and who, after taking office, emphasized slum prevention over slum removal as the centerpiece of America's strategy for urban improvement.[38]

The Illinois and federal laws were not identical, however. There were two key differences between the 1954 Housing Act and the Urban Community Conservation Act on which it was based. First, while Illinois' legislation called for local governments to organize new administrative bodies for community conservation as for land clearance, federal requirements were more flexible. The 1954 act asked city officials to designate a "local public agency"—any agency or commission, public or private, extant or new—to serve as coordinator of redevelopment, rehabilitation, and conservation—or in the act's new language, "renewal." Second, whereas Illinois cities could redevelop blighted areas and conserve areas in decline, the federal law named a third category, "rehabilitation areas," to reflect the influence of Baltimore's work.[39] This new category followed from the now widely accepted belief that the neighborhood life cycle might be slowed, halted, or reversed and permitted all of these strategies to be implemented—to slow blight in the best neighborhoods (conservation), halt it in areas where it had begun (rehabilitation), and reverse it in communities where it was well underway (redevelopment).[40]

A New Vocabulary for the Scientific Management of Cities

How was it that the shift to the language of "renewal" occurred? This term, which became the dominant shorthand reference for federal urban and housing policies after 1954, generally is said to have been coined by former Federal Housing Administration housing economist and presidential advisory committee participant Miles Colean in his book *Renewing Our Cities* (1953), written shortly before the committee assembled for a review of national housing policies.[41] Colean attributed it to Patrick Geddes, Scotland's ecological planner, whose career predated the naturalistic thinking of American urban professionals, yet whose work did not come to their attention until several decades later.[42]

Colean's was by no means the first use of "renewal" in discussions about urban areas in the United States, however. Among the networks of professionals for whom ideas about cities as ecological communities and national resources—and calls for a program of urban conservation—took center stage, the term already was familiar. In a 1946 book on city development, Geddes disciple Lewis Mumford chose as his subtitle "disintegration and renewal."[43] In the 1943 *Master Plan of Residential Land Use of Chicago*, Homer Hoyt envisioned the city "constantly renewing itself."[44] Indeed, Hoyt wrote of renewal as early as 1940, noting in a study of urban decentralization that whereas older cities in Europe "tended to renew themselves and to rebuild any decayed parts," in U.S. cities, the invasion-succession process "was one that did not automatically renew the old parts of the city."[45] As with the appropriation of "blight" and "climax," such talk of renewal appealed to these professionals' belief that patterns in cities shared far more in common with patterns in nature than was popularly perceived. From the 1920s, when natural resource planners came to recognize a distinction between renewable and nonrenewable resources, discussions about "renewal" had preoccupied participants in the conservation field.⁴⁶ These conversations continued through the early 1950s, suggesting how the call for urban renewal expressed the belief that cities were renewable resources as well. Contemporary uses of the term have lost such associations, but when first applied to describing a program for the scientific management of cities renewal evoked the naturalistic traditions of urban interpretation which provided its intellectual rationale.

"A pioneer people . . . is apt to be indifferent to its immediate environment," Colean charged in his 1953 book. "Most of our cities still suffer from the results of this aspect of the pioneer spirit . . . When the mess got too bad, those who could afford to do so generally chose to move, Indian fashion, to a new camp site somewhat distant from the old centers rather than to clean up the one already fouled."⁴⁷ Invoking the Research Committee on Urbanism's 1937 report in both title and message, Colean argued the end of the urban frontier had arrived, and the nation needed to achieve consensus as to what the aims of policies to repair cities should be. At odds with Homer Hoyt's assessment of the "natural" tendencies of U.S. cities, however, this former National Resources Planning Board consultant saw urban areas possessing the capacity to renew themselves — a process that he suggested had been interrupted by previously contradictory policies and programs: "The root of the problem of cities, therefore, lies in the interruption of the renewal process. The solution can be found only when the interruption is prevented or overcome."⁴⁸ Proposing that the "cycle of development in cities" be matched by "a continuous cycle of renewal," he articulated how making the "nature" of cities central to future policies would enable America's urban areas to be returned to a more stable state.⁴⁹

Yet, at the same time Colean insisted the nature of cities become a central preoccupation, he warned of limits to the analogical reasoning he proposed. A literal interpretation of the organic analogy could "cloud the real sources of urban change, confuse thinking and diminish the possibilities of improvement," Colean explained.⁵⁰ In its efforts to follow the founding generation of human ecologists by building comparisons between cities and nature and yet circumscribing these comparisons, Colean's book — and the "renewal" program that followed — marked a turning point for the American urban professions.

By the early 1950s, the vision of cities as ecological communities and national resources these professionals had laid out three decades earlier had led to changes in the policy environment. A broad coalition of individuals and institutions in the public and private sectors who saw continuities in the challenges faced by urban and rural areas had spurred Illinois to codify a vision for its cities' future planned in line with a predictable model of urban growth and decay. This policy, in turn, became a model for the nation in the federal Housing Act of 1954. Passed by a Republican administration seeking to "help the cities help themselves," the act offered financial and technical assistance as local governments intervened in "the deterioration cycle" to "increase municipal revenues at the same time they are reducing the demand for services."[51] Recognizing that these plans for the automatic redevelopment of urban areas would not function if local public agencies and neighborhood groups did not understand the science behind the program, the Housing and Home Finance Agency's new Urban Renewal Administration organized an effort to provide instruction on a massive scale.[52] A close look at this unit's work in the program's early years reveals how, as the dominant language for city problems and solutions shifted away from ecological communities and national resources, agency requirements nevertheless ensured that participants organized their contributions to support renewal's life cycle rationale.

Teaching Local Public Agencies about the Urban Life Cycle

"The tale of slums is the same all across the Nation," Commissioner James Follin told one audience not long after renewal's debut. "The situation in Baltimore and Newark" was "the same in Washington, New Orleans, New York, and Louisville" and "no different either in Miami, Austin, San Antonio, and Durham."[53] Taking the view that a common fate awaited all urban areas, the Urban Renewal Administration invited participating cities to organize their work into two stages: planning and implementation.

In the planning stage, local public agencies, in collaboration with Urban Renewal Administration regional offices, were asked to survey their cities and create a master plan—a *Workable Program for Urban Renewal*—that related the area's past, present, and future, "helping to salvage all that remains sound of the community's past, helping to redevelop its present form into something more serviceable for our own time, and helping, through intelligent planning, to give purposeful shape to its future."⁵⁴ These citywide analyses would designate areas for redevelopment, rehabilitation, and conservation and outline the ideal schedule for neighborhood interventions. In addition to the detailed community analyses necessary for such comprehensive plans, *Workable Program* submissions had several additional criteria to meet: adequate codes and ordinances, adequate administrative organization, financing ability, housing for displaced families, and full-fledged citizen participation.⁵⁵

It would have been difficult for a city planner, real estate appraiser, or academic in the early 1950s—prominent among the membership of the nation's local public agencies—to be unacquainted with the urban life cycle and the importance of accounting for it in a city plan. That residential development extended outward from a downtown core, that blight followed a predictable wavelike pattern, that economic and social problems mapped onto areas of physical decay, and that housing reform would help to cure cities' broader ills—these basic assumptions were alive and well in Philadelphia and Portland, Cleveland and Detroit, Denver and Houston, Los Angeles and beyond, when opportunities to plan for renewal arrived. So too was the sense that, in the postwar period, counteracting the forces of urban decentralization should be a priority. Nonetheless, to remind local officials of the importance of such theories to the process of urban restoration, and to assist them in their work, the Urban Renewal Administration created a variety of instructional materials—manuals, bibliographies, and films.⁵⁶ An Urban Renewal Service and regional Housing and Home Finance Agency offices provided technical assistance on request.

Once the administration approved a city's *Workable Program*, federal funds were disbursed to begin its implementation. Consistent with Colean's observation that "a city can never be completed in the sense that a picture, a statue or a building can be completed . . . there can be no final, once-and-for-all answer to the city's problems," federal officials emphasized to their local counterparts that with the urban life cycle always unfolding renewal efforts ideally would proceed without end.⁵⁷ "In one sense, the goal is never quite achieved," the Housing and Home Finance Agency explained. For "even those cities which have taken all of the steps required for the accomplishment of the 'Workable Program' " would need "to maintain constant follow-up activities to make certain that the improvement is permanent, that the up-grading which is achieved in a particular neighborhood will be sustained, and that new sources of blight will not be permitted to develop."[58] To assist with this process, a Rehabilitation and Conservation Branch was established at Washington headquarters.[59]

Documentation prepared by local public agencies in cities from New York to San Francisco as they organized for renewal finds assessments of residential areas much as in the "types of planning areas" of Hoyt's *Master Plan of Residential Land Use of Chicago*. In their preliminary applications for federal funds, and later, as part of more detailed plans for community improvement, agencies across the nation created statistical maps of their cities to rate the relative health of neighborhood areas on quantitative scales, accompanied by discussions of the ideal schedule for intervention. The maps communicated the belief in predictable patterns of growth, decay, and change in urban areas and that plans should follow this natural life course.[60]

Detroit, whose *Workable Program* was approved by the Housing and Home Finance Agency in June 1955, provides a vivid example. Planners there had already expressed some interest in conservation and rehabilitation in the 1940s, having prepared a citywide plan with postwar redevelopment in mind.[61] Now, renewal funds offered them the opportunity to revisit and rank the city's 155 neighborhoods toward a sequenced program of improvements. The initial resurvey concluded that the city could be "delineated into three major areas of residential patterns which further describe in a general way, the cycle of city growth."[62] Centrally located was the Old City, an area "in need of complete clearance and rebuilding." Surrounding it was the Middle-Aged City, an area "in need of both major public improvements and private rehabilitation." Finally, at Detroit's outer edge lay the New City, an area "in need of minor repairs and protection against blighting influences."[63] Envisioning a three-part program to "a. Preserve, maintain and improve new areas; b. Rejuvenate and halt the decline of sound, middle-aged areas; c. Hold the line in the seriously deteriorating areas and reverse the trend," Detroit officials portrayed the federal government's support for a three-part attack on neighborhood problems as ideally suited to their city.[64] Aware, however, of the political reality that priority determinations would have to be made within as well as among categories, they launched yet another survey to develop more nuanced neighborhood priority classifications on the follow-

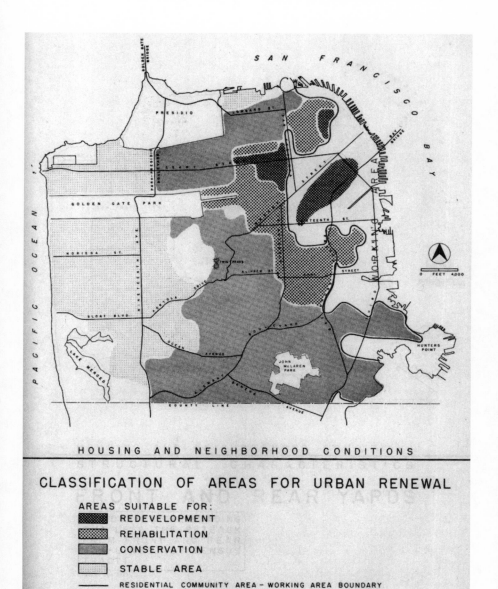

Areas for Conservation, Rehabilitation, and Redevelopment in San Francisco, 1955. From San Francisco Department of City Planning, *Housing and Neighborhood Conditions in San Francisco* (San Francisco: San Francisco Department of City Planning, September 1955), VF NAC 1613.5g27 SanF.

Detroit, Residential Areas by Type of Treatment. From Maurice Parkins, *Neighborhood Conservation: A Pilot Study* (Detroit: Detroit City Plan Commission, 1958). Courtesy of the Burton Historical Collection, Detroit Public Library.

ing scale: redevelopment, rehabilitation (also called "borderline redevelopment"), and conservation (major, medium, and minor).[65] These ratings completed, the city drafted a long-range plan for area improvement, designating "three priority stages: Stage I, to cover a three-year period from 1956 to 1959; Stage II, also for three years, from 1959 to 1962; and Stage III, a four year period, from 1962 to 1966."[66] As Follin put it, "No city in the United States has developed such neighborhood analyses as has Detroit."[67]

Most cities' *Workable Programs* communicated the same key ideas: that plans for intervention should follow the natural laws of neighborhood change; that, depending on a neighborhood's character and the chosen intervention, blight could be slowed, halted, or reversed; and that local public agencies should envision a ten- or twenty-year program for renewal, but maintaining the health of area neighborhoods would be an effort without end.[68] This common message was apparent even though the federal renewal administration, in keeping with the Republican emphasis on local control, gave cities significant flexibility in the rating systems they chose. Taking the view of the Chicago planners who observed in *Ten Square Miles of Chicago* how "the value of the rating schedule presented lies not so much in the absolute numerical scoring, as in the fact that detailed comparative analysis of selected conditions for all residential blocks is facilitated," Illinois lawmakers had not standardized definitions of blighted and conservation areas.[69] "It will be noted that standards by which various needs, deficiencies and problems are to be measured . . . is something which is left to be decided in conference in order to allow adequate leeway for differences between communities," observed the Chicago Community Conservation Board.[70] Similarly, following Robert Mitchell, who had insisted to the Presidential Advisory Committee in 1953 that there could be "no generalized program . . . because conditions differ so widely in varous cities," the Urban Renewal Administration definitions of "standard" or "substandard" housing, "redevelopment," "rehabilitation," or "conservation" areas were not explicitly laid out.[71] Local public agencies were free to borrow rating systems from the American Public Health Association (for example, used in Philadelphia), from other cities (for example, Portland, Maine, which used Cleveland's earlier work), or to devise their own scales (for example, Chicago).[72] Most would follow some variant of the three-class federal rating scale of redevelopment, rehabilitation, and conservation, with a few, including Milwaukee and Tacoma, choosing a more nuanced scale.[73] Program participants were enthusiastic about this opportunity for local control. Thus, compared with earlier efforts at the Federal Housing Administration and Home Owners Loan Corporation to establish a standardized assessment system for housing and neighborhood appraisals so that appraisers' quantitative assessments in Chicago would have meaning in Washington, DC, the renewal administration's push to develop a uniform rating scale was not quite so strong. Yet, as with disputes among scholars about urban models that remained squarely within the ecological tradition, cities as diverse as Detroit, Atlanta, Chicago, and Houston, pursued different assessment schemes with a shared understanding of the overarching forces of urban growth, decay, and change.

Teaching Citizens about Urban Renewal

With prior experiments in urban conservation and rehabilitation demonstrating the need for sustained government action and community involvement in property protection, legislators had made citizen participation a requirement of cities' access to federal funds.[74] Approval of all *Workable Program* submissions was contingent on each local public agency assembling a citizen advisory board, and in turn on organizing property owners for slum prevention in target neighborhoods. Renewal was a collaborative effort, federal documents insisted time and again, "teamwork by citizens and government to end community slums and blight" requiring "the cooperation of and participation by local public bodies, business and professional interests, and civic groups."[75] The expectation was that, once educated about the nature of cities, citizens—like the communities of urban professionals already invested in renewal—would voluntarily join the "groups working toward a common end."[76]

At the same time that the Urban Renewal Administration showed local public agencies how to prepare *Workable Program* maps and documentation, then, it created materials to help local governments mobilize neighborhood participation. Less technical than the educational support provided to city officials, the message was the same: that the urban life cycle was an entirely predictable proposition; that the federal government had identified the required tactics to slow, halt, and reverse the processes of urban decline; and that continuous community action on property maintenance at the local level, rather than piecemeal individual efforts or government action alone, would be critical to the renewal program's success.[77] The seven elements of the *Workable Program* became "seven basic points of community action" as the administration set out to enlist citizens in renewal — for example, by offering advice on how to take advantage of available federal funds.[78]

More numerous than federal brochures for citizens, however, were those produced by local public agencies themselves. That engaging private citizens in the renewal process was not merely a legal but also a logical necessity was a sentiment local officials widely shared. "Urban renewal will *not* get underway until there is an insistent, vocal demand from the community to do so," noted city planners in San Francisco. "The efforts of a small group of civic leaders and dedicated public servants" simply would not be sufficient.[79] Colleagues in Baltimore agreed. "The urban renewal program which to such a large measure consists of conservation and rehabilitation of residential properties, cannot possibly succeed without considerable community participation," observed Albert Rosenberg, director of the Community Organization Division of the Baltimore Urban Renewal and Housing Agency. "The reasons are obvious: many people must be induced to improve their properties . . . to correct code violations . . . to bring neighborhood standards to levels even higher than those required by ordinances."[80] Following the federal government's conceptualization of renewal as a multiparty collaboration — in the words of the San Antonio Department of City Planning, "it is a community-wide program which requires private enterprise, citizen support, individual cooperation, and governmental assistance" — as the program expanded across the country, many local public agencies set out to educate citizens about their central role in renewal's ultimate fate.[81]

As the federal program embraced area improvement strategies that spanned redevelopment, rehabilitation, and conservation, the goal for these educational campaigns varied. Some publicity materials tackled the entire program, but more often, cities provided information about slum prevention activities separately from slum removal. In the case of rehabilitation and conservation, the goal was as it had been in the Waverly and Woodlawn experiments — convincing citizens to intervene in areas that appeared to be in good order.[82] A basic outline of the urban life cycle and its potential interruption through community action frequently appeared. "You live in a neighborhood in a fairly new section," praised a brochure produced in Detroit; yet, it insisted that "no neighborhood, however good or new, will continue that way unless the people who live there — you and your neighbors — work together to maintain and improve it."[83] To this common format cities appended varied messages, reflecting the variations in resources offered to assist residents in their work. Thus, when it came to mobilizing private investments in a property maintenance program many regarded as too costly, Oakland, Cleveland, Baltimore, and Wilmington touted the public and private monies set aside to assist homeowners with modernization, while Boston and Los Angeles, lacking independent modes of financing, emphasized available federal assistance and the notion that "normal maintainance and repair" would have "no effect on valuation for assessment purposes, except to maintain value."[84] Convincing citizens of the necessity of demolishing their homes and relocating them was a trickier proposition, and this genre of brochures followed a standard script. They explained citywide plans for redevelopment, described specific plans for the future of the occupants' neighborhood, answered anticipated questions about the financial and rehousing assistance available, and urged calm in the face of rumors. Vallejo, California, for example, told its residents of clearance areas how "there is no need for alarm as long as you ... Pay your rent ... Cooperate with the agency ... Do not use the premises for unlawful purposes ... DO NOT BELIEVE RUMORS."[85]

Campaigns to promote renewal did not depend on citizens' reading habits alone. Public meetings, demonstration homes, opinion surveys, educational films, and exhibitions of scale models were among the varied tactics that cities, frequently in collaboration with area businesses and nonprofit organizations, used to ensure community-wide familiarity with the renewal task. Chicago, Detroit, Cleveland, Philadelphia, and New York set up information centers in several neighborhoods. (One of Detroit's information centers was a demonstration home; one of Cleveland's, which offered classes on do-it-yourself home repair, was organized in collaboration with local neighborhood improvement associations.) Cincinnati and Dayton sent a publicity trailer throughout their cities. St. Louis used educational television programming. Boston created a short film, *You and Your Neighborhood*, the story of Joe Citizen's efforts to organize his neighborhood. Philadelphia's program was especially comprehensive: the city created an information center, tenant education, television and radio spots, conferences, elementary through high school curriculum, and neighborhood tours.[86]

Several cities, following the precedent established by Waverly and continued by Chicago's Community Appraisal Study, assumed responsibility for forming new block clubs and neighborhood associations in areas targeted for rehabilitation or conservation—often with assistance from local universities, churches, and social service agencies. In Detroit, for example, on the eve of the federal renewal program's creation, the city formalized a neighborhood conservation program by taking three interrelated actions. First, it established a Committee on Neighborhood Conservation comprising heads of city agencies and representatives of area civic, business, and other organizations appointed by the mayor. Second, it named a conservation staff at the city's planning commission. Third, taking the position that "conservation is essentially a grass-roots attempt to maintain, improve or remake our neighborhoods fit for modern living," and that target areas were likely to be disorganized, the city planning commission established a Community Organization Division, hiring sociologist Mel Ravitz to mobilize citizens groups to complement the ongoing efforts of the conservation committee and the planning commission.[87]

Ravitz, observing how many of the "middle-aged" areas that looked fine on the surface soon would be in need of repair—and that "some of the white families are abandoning them for newer outlying neighborhoods as the Negro families move into them from the older, central core of the city," teamed with colleagues in the Conservation Division and the city's Committee on Neighborhood Conservation to teach citizens about how they might "live better where they are" by interrupting the cycle of decay.[88] They started in Boulevard-Gratiot-Mack, a neighborhood "in transition."[89] Emphasizing to community members that "the greatest resource in neighborhood conservation is the people" and that "conservation is the peoples' opportunity to cooperate to secure 'better neighborhoods without moving,'" these collaborators prioritized "helping the people of the pilot area to organize" a local property improvement effort.[90] Even with assistance from volunteers from the Detroit Federation of Settlements and Neighborhood Centers of Metropolitan Detroit the process took six months, but by the end of 1954, "all of the 38 blocks of the neighborhood" had been organized and a Boulevard-Gratiot-Mack Conservation Council created.[91] In early 1955, the council offered feedback to city officials on the plan commission's proposed physical upgrades to the area, and in June 1955, the city sent on to the Housing and Home Finance Agency its application for Detroit's first conservation project under the new federal legislation, "supported by the Boulevard-Gratiot-Mack Neighborhood Council in a letter to the City's Common Council."[92] As Maurice Parkins noted with pride, "This plan was not imposed from above and its acceptance has not relied upon compulsion, but was developed with the full cooperation and participation of both the city department and the residents of the area."[93] Citizen-government consensus on slum prevention was an achievable proposition, he observed, emphasizing the citizenry's "acceptance of the plan was primarily due to the fact that its aim, purposes and goals were recognized and fully understood by the individual residents of the neighborhood and their organizations."[94] Whether this consensus could be translated into action remained to be seen.

Time for Action

Inside the Urban Renewal Administration and local governments, at universities, businesses, and social service organizations, a variety of actors rallied around the idea that, by following a rational plan of action focused on physical improvements, the decline of urban neighborhoods was no longer an inevitable proposition. These extant institutions, however, were not the only ones taking a leadership role to mobilize local public agencies and citizens for the renewal task. One organization founded in the wake of the 1954 Housing Act stands out for its contributions to renewal in cities across the United States: ACTION Inc., the American Council to Improve our Neighborhoods (later renamed ACTION Inc., the National Council for Good Cities), established in 1954 in New York City on the initiative of "distinguished civic leaders and businessmen" as "the national citizens' organization dealing with some of urban renewal's most difficult problems."[95] ACTION provided a new home for many of the figures who had been working since the 1930s to make a nationwide program of urban resource management a reality. With sponsorship from the Ford Foundation to undertake a new generation of urban research and publicity from the Advertising Council for a campaign of "Action on Slums," their continued influences would be widely felt.[96]

Indeed, the cast of characters at this organization included many of the urban professionals from Chicago and other cities who had worked for two decades at the Federal Housing Administration and Home Owners Loan Corporation; at the National Resources Planning Board and the National Association of Real Estate Boards; on the Waverly and Woodlawn conservation experiments; the Community Appraisal Study; and the Subcommittee on Urban Redevelopment, Rehabilitation, and Conservation, to make life cycle planning the basis for federal law. Albert Cole, for example, was hired as ACTION's president immediately after his resignation as administrator of the U.S. Housing and Home Finance Agency; later he became board chair. Chicago mortgage banker Ferd Kramer, advisor to the Woodlawn conservation experiment and Community Appraisal Study participant, served as board chair and subsequently as director of research. Martin Meyerson, previously of the University of Chicago planning program and the Community Appraisal Study, signed on as board vice chair and then vice president for research. Baltimore mortgage banker and former Federal Housing Administration staff member James Rouse, who had headed the Cole Committee's Subcommittee on Urban Redevelopment, Rehabilitation and Conservation, led ACTION from 1958. Hired as consultants or sponsored for further research investigations, planning and real estate experts, including Robert Mitchell, James Downs, Hideo Sasaki, Jack Siegel, Walter Blucher, Miles Colean, John Dyckman, Reginald Isaacs, Edward Banfield, and Paul Ylvisaker, gained access to new funds and new audiences for their ideas.[97]

That these figures would continue their work under the auspices of a private organization was a savvy choice. Even during a decade when urban affairs commanded new levels of federal attention, the renewal policy crafted by a Republican administration stressed the importance of private industry and individual initiative in the program's success as it limited the support that government agencies could provide.[98] Although the organization's original charge was to advise the American citizenry, ACTION affiliates quickly designed for themselves a far broader mission, creating divisions for research (to sponsor academic projects), information (to disseminate publicity), and field service (to offer technical assistance to local public agencies and citizens groups). In short, finding that the Housing and Home Finance Agency would not be following the U.S. Department of Agriculture model they previously had praised—in which the federal government simultaneously sponsored research, provided education and information services, and assisted local governments and citizens groups to implement a set of national goals—ACTION affiliates sought to equip their private organization with the means to do the job.[99] Central to its work was brokering relationships between the many individuals and institutions who needed to come to consensus for renewal to succeed.

ACTION's efforts to reach audiences that spanned urban professionals in planning, real estate, academia, and local government together with renewal participants, including universities, businesses, churches, social service agencies, and citizen organizations, are clearly apparent in the organization's monthly newsletter *The Action Reporter*, which stressed these groups' common interests in improving the future of U.S. cities.[100] This aim of orchestrating government action and citizen initiative is equally apparent in the other activities this organization pursued. Representatives of local public agencies and businesses could attend ACTION-sponsored conferences and workshops and get advice about building momentum for community improvement—for example, with demonstration homes. Block clubs and neighborhood social service organizations could read about the achievements and frustrations of citizens groups such as Philadelphia's Quakers and Chicago's Hyde Park–Kenwood Community Conference in stimulating property protection and watch the movie *Man of Action*, the story of John Q. Public, who, recognizing local officials could accurately predict the city's future but could not reverse the cycle of decay and decline without community involvement, worked to organize his neighbors to combat urban blight.[101] ACTION's upbeat message emphasized how "urban renewal is achievable if it is kept in mind that every public and private building project makes its contribution to the sound maintenance and development of the urban area," and repeated Eisenhower's observation that "the Federal Government, or any government, can succeed only as each individual citizen assumes his own local responsibility." Thus, as they heard in a variety of federally produced materials about how urban renewal was an opportunity to "help communities to help themselves," and how it was "teamwork by citizens and government to end community slums and blight," the diverse participants in the renewal effort heard nearly the identical message from this national organization—a turn of events perhaps unsurprising given the central role of ACTION affiliates in crafting the federal policy itself.[102]

From Planning to Implementation

Scholars who study the construction of public problems recognize the rarity of occasions when diverse actors come together to speak a common language and rally in support of a single issue. This book has documented how on account of the persistence of a community of enthusiasts working for several decades, by the late 1950s, a consensus had emerged that urban areas possessed predictable life cycles and that physical planning to slow, halt, or reverse the cycle of decay and decline, rather than allowing "nature" to take its usual course, would save postwar American cities.[103] Urban renewal "represents a complete change from the concept that a city must, of necessity, decay, wear out, and then be abandoned to blight," observed scholars in the *Indiana Law Journal*. "It asserts the belief that cities can be brought and kept up-to-date—made and kept livable," wrote the Saint Paul City Planning Board. "We do not have to capitulate to a cycle of *growth* and *decay*—help us to *conserve* and *renew*," declared Hyman Levine of Chicago's Greater Lawndale Conservation Commission.[104] From the late 1930s, when the Home Owners Loan Corporation first suggested that it would benefit from conservation "by the increased value of the properties it owns, just as would every private homeowner; that mortgage companies would be making their mortgages more secure and find an outlet for funds now idle; that the city would be rewarded for improvements by sounder technical values and the avoidance of future slums," to the suggestion of National Association of Real Estate Boards in the 1950s that conservation "*benefits everybody*," slum prevention increasingly became the focal point of this broad consensus.[105] Although there were still a few ongoing disputes between those "who see conservation primarily in terms of volunteer community organizations, as a matter of community cooperation at the 'grass roots' level, and those who regard it primarily as a governmental problem arising out of a breakdown of municipal housekeeping and services, and faulty municipal administration," most agreed with how the renewal program framed matters—insisting on contributions from both.[106]

Yet, as Chicago and other cities made the transition from planning to implementation, the apparent consensus began to break down. "The happy unanimity over the general goal faded," Peter Rossi and Robert Dentler observed of Hyde Park, "when the particulars of renewal planning began to be spelled out."[107] Teamwork between local governments and city residents, a centerpiece of the program, soon faltered on numerous counts.

Disputes over slum clearance were especially rancorous. Community groups organized to aid their cities turned on their sponsors to protest the definition of "blighted areas" and challenge forced removal, stalling the implementation of some cities' plans.[108] In Washington, DC, for example, even before the passage of the 1954 Housing Act, the first redevelopment project had to be abandoned because residents refused to leave.[109] Charges of "Negro removal" confronted program organizers as it became apparent that neighborhoods targeted for demolition and citizen relocation were largely African American.[110] Representing a group of African American Chicagoans, Roosevelt University professor St. Clair Drake, who had served on the Community Survey Committee of the Hyde Park–Kenwood Community Conference, went head-to-head with University of Chicago administrators at a city hearing to dispute their classification of his neighborhood as a slum and suggested it instead should be targeted for conservation.[111] Residents complained that local public agencies, as unelected bodies, did not represent the "public interest," and filed lawsuits in numerous jurisdictions to dispute the legality of renewal plans.[112] Los Angeles's plans for its Bunker Hill area, designated for redevelopment as early as 1951, were delayed for years.[113]

As local public agencies identified more projects for spot demolition and relocation than originally anticipated, citizens' early enthusiastic embrace of rehabilitation and conservation also began to wane. In Detroit, for example, the planning commission's enthusiasm about the "high level of citizen participation and cooperation" indicated by "actual improvements in private property" was replaced with charges that citizens were "untrustworthy" when staff discovered that action on property maintenance in the city's pilot conservation neighborhood was no greater than in a similar control community nearby.[114] Neighborhoods' commitment to slum prevention eroded still further as city officials asserted control over community organizations, expecting that residents' participation would be restricted to creating "a general awareness of the importance of urban renewal" and providing "the understanding and support which is necessary to insure success," while giving business interests the upper hand in urban redesign.[115] In Chicago, for example, Mayor Richard Daley appointed local Community Conservation Councils, insisting that in the effort to promote community property standards landlords, businesspeople, and local officials be represented alongside social welfare and neighborhood groups.[116] The anticipated costliness of slum prevention, together with concerns that such maintenance might raise property values and push residents out, became yet another obstacle to participation—even in relatively affluent neighborhoods.[117] Whereas Detroit, Oakland, Baltimore, and Wilmington provided funds to reduce individual homeowners' expenses, more cities—from Boston to Little Rock—did not.[118] Elsewhere the problem of relying on voluntarism for a program of continuous city improvement became broadly apparent when residents lost interest in the program. Property owners in some rehabilitation areas, for example, in Philadelphia, more apathetic than hostile, simply dropped out.[119]

These problems were compounded when, in the communities where residents did embrace rehabilitation and conservation, local governments did not live up to their promised obligations. The North Lawndale Citizens Council, unable to persuade Chicago's Buildings Department to enforce building codes and ordinances in their neighborhood, leaked the back-and-forth correspondence to the *Chicago Daily News* to gain publicity.[120] Similarly, in Detroit, residents of the city's flagship conservation neighborhood "expressed a good bit of resentment toward the city in connection with code enforcement," reporting complaints to the frequent surveyors who studied their community in the hope that these would make it to city leaders' ears.[121]

Although city residents' frustrations with "the government" in this program generally referred to local governments, federal officials shared blame for this state of affairs. That the technical advice provided by the Urban Renewal Administration did not extend to defining standard and substandard housing gave local public agencies the opportunity to appease mayoral and business interests by targeting neighborhoods for redevelopment and conservation at whim.[122] In New York, for example, charges of corruption and favoritism followed the application of eminent domain to create public facilities such as Lincoln Center and the New York Coliseum, sited in areas not obviously classified as "slums."[123] With congressional appropriations limited to two- and three-year increments, and with highway and public housing programs competing for attention and funding, cities were stymied in their efforts to develop long-term comprehensive renewal plans.[124] In Detroit, for example, despite the recognition that "the matter of timing" would be "of crucial importance," four years after Urban Renewal Administration Commissioner James Follin's praise for the city's *Workable Program,* only one of the more than fifty conservation areas had advanced to "the project execution stage," and once enthusiastic citizens now faced a "demoralizing hiatus."[125] Even for those projects that tempted developers away from suburban areas to build new city homes, administrative red tape lengthened time to completion, particularly following building demolition.[126] In Buffalo, for example, a decade of work on one project yielded only six new homes.[127] The Housing and Home Finance Agency, while making voluntary community action a requirement for cities' access to federal funds, and even commissioning research on block clubs' best practices, offered its resources only to individual homeowners rather than community groups.[128] With cities like Chicago refusing to certify conservation without demonstrated community-wide investment, this further slowed the renewal task.[129] Despite the Presidential Advisory Committee's recognition that "slum prevention is cheaper than slum clearance," and that "rebuilding our present housing supply" would provide new business opportunities for the housing industry, in an era in which other Housing and Home Finance Agency programs celebrated the pastoral ideal that growing suburbs represented, the Urban Renewal Administration failed to launch a rehabilitation and conservation industry for urban housing or to publicize available resources for individual property owners.[130] Although federal officials had hoped to create a renewal program dominated by conservation and rehabilitation and "to use the clearance process selectively like a scalpel rather than indiscriminately like a bulldozer," in the words of Philadelphia city officials, in the end, the "federal bulldozer" prevailed.[131] By 1960, the National Housing Inventory had documented how the overall supply of housing across the United States had increased faster than the U.S. population—but not in central cities where most renewal projects occurred.[132]

Questioning the Theory behind the Program

Compounding these problems was the evidence, as renewal went on, that the program's linear life cycle model of neighborhood growth, decay, and change—with its expectations about how this predictable pattern of events might be interrupted—was ill-equipped to anticipate the effects that the complex interplay among urban physical, economic, and social forces actually produced. Even though calls for a program of human conservation in cities had long been discarded, public and private organizations touted the many economic and social benefits that would follow directly from urban physical transformations. Yet, these linked improvements did not arrive. By destroying the informal economy and social order of communities that existed in blighted areas, and reducing the supply of affordable housing, redevelopment exacerbated the problems many of these neighborhoods' residents faced. Officials occasionally claimed that replacement homes were available when none existed; persistent racial segregation and a lack of low-cost housing alternatives put population pressures on some conservation areas, creating new slums.[133] The widely publicized campaign for "no slums by 1960" fell short of expectation.[134] As Walter Blucher noted in 1960, "The slums and blighted areas of America are growing faster than they are being destroyed."[135] "Urban renewal —for all its promise and its valid physical accomplishments—has barely touched America's urban chaos," James Scheuer of the New York City Renewal and Development Corporation admitted to the 1964 meeting of the American Society of Planning Officials.[136] Postwar cities remained a "mess" in the minds of many observers.[137] "The slum has too often merely been pushed next door," concluded one real estate developer. "The real needs of humans have been subordinated to concepts."[138] Despite the belief expressed by so many participants that, in the words of the University of Chicago's Jack Meltzer, urban renewal reflected the latest "phase in our knowledge of the science of cities and their people," aggregating these and other problems critics pointed to the limits of this scientific tradition.[139]

Perhaps the most significant weakness of this theory of city improvement was its failure to recognize that physical change would not be the deciding factor for city residents contemplating whether to remain in their homes. Among property owners in communities in "transition" it was the character of owners more than property that mattered. As the renewal program continued to unfold, a few white neighborhoods would use rehabilitation and conservation as techniques for neighborhood stabilization in the tradition of the Waverly and Woodlawn experiments, seeking to prevent shifts in racial balance with implications for property valuation. Thus, in Chicago's Back of the Yards neighborhood, best known as a site for community organizer Saul Alinsky's work, the conservation program brought together previously antagonistic groups of Serbs, Croatians, Czechs, Slovaks, Lithuanians, and Poles on matters of common interest, including the desire to hold off a pending influx of racial minorities. The program's first year saw 560 loans granted for home improvement.[140] Buffalo's Fruit Belt, a community poised to transition from Italian American to African American, organized its conservation program toward a similar goal.[141] City officials reported that "every single dwelling of the 1,461 in the project area was thoroughly inspected ... The cooperation of the homeowner and others concerned with the program was good and the rate of compliance was high."[142] In these neighborhoods, as Herbert Thelen had feared, the quest for community improvement was motivated by the desire for racial homogeneity.[143]

More often, however, in a context where the restrictive covenants that organized earlier residential segregation had been declared illegal, yet real estate appraisers continued to value homogenous communities, white property owners who witnessed neighborhood change chose to depart for suburban alternatives, leaving the conservation and rehabilitation effort to African American city residents. Thus, surveys of Detroit's pilot conservation neighborhood found that longtime resident whites ranked "the type of neighbors moving in" over property deterioration as a matter for concern, telling surveyors they were less eager to participate in community improvement than they were to leave.[144] African American residents, by contrast, still held out hope for the city-organized conservation effort. "Negroes in the neighborhood are more willing to join such an organization and work in it," wrote observers from Wayne State University's Sociology Department in the mid-1950s. "These facts represent a fundamental challenge to efforts at organizing the residents of the area into effective block units."[145] Later analyses by the Detroit Urban League and the city's Plan Commission confirmed these divergent feelings about slum prevention persisted among the property owners who remained. As the Urban

League reported on African Americans' continued interest in neighborhood improvement, noting that "most of the block clubs (99.2 percent) had a majority of non-white members," the Plan Commission found growing hostilities among area whites who "did not temper their criticism of Negroes."[146] A typical comment: "All these colored moved in, each with 15 or 20 kids; they just rant all day long. Noisy! (Female single-family owner, 14 years in Conservation Neighborhood.)"[147]

Similar attitudes prevailed in other cities. In Philadelphia, for example, where local government officials took an active role "to acquaint minority leaders with a knowledge of renewal problems and methods of developing local organization and citizen participation," surveyors found "great resources of initiative within its Negro population." More specifically, "over 1200 block improvement groups" were organized "in older largely Negro neighborhoods... In some of the most active urban renewal areas Negro leaders are helping to further pioneering rehabilitation programs."[148] At the *Chicago Defender*, which ran numerous articles critical of slum clearance — and indeed was so concerned about the negative effect of these policies that it had established a public service bureau to mediate between city officials and residents likely to be affected by the 1947 Blighted Areas Rehabilitation Act — staff seconded this support for slum prevention.[149] They did so through not only publicity in news and opinion articles but also by editor and publisher John Sengstacke and his colleague journalist Enoc Waters taking on leadership roles in citywide and local conservation projects. Sengstacke became vice chair for the Chicago branch of Operation Home Improvement, and Waters served on the board of the Greater Lawndale Conservation Commission.[150] The enduring commitment to conservation and rehabilitation in mostly African American neighborhoods was consistent with earlier property improvement efforts among middle-class African Americans as Thomas Philpott has described. While whites did not see gradations of more or less desirable African Americans with whom to associate, middle-class African Americans considered themselves to live in comparatively desirable neighborhoods they hoped could be maintained.[151] The appeal was straightforward: As the Detroit conservation program motto "live better where you are" implied, success in slum prevention would forestall "Negro removal" in these communities. In fact, in several cities, including New York and St. Louis, African Americans, judging rehabilitation and conservation as appealing alternatives to the clearance they vehemently protested, pushed for the expansion of these measures.[152] In many such efforts, as in Chicago, local Urban League chapters—seeing political opportunities in block organizations—provided administrative and financial support.[153] "A block club can be more than an acceptable means of maintaining a physically attractive neighborhood; a congeries of busy-bodies; an extremely vigilant pressure group; so many votes in the next election; or only a vague term," noted the League's George Henderson in 1960.[154] Henderson, like John Dyckman, believed conservation was democracy in action; he concluded that the tool earlier used to promote racial separation was just what the nation needed to maintain civic spirit in its cities. "That which encompasses the concept of 'block club' may be a means of bringing about a more stable and democratic society."[155]

At their debut, theories of the urban life cycle had recognized the ways in which the movements of antagonistic population groups were intertwined. Race and nationality had been central in the earliest ecological studies of U.S. cities and a focal point of the first experiments with neighborhood property conservation in Baltimore and Chicago. The Community Appraisal Study, recognizing the need to reduce tensions in areas experiencing demographic change, had made training in human relations central to its mission. Yet scaled up to the national level the urban renewal program banked on the expectation that forces of urban decline might be interrupted through physical change alone. Like Homer Hoyt's *Master Plan of Residential Land Use of Chicago*, with its scientific language of statistics and "types of planning areas," reflections on race largely disappeared from the documents, maps, and brochures that cities produced. This absence was especially striking in the materials distributed to residents of redevelopment areas: The populations soon to be displaced were depicted as white (as examples from Pittsburgh, Sacramento, Philadelphia, Marina Vista, and Vallejo demonstrate) when in fact the majority relocated were not.[156]

Although educated urban professionals could pontificate about how racial minorities were not the cause of urban blight, and propose how block organizations for property improvement would "reduce racial tensions if the area is mixed," by involving "individual members in problem solving that transcends race, color, or creed," such observations had little influence on the behaviors of white families, an effect exacerbated by persistent real estate valuation practices at federal and local levels.[157] Restrictive covenants were now illegal, but associations between heterogeneous racial populations and declining property values remained.[158] Even in the few neighborhoods where community leaders explicitly espoused integration among "discordant ethnic, racial, or economic groups," racial tensions flared.[159] Such was the case in Chicago's Lawndale, where a coalition of African American and Jewish residents ran the Greater Lawndale Conservation Commission.[160] For example, prominent local white businesspersons A. R. Cox (from Sears) and William G. Dooley (president of the Sears Community Bank), when invited to join its board, at first demurred saying neither "could accept the posts offered them on the commission because of the attacks made on them personally and on their businesses by groups in the community" across the racial spectrum "opposed to a Conservation program."[161] Similarly, while neighborhood surveys found most African Americans to be enthusiastic about conservation and rehabilitation — "the interest of Negro homeowners has been high in terms of local improvement" — simultaneously from within the community, "a tiny, though vociferous segment" remained committed to "spreading misinformation and fear about the program."[162] (Speculation was that these were most likely South Central Association executive director and local African-American political leader "John Ragland and his supporters.")[163] This included one episode in July 1955, when, pretending to represent the NAACP, these conservation opponents sought to erode African American support by distributing leaflets across the area equating the local conservation plan with Jim Crow.[164]

A few cities, like Chicago, chose to assign their human relations commissions to the job of recruiting broader support for slum prevention in neighborhoods in transition. Project leaders were daunted by the challenge. "Changing neighborhoods" remained "likely places for outbreaks of extreme tension, hostility, and even violence," noted John McDermott of the Philadelphia Commission on Human Relations, skeptical about the likelihood of bringing white residents into the renewal effort. "Efforts to stabilize the changing neighborhood are very important even if they eventually fail and the neighborhood becomes incorporated into an advancing ghetto," McDermott offered, seeking to scale back public expectations. "Such programs can have the important by-products of first, a positive educational effect on those who eventually move away; and second, an orderly transition in the life of a community."[165] In Philadelphia and other urban areas, McDermott's predictions came true as the white flight beyond city limits that had accelerated in the late 1940s showed no signs of stopping by decade's end.[166]

Academic critiques of human ecology in the 1930s had not ended the enthusiasm for its applications to explaining urban life, but disappointments with its abilities to predict the outcome of urban renewal in the 1960s largely did. The voices of caution that had been silenced as the momentum for life cycle planning grew — for example, real estate professionals who questioned the efficacy of block clubs, local officials anxious about the political consequences of a city improvement program with no yardstick of success, and citizens frustrated by the apparent gains for mayors, businesses, experts, and the middle classes in the face of little progress on housing for the urban poor — found themselves attracting new audiences as they assessed the outcomes of the renewal effort.[167]

At first, pointing to improved citizen-government relations in some cities, and the decreasing rate of slum creation, diehard supporters mounted a countercampaign. ACTION, for example, which insisted that the urban renewal program was "still experimental" and might be refined, used its connections to place positive articles in the national press.[168] "Slums are decreasing in this country at the rate of about a half-million dwelling units per year — largely due to rehabilitation," Urban Renewal Administrator Richard Steiner told the National Association of Housing and Redevelopment Officials (formerly the National Association of Housing Officials).[169] His successor David Walker suggested the "growing recognition of the permanent or continuous nature of the program" — the same factor that made it difficult to sustain citizen interest — should make it "as much a part of municipal life as zoning, traffic engineering, or fire prevention."[170] Yet as initially encouraging reports on Hyde Park — the nation's first official renewal project and a media magnet for positive portrayals — gave way to less flattering depictions of a neighborhood area experiencing the same tensions as so many other communities, these damage control efforts could not be sustained for the longer term.[171] From Walter Blucher, Martin Meyerson, John Dyckman, and Jack Meltzer to the Urban League and the American Institute of Planners, in the face of mounting critiques, many of the individuals and institutions that only a few years earlier had enthusiastically rallied around a national renewal effort would repudiate the program.[172]

CONCLUSION

From Ecology to System

Historians have long recognized how Americans' enduring preference for nature over cities shaped the history of urban areas in the United States from the nation's origins to its recent past. In their accounts of Garden Cities and New Towns, the public park and antipollution movements, and the rise of suburbia and exurbia, stories abound of the individuals and institutions who, by bringing nature to cities, saving nature in cities, or fleeing cities for nature sought to deliver to property owners a closer relationship with the natural world. Yet, during the twentieth century, nature's priority in the American mind affected the fate of U.S. urban history in another important way. This book has documented how, over four decades, the nation's approach to the scientific understanding and management of nature provided a conceptual and an organizational model for three communities of urban experts to emulate as they sought to push the needs of cities into more prominent public view. Eschewing traditional oppositional frames for relationships between urban and rural areas, they focused, for a time, on continuities instead.

The Nature of Cities takes seriously the analogical thinking that dominated so many discussions about cities between 1920 and 1960 to argue that a new narrative should assume its place in urban histories of the period. This narrative reveals how claims about urban problems and solutions mapped onto the changing political position of the nation's movement for natural resource conservation. Bringing perspectives from the history of environmental science to trace the rise and fall of a dominant understanding of cities as "ecological" communities and national "resources," the book offers an alternative historiographic context for assessing the work of many well-known urban actors. It suggests that we cannot understand the broad popularity of ecological models of cities from the 1920s, nor the decades-long mobilization for a city improvement program pairing technocratic expertise and citizen action that culminated in urban renewal in the 1950s, without attending to these connections. And it points to a broader legacy for Frederic Clements's ideas.

Environmental History Meets Urban History

Historians of environmental science have characterized the period between 1920 and 1940 as a time when ecology and natural resources planning garnered substantial national attention. In the 1920s, following the close of the frontier and the recognition that cities were expanding into rural areas, land use issues became matters of growing expert concern. Environmental problems during the depression of the 1930s brought a new level of public attention to the scientists and scientific managers seeking to address the nation's resource problems. From range management to soil conservation, central in the policy planning was Frederic Clements's contention that, if subjected to scientific analysis and control, present environmental problems might be reversed and future ones prevented. Through public awareness campaigns and work-relief programs, the importance of an ecological approach to natural resources planning became known to citizens across the United States.

These developments lay the foundations for the rise of naturalistic thinking in the urban professions. In a period of disciplinary formation in American social science, sociologists, geographers, and economists were attracted to ecological modes of interpretation because of the scientific direction such ideas represented for the nascent study of cities. Drawing on a range of competing traditions in plant and animal science to emphasize the similarities between human communities and associations in the plant and animal world, they found in Clementsian theories of succession to climax the suggestion of order within the disorder of U.S. cities and a template for explaining and forecasting urban patterns. Such appeals to nature provided a starting point for professionals in two fields — city planning and real estate — seeking an applied science for their work on the urban "frontier" in the wake of a city-building boom and subsequent economic depression. City planners seeking to remedy the growing physical, social, and economic disrepair in urban areas saw great promise in the thinking that united predictive social science with the prestige of natural resources planning. Real estate appraisers eager to maintain the security of their investments discovered in these city models valuable knowledge about the predictable cycles of neighborhood physical, economic, and social change. For these two expert communities concerned about the acceleration of urban decentralization, social scientists' "scientific" visions of cities as ecologies offered new arguments for intervention toward longstanding professional aims—and scholars' appropriations of Clements's ideas in particular helped pave the way for associating the urban professions with the scientific management of nature that already had captured the public's attention.

The vogue for a general scientific management in the urban professions that historians already have described, in other words, had important links to more specific matters of pressing national concern.[1] Following natural resources planning colleagues such as Hugh Bennett and Rexford Tugwell, for whom ecological models lay the foundation for restoring nature's balance, among these analogy-building actors similar hopes swelled for the urban sphere. City planners, calling cities "national resources"— the category President Franklin Roosevelt used to refer to farms and forests—and explaining how the pioneer mentality that had caused the environmental and human degradation of the Dust Bowl equally plagued urban areas, petitioned for a comprehensive program for "the conservation of physical and human resources."[2] Real estate appraisers, observing how the Clementsian school offered not only a theoretical paradigm for explaining the urban life cycle but also a rationale for sequenced intervention to return disordered communities to more stable states, set out to demonstrate in Baltimore and Chicago how collaborative efforts from experts and citizens on the more narrow matter of physical property protection might interrupt the otherwise "inevitable cycle of urban birth, life and death."[3]

The outbreak of World War II amplified the country's attention to managing nature as the continued availability of resources became a national security issue. Historians of environmental science have documented how, with the United States' entry into the war, President Roosevelt reoriented discussions of conservation to address these new concerns. As programs for the conservation of human resources such as the Civilian Conservation Corps were dismantled to mobilize citizens for military participation, the efforts to conserve physical resources that had enlisted the support of so many urban and rural citizens now garnered new rationales — patriotism and the national defense. In calls to cultivate victory gardens and gather scrap metals to enable maximum wartime production, the home front campaign kept national resources planning prominent in the public eye.

With security issues at the forefront of the American psyche, momentum for urban revitalization dwindled, and the links to the natural resources conservation movement that earlier had provided a compelling rationale for insisting that cities receive more substantial public consideration now lost some of their powers of persuasion. This new context for conservation nevertheless paved the way for the continued influences of naturalistic thinking in the urban professions. Early perceptions of success with community conservation in Baltimore and Chicago — the belief that a "simple formula" was "now available" for saving neighborhoods — were a point of pride that few urban professionals were prepared to ignore.[4]

Confident that cities were "ecological" communities and national "resources" whose life cycles might be interrupted, a few impatient advocates followed their natural resources planning colleagues to suggest how the continued conservation of urban neighborhoods might be redefined as an urgent matter of national security. Others preferred to wait; they hoped that the momentum for natural resources planning building before the conflict might be applied to a more expansive vision of the conservation of urban resources as part of later postwar demobilization instead. Homer Hoyt's wartime work as research director at the Chicago Plan Commission exemplified this latter approach. Turning his prior contributions to models of the inevitability of urban decline on their heads, he insisted these visions of cities' natural laws offered inspiration for building on previous efforts at neighborhood conservation, expanding tactics of scientific management to encompass both clearance and conservation and taking strategies to interrupt the urban life cycle citywide.[5] Finding a middle ground between city planners' visions of conservation as master planning for the rehabilitation of property and people and real estate professionals' hopes for conservation as neighborhood property protection, Hoyt suggested in his *Master Plan of Residential Land Use of Chicago* (1943) how, by undertaking a continuous program of urban physical revitalization in which both government and citizens did their part, the city in a garden would be "constantly renewing itself."[6]

The postwar period witnessed the beginnings of an active questioning of the approaches to engaging with nature that had dominated earlier decades. Historians of environmental science have chronicled how, from the wisdom of organizing resource planning toward business development to the value of Clements's equilibrium paradigm as the basis for land use interventions, a range of past assumptions came under assault before ceding to a new environmentalism that characterized the years after 1960.[7] As the United States faced the Korean conflict and the beginnings of the Cold War, for a time under President Harry Truman resource planning policies continued to be aligned with national security concerns. Yet, from the 1950s, the opportunity to assess the payoffs from years of massive governmental investments in conservation revealed results that did not deliver on their ambitious promises. Many of the short-term outcomes so widely praised were erased in the longer term as environmental problems returned. With the arrival of Truman's Republican successor Dwight Eisenhower, whose limited patience for resource planning was apparent even before his ascent to the White House, public concerns about conservation soon took a backseat to the nation's security needs.

This new political context set the stage for the final flowering of explicitly naturalistic thinking in the urban professions and the irony of the subsequent disappearance of this language even as a life cycle template became the basis for federal urban policy. In the immediate aftermath of World War II, a Chicago-based community of experts from academia, planning, and real estate, led by Michael Reese Hospital planning director and Harvard planning professor Reginald Isaacs, rallied to realize in their city the vision Homer Hoyt had laid out during the conflict. These parties, successful in achieving legislation for redevelopment and conservation at the state level and in recruiting a broad base of citizens to participate in the improvement effort, pressed to expand the potential applications of their work. Public confidence in the Chicago experiment helped to inspire a program for nationwide urban renewal under the federal housing legislation of 1954.

The juxtaposition of government messages to participants in urban renewal with government messages of past federal programs for resource conservation underscores a remarkable continuity of themes: from the notion that land can wear out to the pairing of economic and moral arguments about waste, from the claims that scientific study of land made it possible to forecast the rate and direction of future decline to the insistence that protection be organized around this predictable cycle, from the urgings for comprehensive and continuous planning to the call for citizen-government cooperation. In its initial usage the term "renewal," with its sense of cities as renewable resources, served as a reminder of the campaign to fashion urban policy in the image of natural resources policy. Yet with its supporters' legislative successes arriving as conservation was dwindling in the public mind-set as a cultural repository of scientific management expertise, this metaphor quickly died.[8] As the effort to hold off cities' further expansion through central area improvement continued to unfold, the explicit suggestion of their commonalities with nature so central to earlier rationales for intervention—the language of city as garden and conservation of urban resources that had helped to enshrine a life cycle template in federal urban policy—quietly disappeared.

Certainly, there were some exceptions. Local public agencies in cities, including Detroit, Chicago, Cleveland, and Baltimore, all of which had undertaken experiments with urban community conservation and rehabilitation before the 1954 Housing Act made slum prevention a national priority, occasionally bucked this trend in their work to implement the federal program. Thus, in Detroit, planning commission conservation staff member Maurice Parkins sent to the U.S. Housing and Home Finance Agency his interpretation of the "birth," "growth," "maturity," and "middle age and decline" of varied neighborhoods together with his suggestion that with the designation of "conservation areas" the "life cycle had been completed" and "rebirth" was beginning.[9] "Neighborhood decay is often called blight—a spreading disease in plants or trees that eventually destroys them," Chicago's Community Conservation Board explained in a brochure for citizen participants in the slum prevention effort. "The choice is not between conservation, and letting things remain as they are. Conditions won't stay as they are. The only choice is—Fight or Blight. How to fight is the subject of this booklet."[10] In Cleveland, leaflets distributed to property owners compared neighborhood blight with rotting apples in language nearly identical to that the Federal Home Loan Bank

Board had used in the Waverly experiment more than a decade earlier. "It's a fact that one bad apple will ruin the whole barrel if not removed. On a street or block the 'bad apple' usually stands out like a sore thumb. Be sure it isn't yours."[11] And Baltimore officials, reflecting on how "the United States Department of Agriculture during the last century established an agricultural extension service to help farmers improve their farms," insisted that municipalities now "must establish an urban renewal extension service to help people save and improve their neighborhood." For "grass roots participation in urban renewal" would "not be secured by accident." Rather it would require "organized, systematic, daily professional liaison between the many departments of City Hall and each neighborhood."[12]

The Ford Foundation, too, briefly mobilized for an urban Morrill Act as the centennial establishing America's land-grant institutions approached. "Urbanites no less than their rural predecessors, need help with family budgets, nutrition, maintenance, land use, housing, vocational guidance, credit, and conservation," noted public affairs director Paul Ylvisaker, the concept's chief spokesperson, calling for the creation of "urban counterparts of the agricultural research, education, and extension programs of the land-grant colleges."[13] Ford, sponsor in the early 1950s for an urban division at Resources for the Future as well as for ACTION, Inc., and increasingly aware of the need for policies of human renewal to accompany those of physical renewal, saw inspiration in the human conservation efforts of an earlier generation.[14] The foundation seeded eight universities, National 4-H, and ACTION-Housing with $4.5 million to pilot test the possibilities for such work, catalyzing discussions in federal circles and complementing President Kennedy's call to revive the Civilian Conservation Corps.[15]

Such representations of urban problems and solutions, however, were increasingly uncommon. By the middle of the 1960s, characterizations of cities that emphasized their continuities with rural areas, already dwindling in popularity, had all but disappeared. Three decades earlier, social scientists studying St. Louis and members of the American City Planning Institute evaluating a proposed statute from the National Association of Real Estate Board, would see models of cities as ecological communities and national resources in places these interpretations did not apply. Now, when such models actually undergirded federal urban policy, their legacy was not the subject of extended examination. Instead, at the same time that urban professionals recognized the waning cultural influences of the conservation movement required them to find a new rhetorical focus to recruit support to the city improvement cause, their basic view of urban problems and solutions did not significantly change.[16] Even Ylvisaker's own Ford Foundation work on "grey areas" and his research at ACTION Inc. took slum prevention away from associations with the conservation movement by giving this technique a new name.[17]

The decline of explicit analogizing, and the disappearance of scrutiny for the analogical reasoning that had helped to establish the legitimacy of ecological thinking in the urban professions, had important consequences for the history of U.S. cities.[18] Although urban affairs in the four decades to 1960 garnered neither public attention nor financial backing on par with natural resources concerns, renewal promoters implementing a program whose conceptual and organizational foundation was the earlier conservation movement encountered several of the challenges that their colleagues in ecological science and natural resources planning faced. Despite a shared language of cities as ecological communities and national resources, and common interest in designing federal policies for urban restoration, conflicts erupted among urban professionals over the scope and meaning of the ideal interventions. With no nationally sanctioned standards for housing quality assessment, the notion of a "scientific" program of land use, open to local manipulation, was undermined from the outset. Apathy set in among communities whose initial burst of enthusiasm for neighborhood action, greeted by few immediately visible results, could not be sustained for the ongoing and long-term program these experts demanded. The work of individuals and institutions, including Homer Hoyt, Reginald Isaacs, and ACTION Inc., to broker relationships among the varied participants eager to place city problems and solutions higher on the national agenda could only go so far, and the initial consensus among a diversity of actors about the responsibilities of government and citizens to collaborate on slum removal and slum prevention rapidly disintegrated. With many cities opting for plans whose benefits accrued to developers and private property owners, the longer-term picture for urban renewal echoed the fate of conservation in the natural resources planning domain: a program that, in theory, was to offer a scientific and comprehensive attack on the problems of impoverished land areas and by extension their inhabitants became, in implementation, chiefly a business development scheme. Confidence that the future of land capacity could be easily predicted, and that upgrading the physical status of land would upgrade the economic and social status of its residents, gave way to anxieties about the difficulties of both tasks. The life cycle theory of human ecology, like Frederic Clements's organismic theory of succession to climax occasionally opposed from its debut, attracted the harshest criticisms following apparent failure to transform the American landscape as predicted. Faced with these and other disappointments, the renewal organizers who, like their natural resources planning colleagues declared their program an immediate success, were forced to revise their assessments in the longer term. Few of these developments would have been surprising if, in the press to garner new public support to finance a program of nationwide city improvement, the fate of the New Deal era conservation program that had substantially influenced their views about cities had not been overlooked.[19]

Facing the Future

All paradigms are by definition temporary, and the one described in this book was no different. As they witnessed the limitations of the nation's program for physical renewal, together with the forces of suburbanization and highway development that continued to transform the American landscape, many of the parties to the conversation about the future of urban areas nevertheless remained optimistic that a key to understanding and managing the orderly disorder of Cold War cities might be found. By the 1960s, these diverse actors had discovered an alternative model for scientific urban analysis and reform with equally broad appeal. Like the ecologists and natural resources planners who found new life for their work in the nonlinear systems sciences, individuals and institutions who earlier embraced notions of cities as ecological communities and national resources now turned to this interpretive tradition as well.[20] City planners had a troubling "tendency to cling to old, outmoded analogies when contemporary scientific concepts provide more useful notions," John Dyckman complained in 1961, noting how "organic analogies, for example, persist in planners' thinking about 'blight,' slums, and urban ecology, despite evidence that these models have frequently been misleading." The systems sciences offered "potentially very fruitful concepts," he explained, ideas such as "*channel capacity*, *storage*, and *overload*," that had been "little exploited" to date.[21]

Adopting and adapting the flow charts, simulation games, and computer models of systems scientists, urban professionals in the social sciences, real estate, and city planning; federal, state, and local officials; and representatives of a range of business, nonprofit, and citizen associations came to recognize in the nation's cities "subsystems," "feedback loops," and "homeostatic mechanisms" — many of the same patterns regulating behavior in humans, animals, machines, and organizations.[22] With the shift in analytical approaches came alternative proposals for reform, reorienting urban programs around the nonlinear interactions of physical, economic, and social forces in city systems. Experimental modifications to renewal pioneered in the Community Renewal Program under John F. Kennedy became codified in President Lyndon Johnson's Great Society initiatives, such as Model Cities, which brought "the unique analytical capability of the cybernetics fraternity . . . to restore the options and the beauty to urban life."[23]

In urban theory and practice, as in ecological science and resource planning, this new focus did not spell the total disappearance of life cycle ideas.[24] Instead, they were subsumed by an interpretive tradition that, like its predecessor, sought insights about cities, and a rationale for urban intervention, in the era's preeminent tradition of scientific analysis and management: systems-inspired military planning in the initial decades of the Cold War. Casual associations with militarism were apparent in earlier generations of discussions about cities as human ecologists spoke of "invasion" and "occupation," federal housing agencies created residential "security" maps as part of their "united front" against urban decay, real estate appraisers praised vigilant homeowners associations for "property protection," and city governments organized citizens for a "fight on blight" and a "war against slums."[25] As urban analysts and reformers continued the press to make cities national priorities in the 1960s, their new scientific language built on these earlier associations to align urban affairs with America's most urgent political concerns. Now, they cultivated closer relations with the scientific community on whose ideas they came to depend.[26] At the center of these city programs, like the Cold War military efforts that inspired them, was a vision for the comprehensive management of impoverished landscapes and impoverished populations — a shift away from the assumption that physical changes alone would lead to cities' salvation.[27] Calls for physical and human renewal in a "war on poverty" now superseded calls for the paired conservation of physical and human resources, but the idea was much the same.[28]

The dreams of the National Resources Planning Board for a cabinet-level urban department would finally be realized with the creation of the U.S. Department of Housing and Urban Development under President Johnson in 1965. Robert Weaver, former aide to Harold Ickes, became its first secretary. Yet, the model for the new agency was no longer the U.S. Department of Agriculture. Instead, it was the U.S. Department of Defense.

Abbreviations

AAAAS	*Annals of the American Academy of Arts and Sciences*
AAAG	*Annals of the Association of American Geographers*
AAAPSS	*Annals of the American Academy of Political and Social Science*
AAAS	American Association for the Advancement of Science
AC	*American City*
ACPI	American City Planning Institute
AIP	American Institute of Planners
AIREA	American Institute of Real Estate Appraisers
AJ	*Appraisal Journal*
AJS	*American Journal of Sociology*
APCA	*American Planning and Civic Annual*
AREP	*Annals of Real Estate Practice*
ASPO	American Society of Planning Officials
ASR	*American Sociological Review*
BG	*Botanical Gazette*
CCB	Community Conservation Board of Chicago
CCHR	Chicago Mayor's Commission on Human Relations
CCI	Chicago Community Inventory
CD	*Chicago Defender*
CDD	*Chicago Daily Defender*
CDT	*Chicago Daily Tribune*
CDUR	Chicago Department of Urban Renewal
CHA	Chicago Housing Authority
CJEPS	*Canadian Journal of Economics and Political Science*
CLCC	Chicago Land Clearance Commission

CPC	Chicago Plan Commission
CST	*Chicago Sun-Times*
CT	*Chicago Tribune*
CUL	Chicago Urban League
DCNC	Detroit Committee for Neighborhood Conservation and Improved Housing
DCPC	Detroit City Planning Commission
DCPCR	Records of the Detroit City Planning Commission, Burton Historical Collection, Detroit Public Library
DHC	Detroit Housing Commission
DJBC	Detroit Junior Board of Commerce
DUL	Detroit Urban League
EG	*Economic Geography*
EM	*Ecological Monographs*
FHA	United States Federal Housing Administration
FHAR	Records of the Federal Housing Administration, Record Group 31, U.S. National Archives and Records Administration
FHLBBR	Records of the Federal Home Loan Bank Board and Predecessor Agencies, Record Group 195, U.S. National Archives and Records Administration
FHLBR	*Federal Home Loan Bank Review*
GLCCR	Records of the Greater Lawndale Conservation Commission, 1950–1967. NUCMC MS 71-899, Chicago Historical Society
GR	*Geographical Review*
HHFA	U.S. Housing and Home Finance Agency
HOLC	U.S. Home Owners Loan Corporation
HPD	*Housing Policy Debate*
HPKCC	Hyde Park–Kenwood Community Conference
HUDR	Records of the Department of Housing and Urban Development and Predecessor Agencies, Record Group 207, U.S. National Archives and Records Administration
IMP	*Insured Mortgage Portfolio*
ISAS	Illinois State Academy of Science

ISRR	Institute of Social and Religious Research
JAH	*Journal of American History*
JAIP	*Journal of the American Institute of Planners*
JAIREA	*Journal of the American Institute of Real Estate Appraisers*
JAPA	*Journal of the American Planning Association*
JE	*Journal of Ecology*
JES	*Journal of Educational Sociology*
JFE	*Journal of Farm Economics*
JH	*Journal of Housing*
JHB	*Journal of the History of Biology*
JLPUE	*Journal of Land and Public Utility Economics*
JPE	*Journal of Political Economy*
JPER	*Journal of Planning Education and Research*
JUH	*Journal of Urban History*
LACRA	Community Redevelopment Agency of the City of Los Angeles
LAT	*Los Angeles Times*
LCP	*Law and Contemporary Problems*
LCRC	Local Community Research Committee of the University of Chicago
LE	*Land Economics*
LPNDC	Lincoln Park Neighborhood Collection, Digital Collection, DePaul University Libraries.
MHPC	Metropolitan Housing and Planning Council of Chicago
NAHB	National Association of Home Builders
NAHO	National Association of Housing Officials
NAHRO	National Association of Housing and Redevelopment Officials
NAREB	National Association of Real Estate Boards
NCCC	National Conference of Catholic Charities
NHA	National Housing Agency
NORC	National Opinion Research Center
NPC	*New Pittsburgh Courier*
NRC	U.S. National Resources Committee
NREJ	*National Real Estate Journal*

NRPB	U.S. National Resources Planning Board
NRPBR	Records of the National Resources Planning Board and Predecessor Organizations, Record Group 187, U.S. National Archives and Records Administration
NYT	*New York Times*
OHRC	Office of the Housing and Redevelopment Coordinator, Chicago
PAR	*Public Administration Review*
PASS	*Publications of the American Sociological Society*
PCHR	Philadelphia Commission on Human Relations
PCPC	Philadelphia City Planning Commission
PQ	*Phylon Quarterly*
PRA	Philadelphia Redevelopment Authority
QJE	*Quarterly Journal of Economics*
QRB	*Quarterly Review of Biology*
RE	*Real Estate*
RER	*Real Estate Record*
RES	*Review of Economic Statistics*
SACRPH	Society for American City and Regional Planning History
SCS	U.S. Soil Conservation Service
SF	*Social Forces*
SM	*Scientific Monthly*
SSH	*Social Science History*
SSPB	South Side Planning Board
SURRC	Subcommittee on Urban Redevelopment, Rehabilitation and Conservation of the U.S. Presidential Advisory Committee on Government Housing Policies and Programs
TCR	*Teachers College Record*
UAQ	*Urban Affairs Quarterly*
UCLR	*University of Chicago Law Review*
ULI	Urban Land Institute
URA	U.S. Urban Renewal Administration
USDA	U.S. Department of Agriculture
USRA	U.S. Resettlement Administration

VF	Vertical Files, Harvard University Graduate School of Design Library
WP	*Washington Post*
WPA	U.S. Works Projects Administration (later Works Progress Administration)
WPTH	*Washington Post and Times Herald*
WUL	Washington Urban League

Notes

Introduction: Revisiting American Antiurbanism

1. John T. Woolley and Gerhard Peters, "Speech by Senator John F. Kennedy Delivered before the Urban Affairs Conference," Pittsburgh, PA, October 10, 1960, The American Presidency Project (Santa Barbara: University of California [hosted]; Gerhard Peters [database]), www.presidency.ucsb.edu/ws/?pid=25758.

2. Kennedy employed this imagery on numerous occasions. See, for example, "Kennedy's Pitch for the City Vote," *Business Week*, February 10, 1962, 32–33.

3. Woolley and Peters, "News Release on Conference on Urban Affairs from the Democratic National Committee Publicity Division," Washington, DC, October 20, 1960, The American Presidency Project. www.presidency.ucsb.edu/ws/print.php?pid=74130.

4. Of course, these were not the only three urban professions as others, for example, the actors Peter Dreier refers to as "housers" (see Dreier, "Labor's Love Lost? Rebuilding Unions' Involvement in Federal Housing Policy," *HPD* 11, no. 2 [2000]: 327–391) played a role in how the history of U.S. cities unfolded. This book focuses on professionals in social science, city planning, and real estate because of their mutual interest in building a science of urban understanding and a shared language of conservation, despite the differences in visions of city problems and solutions they espoused.

5. Of course, this conceptual divide is not unique to the United States, as Raymond Williams elegantly reminds us in *The Country and the City* (London: Chatto & Windus, 1973).

6. Ralph Waldo Emerson, "Nature," in *Essays: Second Series* (1844).

7. Several literary scholars, examining paradigmatic antiurban texts such as Emerson's, have taken note of this alternative view. See James Machor, "Pastoralism and the American Urban Ideal: Hawthorne, Whitman, and the Literary Pattern," *American Literature* 54, no. 3 (1982): 329–353. Morton and Lucia White

mention a slightly earlier version of this tradition, noting nineteenth-century "sociologists, philosophers, and social workers who refused to call the city unnatural, and who insisted that its development was governed by empirical regularities and hence subject to scientific study." Morton White and Lucia White, *The Intellectual versus the City: From Thomas Jefferson to Frank Lloyd Wright* (Cambridge, MA: Harvard University Press, 1962), 234.

8. I refer to studies by scholars such as Emanuel Gaziano, Henrika Kuklick, Amy Hillier, and John Metzger; see the "Essay on Sources."

9. See the "Essay on Sources" for the rich literature Donald Worster, Ronald Tobey, Christopher Masutti, and others have produced on Frederic Clements's influences on ecological theory and resource planning policy.

10. These included the American Society of Planning Officials, National Association of Housing Officials, National Association of Assessing Officers, Institute of Real Estate Management, American Institute of Real Estate Appraisers, American Municipal Association, National Association of Real Estate Boards, Urban Land Institute, International City Managers' Association, and Society of Residential Appraisers. Many of these organizations shared a single building operated by the Public Administration Clearinghouse next to the University of Chicago. See Herbert Emmerich, "Cooperation among Administrative Agencies," *American Journal of Economics and Sociology* 15, no. 3 (1956): 237–244.

11. William Cronon, *Nature's Metropolis: Chicago and the Great West* (New York: W. W. Norton, 1991).

12. George Lakoff and Mark Johnson, *Metaphors We Live By* (Chicago: University of Chicago Press, 1980).

Chapter One: The City Is an Ecological Community

1. Robert E. Park, "Physics and Society," *CJEPS* 6, no. 2 (1940): 142.

2. Robert A. Woods et al., *The City Wilderness: A Settlement Study* (Boston: Houghton Mifflin, 1898); and Miriam Beard, "Wild Urban Jungle Has No Terrors for Man," *NYT*, March 14, 1926, SM2.

3. For a concise summary of the specific plant and animal ecologists most frequently cited in sociological research, see James Quinn, "Topical Summary of Current Literature on Human Ecology," *AJS* 46 (1940): 191–226. Human ecologists also cited a number of other scientists in their work, especially sources on the scientific method and the unity of knowledge.

4. Exceptions include J. Nicholas Entrikin, "Robert Park's Human Ecology and Human Geography," *AAAG* 70, no. 1 (1980): 43–58; Matthias Gross, "Human Geography and Ecological Sociology: The Unfolding of a Human Ecology, 1890 to 1930—and Beyond," *SSH* 28, no. 4 (2004): 575–605; and Gerhard

Fuchs, "Das Konzept der Ökologie in der amerikanischen Geographie," *Erkunde* 21 (1967): 81–93.

5. George Renner, "Human Ecology—a New Social Science," *TCR* 39 (1938): 488–493.

6. Of course, Europe already possessed some tradition of urban studies, given its longer history of urbanization, and research from figures such as Johann von Thünen and Ada Weber was known to U.S. scholars. Many of the earlier U.S.-based urban studies had been conducted in connection with late nineteenth- and early twentieth-century reform efforts—from sanitary to social surveys. The figures described in this chapter were self-consciously seeking to create a more theoretical tradition for the study of American cities with an urban "science" as their intellectual aim.

7. David Livingstone, "Evolution, Science, and Society: Historical Reflections on the Geographical Experiment," *Geoforum* 16 (1985): 119–130; and Robert Nisbet, *Social Change and History* (New York: Oxford University Press, 1969).

8. Stephen J. Cross and William R. Albury, "Walter B. Cannon, L. J. Henderson, and the Organic Analogy," *Osiris* 3 (1987): 165–192.

9. Although eugenics was occasionally termed *human ecology*, the idea of an urban ecology—in contrast to eugenics—suggested that it was place, more than people, that was responsible for the fate of a particular racial or nationality group. A few scholars, including Griffith Taylor and Ellsworth Huntington, were closer to the eugenical viewpoint in their interpretations of social phenomena. Examples of literature explicitly aligning the two fields include Roswell Johnson, "Eugenics of the City," in *The Urban Community: Proceedings of the American Sociological Society, 1925*, ed. Ernest Burgess (Chicago: University of Chicago Press, 1926); Horatio Pollack, Benjamin Malzberg, and Raymond Fuller, "Child Hygiene in Human Ecology," in *A Decade of Progress in Eugenics; Scientific Papers of the Third International Congress of Eugenics*, ed. Harry Perkins and Harry Laughlin (Baltimore: Williams and Wilkins, 1934); Samuel Holmes, *The Negro's Struggle for Survival: A Study in Human Ecology* (Berkeley: University of California Press, 1937); and C. G. L. Bertram, "Eugenics and Human Ecology," *Eugenics Review* 43 (April 1951): 11–18. Others, including George Renner, by identifying the value of human ecology for understanding population problems in the United States, addressed a number of overlapping themes.

10. Charles Abrams, *Revolution in Land* (New York: Harper and Brothers, 1939); *Distribution of Population of Chicago by Square Mile Sections* (Chicago: n.p., 1920); and CHA, *Information in Regard to the Proposed South Park Gardens Project* (Chicago: CHA, February 17, 1938), 4. The population would be at 3,376,438 in 1930.

11. Robert Kohler, *All Creatures: Naturalists, Collectors, and Biodiversity, 1850–1950* (Princeton, NJ: Princeton University Press, 2006); C. W. G. Eifrig, "Ornithological Notes from the Chicago Area," *American Midland Naturalist* 5, no. 2 (1917): 50; Henry C. Cowles, "The Physiographic Ecology of Chicago and Vicinity: A Study of the Origin, Development, and Classification of Plant Societies," *BG* 31, nos. 2–3 (1901): 73–108, 145–182; Henry C. Cowles, *The Ecological Relations of the Vegetation on the Sand Dunes of Lake Michigan* (Chicago: University of Chicago Press, 1899); and Hazel Schmoll, *Ecological Survey of Forests in the Vicinity of Glencoe, Illinois* (Chicago: Illinois State Academy of Science, 1919).

12. Herbert Hanson, "Ecology in Agriculture," *Ecology* 20, no. 2 (1939): 114.

13. Charles Adams, *Guide to the Study of Animal Ecology* (New York: Macmillan, 1913), 27. These developments were especially relevant to student studies, which often depended on sources close to campus.

14. Cited in Jesse Steiner, *Readings in Human Ecology* (Seattle: University of Washington Bookstore, 1939), 17.

15. Adams, *Guide to the Study of Animal Ecology*, 11–13. That individual scholars contradicted their own position on the extent to which urbanization was a natural process was quite standard in this period, reflecting the ambivalence toward nature-city relations.

16. In their pleas that the nation appreciate the importance of wilderness, ecologists occasionally drew analogies to human society to express the beauty and complexity of the natural world. As University of Illinois plant ecologist William McDougall observed in 1927, "When an ecologist goes into a forest he does not see merely a number of trees, shrubs, and herbaceous plants with no relations to one another except that they happen to be growing in close proximity." Instead, he sees a community whose "succession of life problems" is "just as intensely interesting as any city or other community dominated by the genus of bipeds to which we belong." Excerpted in Steiner, *Readings in Human Ecology*, 1.

17. On social scientists seeking to be more scientific, see Morris Cohen, "The Social Sciences and the Natural Sciences," in *The Social Sciences and Their Interrelations*, ed. William Ogburn and Alexander Goldenweiser (Boston: Houghton Mifflin, 1927).

18. Ellsworth Huntington, "Geographical Record," *GR* 15, no. 2 (1925): 316; and Robert Park, "The City as a Social Laboratory," in *Chicago: An Experiment in Social Science Research*, ed. Thomas V. Smith and Leonard D. White (Chicago: University of Chicago Press, 1929), 1.

19. Louis Wirth, *The Ghetto* (Chicago: University of Chicago Press, 1928); F. Stuart Chapin, *Experimental Designs in Sociological Research* (New York:

Harper, 1947; much of this book was published earlier, as articles); Pauline Young, *Scientific Social Surveys and Research* (New York: Prentice Hall, 1939). Human ecologists would not be the first observers to call the city their laboratory; John Jordan has commented on the frequent use of the term "social laboratory" in the 1910s. Jordan, *Machine Age Ideology* (Chapel Hill: University of North Carolina Press, 1995), 83.

20. Ernest Burgess, "Basic Social Data," in Smith and White, *Chicago*, 47.

21. Plant and animal ecologists adopted some experimental techniques with varying success during the late nineteenth and early twentieth century, a negotiation that was part of larger tensions between the "landscape" and "labscape" as central space for scientific practice as described by Robert Kohler, *Landscapes and Labscapes* (Chicago: University of Chicago Press, 2002). See, for example, Frederic Clements and John Weaver, *Experimental Vegetation* (Washington, DC: Carnegie Institution, 1924); and Frederic E. Clements, "Experimental Methods in Adaptation and Morphogeny," *JE* 17, no. 2 (1929): 356–379. Further discussion of the rationale behind Clements's interest in experimentalism can be found in the many studies in the "Essay on Sources" that treat the history of Clementsian ecology.

22. According to geographer Carl Sauer, by 1924, the scientific survey methods of geographers, economists, and sociologists had grown close. See Carl O. Sauer, "The Survey Method in Geography and Its Objectives," *AAAG* 14, no. 1 (1924): 17–33. Ecologists' methods are described in Frederic Clements, *Research Methods in Ecology* (Lincoln, NB: University Publishing, 1905); Stephen Forbes, "On the Local Distribution of Certain Illinois Fishes: An Essay on Statistical Ecology," *Bulletin of the Illinois State Laboratory* 7 (April 1907); Arthur Tansley, *Practical Plant Ecology* (New York: Dodd and Mead, 1923); Daniel Schneider, "Local Knowledge, Environmental Politics, and the Founding of Ecology in the United States," *Isis* 91 (2000): 681–705; Henrika Kuklick and Robert E. Kohler, eds., "Science in the Field," *Osiris* 11 (1996): entire issue; Sharon Kingsland, *Modeling Nature* (Chicago: University of Chicago Press, 1985); and Frank Egerton, "The History of Ecology: Achievements and Opportunities," *JHB* 16 (1983): 259–310, and 18 (1985): 103–143.

23. Arthur G. Tansley, "British Ecology during the Past Quarter Century: The Plant Community and the Ecosystem," *JE* 27, no. 2 (1939): 513–530.

24. Charles Redway Dryer, "Genetic Geography: The Development of the Geographic Sense and Concept," *AAAG* 10 (1920): 15. In short, rather than depend only on earlier traditions of organicist thought within their own fields, social scientists explicitly borrowed from ecologists' appropriations of these ideas because of the scientific imprimatur they delivered. Gregg Mitman, *The State of Nature: Ecology, Community, and American Social Thought, 1900–1950* (Chi-

cago: University of Chicago Press, 1992), has documented some of the specific aspects of social theory on which American ecologists relied.

25. Among them were Columbia University; University of Wisconsin; Tulane University; Fisk University; University of North Carolina; University of Washington; University of Nebraska; University of Virginia; University of Southern California; University of Michigan; University of California, Los Angeles; Northwestern University; Washington University in St. Louis; and University of Pennsylvania. Reflecting close ties to scholarship in the United States, Canadian universities, most notably McGill, also developed traditions of human ecology research. Andrew Abbott, *Department and Discipline* (Chicago: University of Chicago Press, 1999), examines the history of the University of Chicago Sociology Department.

26. E. Franklin Frazier, "Negro Harlem: An Ecological Study," *AJS* 36, no. 2 (1930): 72–88; Cloyd V. Gustafson, "An Ecological Analysis of the Hollenbeck Area of Los Angeles" (MA thesis, University of Southern California, 1940); Kimball Young, John Gillin, and Calvert Dedrick, "The Madison Community," *University of Wisconsin Studies in the Social Sciences and History* 21 (1934); and Denis Charles McGenty, "A Natural History of the Central Business District of Minneapolis: A Study in Human Ecology" (MA thesis, University of Minnesota, 1935).

27. Although the focus of this book is ecological studies of U.S. cities, the power of ecological approaches also extended to studies of non-U.S. cities, small towns, suburbs, regions, and rural areas. These include Kathren Pearl McKinney, "An Ecological Study of Paris, France 1911, 1921, 1931" (MA thesis, Washington University, 1939); Asael T. Hansen, "The Ecology of a Latin-American City," in *Race and Culture Contacts*, ed. E. B. Reuter (New York: McGraw-Hill, 1934), 124–142; Carl Dawson, *The City as an Organism, with Special Reference to Montreal* (Montreal: McGill University, 1926); Donald Pierson, "The Negro in Bahia, Brazil," *ASR* 4, no. 4 (1939): 524–533; Selden Cowles Menefee, *A Plan for Regional Administrative Districts in the State of Washington: An Ecological Study* (Seattle: University of Washington, 1935); John Adrian Rademaker, "The Ecological Position of the Japanese Farmers in the State of Washington" (PhD diss., University of Washington, 1939); T. Lynn Smith, "An Analysis of Rural Social Organization among the French-Speaking People of Southern Louisiana," *JFE* 16, no. 4 (1934): 680–688; Laura Frances Hildreth Hoffland, "Seattle as a Metropolis: The Integration of the Puget Sound Region through the Dominance of Seattle; An Ecological Study" (MA thesis, University of Washington, 1933); Andrew Lind, *An Island Community: Ecological Succession in Hawaii* (Chicago: University of Chicago Press, 1938); Jean Hunter, "The French Canadian Invasion of the Eastern Townships" (MA thesis, McGill University, 1939); Norman Hayner, "Delin-

quency Areas in the Puget Sound Region," *AJS* (1933): 314–328; Clarence Glick, "Winnetka: A Study of a Residential Suburban Community" (MA thesis, University of Chicago, 1928); and A. B. Hollingshead, "Changes in Land Ownership as an Index of Succession in Rural Communities," *AJS* 43, no. 5 (1938): 764–777.

In addition, during this period, plant and animal ecology provided a template for understanding phenomena that ranged from occupational change, to mental health, to race relations, to the influences of religious institutions. See, for example, Everett Hughes, "Personality Types and the Division of Labor," *AJS* 33, no. 5 (1928): 754–768; Earl Johnson, "A Study in the Ecology of the Physician" (MA thesis, University of Chicago, 1932); Erdmann Doane Beynon, "Occupational Succession of Hungarians in Detroit," *AJS* 39, no. 5 (1934): 600–610; Horace R. Cayton, "The Negro Police Officer: A Study in Occupational Succession," cited in "Students' Dissertations in Sociology," *AJS* 39, no. 1 (1933): 87; Wesley Clair Mitchell, "Occupational Succession," in *Recent Economic Changes in the United States*, ed. C. R. Hoffer (Washington, DC: Government Printing Office, 1929), vol. 2; and A. E. Holt, "The Ecological Approach to the Church," *AJS* 33, no. 1 (1927): 72–79.

28. I refer to the work of Sharon Kingsland and others on the historiography of American ecology; see "Essay on Sources."

29. Adams, *Guide to the Study of Animal Ecology*, 5–6. These "units of communal life" were also called "formations, to use the term of the plant ecologist." Roderick McKenzie, "The Ecological Approach to the Study of the Human Community," *AJS* 30 (1924): 300.

30. Richard Ely and Edward Morehouse, *Elements of Land Economics* (New York: Macmillan, 1924), 85.

31. William Jones, *The Housing of Negroes in Washington DC: A Study in Human Ecology* (Washington, DC: Howard University Press, 1929), 85.

32. Ernest Shideler, "The Chain Store: A Study of the Ecological Organization of a Modern City" (PhD diss., University of Chicago, 1927), abstract, p. 3. On the tendency of "certain forms of enterprise to segregate and seek company with related organizations," see also William Wallace Weaver, "West Philadelphia: A Study of Natural Social Areas" (PhD diss., University of Pennsylvania, 1930), 109.

33. Robert Park, "The Concept of Position in Sociology," *PASS* 20 (1925): 4; and Nels Anderson, "The Slum: A Project for Study," *SF* 7, no. 1 (1928): 87.

34. George Renner and C. Langdon White's textbook *Geography: An Introduction to Human Ecology* (New York: D. Appleton-Century, 1936) later popularized the term "industrial ecology."

35. Harlan Paul Douglass, *The Suburban Trend* (New York: Century, 1925), 34.

36. Harvey Zorbaugh, "The Natural Areas of the City," *PASS* 20 (1926): 192.

37. Herbert Dorau and Albert Hinman, *Urban Land Economics* (New York: McGraw-Hill, 1928), 223.

38. Roderick McKenzie, "The Concept of Dominance and World Organization," *AJS* 33, no. 1 (1927): 29.

39. Robert Park, "Human Ecology," *AJS* 42 (1936): 8.

40. Wirth, *The Ghetto*, 282–284.

41. Robert M. Haig, "Toward an Understanding of the Metropolis," *QJE* 40 (May 1926): 403–404. This was the second article in a two-part series republished the following year in *Regional Plan of New York and its Environs, Regional Survey*, Vol. 1 (New York: RPA, 1927). "Accommodation" was another term used in plant and animal studies.

42. H. Paul Douglass, *St. Louis Church Survey* (New York: George H. Doran, 1924), 4.

43. Marvin Mikesell, "The Rise and Decline of Sequent Occupance," in *Perspectives on Geography 3: The Nature of Change in Geographical Ideas*, ed. Brian Berry (Dekalb: Northern Illinois University Press, 1978). Numerous other geographers working on cultural aspects of sequent occupance are noted in Richard Ellwood Dodge, "The Interpretation of Sequent Occupance," *AAAG* 28, no. 4 (1938): 233–237. Even those scholars who identified a difference between the processes (such as George Renner and C. Langdon White) nevertheless often referenced them together.

44. Derwent Whittlesey, "Sequent Occupance," *AAAG* 19, no. 3 (1929): 162.

45. Dryer, "Genetic Geography," 15.

46. Norman Scott Brien Gras, "The Rise of the Metropolis," *PASS* (1926). Commenting on Gras's *An Introduction to Economic History* (New York, 1922), Chicago graduate student Ernest Shideler had observed in his dissertation's annotated bibliography how "this study might have been entitled 'A Study in Ecologic Succession' "; Shideler, "The Chain Store," Bibliography, 3.

47. Ernest Burgess, "Residential Segregation in American Cities," *AAAPSS* 140 (1928): 112.

48. McKenzie, "The Ecological Approach to the Study of the Human Community," 298.

49. Martin Bulmer, *The Chicago School of Sociology* (Chicago: University of Chicago Press, 1984), notes that, during the 1920s, Henry Cowles and Charles Child addressed the Society for Social Research. Among the numerous reviews of plant and animal studies presented in AJS was L. L. Bernard, Review of Charles M. Child, *Physiological Foundations of Behavior*; C. Judson Herrick, *Foundations of Animal Behavior*, *AJS* 31, no. 1 (1925): 95–96. Readings in plant and animal

ecology appeared in the mimeographed course reader that Park and Burgess assembled for the course they taught. Later it was published as *Introduction to the Science of Sociology* (Chicago: University of Chicago Press, 1921).

50. Albion Small previously had taught a class with this name. See Albion W. Small, *Syllabus: Introduction to the Science of Sociology* (Waterville, ME: Mail Office, 1890).

51. Vivien Palmer, *Field Studies: A Student's Manual* (Chicago: University of Chicago Press, 1928), xv. Palmer was a former graduate student; her book was among the first in the social sciences to bear the title "field studies," although department faculty Albion Small and George Vincent had called their earlier textbook, *An Introduction to the Study of Society*, a "laboratory guide."

52. These texts included the journal *Ecology*, about which Nels Anderson and Eduard Lindeman observed, "This magazine frequently contains articles treating the subject of ecology as it relates to sociology," as well as Charles Adams's *An Ecological Survey in Northern Michigan*, as library records show. Nels Anderson and Eduard Lindeman, *Urban Sociology* (New York: Alfred Knopf, 1928).

53. "AAG Titles and Abstracts of Papers Presented to the Association from 1904 to 1910, Inclusive," *AAAG* 1 (1911): 101–150. Goode's PhD was in economics. The term o*ntography*, coined by Harvard geographer and AAG founder William Davis in the early 1900s, referred to the intersection between the physical environment and the environmental organism.

54. "Henry Chandler Cowles, Physiographic Plant Ecologist," *AAAG* 30, no. 1 (1940): 41; May T. Watts, *Reading the Landscape of America* (New York: Macmillan, 1957). Campus plant and animal specialists such as Cowles were affiliated with the departments of botany, zoology, and physiology; despite the university's prominence in ecological research, no department was devoted to the field. However, the three departments collaborated on a teaching sequence. See *Second Year College Sequence in Botany, Zoology and Physiology. Syllabus* (Chicago: University of Chicago, 1931).

55. "News Items," *EG* 1, no. 1 (1925): 131. The inaugural issue of *Economic Geography*, which listed information from around the nation's Geography departments, noted, "An interesting course in addition to the many given in the Department of Geography is the course in human ecology given under Sociology and Anthropology entitled 'Human Ecology' by Associate Professor McKenzie. It is a study of the geographic and economic factors which determine the location and growth of communities." Robert Platt, too, studied communities in organismic terms, for example, in "A Detail of Regional Geography: Ellison Bay Community as an Industrial Organism," *AAAG* 18 (1928): 81–126.

56. Harlan Barrows, "Geography as Human Ecology," *AAAG* 13 (1923): 1–14. This claim would be the subject of much discussion as in Clarence Jones,

"Trends in Modern Geography," *Science* 60, no. 1556 (October 24, 1924): 374–376. Other scholars had made similar observations, for example, Barrington Moore, "The Scope of Human Ecology," *Ecology* 1 (1920): 3–5.

57. Smith and White, *Chicago*, 24, 27. When the site opened, one of the human ecologists' key sponsors suggested that, in their emphasis on fieldwork, these scholars' efforts to build a science of society had been achieved. See Beardsley Ruml, "Recent Trends in Social Science," in *The New Social Science*, ed. Leonard D. White (Chicago: University of Chicago Press, 1929), 102–103.

58. Emory Bogardus, "Cooperative Research on the Pacific Coast," *JES* 4, no. 9 (1931): 567.

59. Entrikin, "Robert Park's Human Ecology and Human Geography," 55.

60. Dryer, "Genetic Geography," 16.

61. Andrew Lind, *A Study of Mobility of Population in Seattle* (Seattle: University of Washington Press, 1925). This point was made frequently, for example, by Everett Hughes, *The Chicago Real Estate Board* (Chicago: University of Chicago Press, 1931), 11.

62. Whittlesey, "Sequent Occupance," 164.

63. Anderson and Lindeman, *Urban Sociology*, 44n1. Herbert Spencer had made a similar point about the functions of analogies in his own work at the intersections of science and social thought, noting how analogies were "a scaffolding to help in building up a coherent body of sociological inductions" that should "stand by themselves." Cited in Leo Schnore, "The City as a Social Organism," *UAQ* 1, no. 3 (1966): 69.

64. Charles Child, *The Physiological Foundations of Behavior* (New York: Henry Holt, 1924); Frederic Thrasher, "The Study of the Total Situation," *JES* 1, no. 8 (1928): 487.

65. Clements, "Experimental Methods in Adaptation and Morphogeny," 376.

66. Palmer, *Field Studies*, 168. Here she was commenting on the life history, a field study technique used in ecology as well as social research. See Steven C. Stearns, "Life History Tactics: a Review of the Ideas," *QRB* 51 (1976): 3–47.

67. Weaver, "West Philadelphia," 11.

68. Dorau and Hinman, *Urban Land Economics*, 125.

69. Park, "The City as a Social Laboratory," in Smith and White, *Chicago*, 8.

70. Burgess, "Residential Segregation in American Cities," 108. The zonal map, while formally based on Chicago, was schematized to represent the patterns organizing life in all cities. This idealized template was based on the wealth of faculty and student research at the University of Chicago's Sociology Department — the storehouse of statistical data at the program's Sociological Laboratory and

in the *Social Base Map*, which provided Burgess with the details from which to abstract a general theory of the American city. See LCRC, *Social Base Map of Chicago* (Chicago: University of Chicago Press, 1926).

71. Gras, "The Rise of the Metropolis," 158.

72. Ernest Burgess, "The New Community and Its Future," *AAAPSS* 149, no. 1 (1930): 162.

73. Because of both men's ties to the Behavior Research Fund, when psychology professor Heinrich Klüver published a book on behavior mechanisms in monkeys—*Behavior Mechanisms in Monkeys* (Chicago: University of Chicago Press, 1933)—for example, Burgess wrote its foreword.

74. Cowles, "The Physiographic Ecology of Chicago and Vicinity," *BG* 31, no. 2 (1901): 79; Schmoll, *Ecological Survey of Forests in the Vicinity of Glencoe, Illinois*, 220. Concentric zonation in nature is also discussed in John W. Harshberger, "Suggestions toward a Phyto-Geographic Nomenclature," *Science* 21, no. 542 (May 1905), 789.

75. H. W. Graham and L. K. Henry, "Plant Succession at the Borders of a Kettle-Hole Lake," *Bulletin of the Torrey Botanical Club* 60, no. 4 (April 1933): 301–315.

76. Ernest Burgess, "The Growth of the City," in *The City*, ed. Robert Park, Ernest Burgess, and Roderick McKenzie (Chicago: University of Chicago Press, 1925).

77. Cowles, "The Physiographic Ecology of Chicago and Vicinity," 79; and idem, *The Ecological Relations of the Vegetation on the Sand Dunes of Lake Michigan*.

78. Frederic Clements, *Plant Succession* (Washington, DC: Carnegie Institute, 1916); and Frederic E. Clements, "Nature and Structure of the Climax," *JE* 24, no. 1 (1936): 252–284.

79. The life cycle concept is explored in Hughes, *The Chicago Real Estate Board*; Paul H. Landis, "The Life Cycle of the Iron Mining Town," *SF* 13, no. 2 (1934): 245–256; and the many excerpts from plant and animal scientists in Steiner, *Readings in Human Ecology*. This idea of the repetition of phenomena at different levels built on Frederic Clements's notion of plant communities as superorganisms that were more than the sum of their parts.

80. Angela O'Rand and Margaret Krecker, "Concepts of the Life Cycle: Their History, Meanings, and Uses in the Social Sciences," *Annual Review of Sociology* 16 (1990): 241–262.

81. Ernest Burgess, "Is Prediction Feasible in Social Work?" *SF* 7, no. 4 (1929): 533. Steiner's *Readings in Human Ecology* provides several readings on cyclical behavior in society.

82. Ernest Burgess, "Can Neighborhood Work Have a Scientific Basis?" in

Park et al., *The City*, 148. Certainly, as social scientists set out to analyze urban patterns in terms of climax communities and urban life cycles, there were some exceptions to this interpretative framework. As geographer Charles Redway Dryer suggested of the macro-phenomenon of urbanization, "a climax of stability" could be seen in the mere existence of urban industrial society. Dryer, "Genetic Geography," 15. At a more micro level, some stability was apparent, some observers of the residential succession process suggested. For example, Burgess and McKenzie proposed that a temporary climax community, in Burgess's terms an "equilibrium of community stability," was achieved at the end of each invasion-succession cycle. Burgess, "Residential Segregation in American Cities," 112.

83. Burgess, "The Growth of the City."

84. McKenzie, "The Ecological Approach to the Study of the Human Community," 293. There were some debates within scientific circles about the stability of the climax stage, which spurred Clements to revise his theories and to generate new terms such as subclimax and disclimax. Yet, as historians of environmental science have documented, Clements's overarching theory of succession to equilibrium climax remained the dominant mode of understanding the natural environment in this period.

85. Anderson and Lindeman, *Urban Sociology*, 87; Burgess, "The New Community and Its Future," 161–162.

86. For example, University of Chicago graduate students conducting studies of suicide, family types, and gangs in Chicago spotted these data on city maps to find the zones that divided types of family areas, types of social disorganization, and neighborhoods populated by specific racial or nationality groups. See Clifford Shaw and Frederick Zorbaugh, *Delinquency Areas* (Chicago: University of Chicago Press, 1929); Frederic Thrasher, *The Gang* (Chicago: University of Chicago Press, 1927); Walter Reckless, *The Natural History of Vice Areas in Chicago* (Chicago: University of Chicago Press, 1925); and Ernest R. Mowrer, *Family Disorganization* (Chicago: University of Chicago Press, 1927).

87. Weaver, "West Philadelphia," 25–26.

88. Calvin F. Schmid, *Suicides in Seattle, 1914 to 1925: An Ecological and Behavioristic Study* (Seattle: University of Washington Press, 1928), 13.

89. Weaver, "West Philadelphia," 22.

90. President's Conference on Home Building and Home Ownership, 1931, *Housing Objectives and Programs* (Washington, DC: The Conference, 1932). The Interior Department jointly sponsored this meeting. Hoover had a long interest in planning issues. As secretary of commerce, he established the Better Homes in America campaign and later advocated for a model zoning ordinance. This position placed him at the center of discussions about cities' land use policies because,

lacking a federal housing agency, the era's research and policy discussions about urban issues chiefly took place at the Department of Commerce's Division of Building and Housing.

91. Roderick McKenzie, ed., *The Metropolitan Community* (New York: McGraw-Hill, 1933).

92. National Resources Committee, *Our Cities: Their Role in the National Economy* (Washington, DC: NRC, 1937). This organization was created in 1933 as the National Planning Board and reported directly to the president for most of its decade-long tenure. Its name changed several times: National Resources Board (1934–1935), National Resources Committee (1935–1939), and National Resources Planning Board (1939–1943). National Resources Planning Board is used throughout this book for simplicity.

93. William E. Leuchtenburg, *Franklin D. Roosevelt and the New Deal* (New York: Harper, 1963).

94. Elsa Schneider Longmoor and Erle Fiske Young, "Ecological Interrelationships of Juvenile Delinquency, Dependency, and Population Mobility," *AJS* 41, no. 5 (1936): 598–610; Calvin Fisher Schmid, *Social Saga of Two Cities: An Ecological and Statistical Study of Social Trends in Minneapolis and St. Paul* (Minneapolis, MN: Minneapolis Council of Social Agencies, 1937); T. Earl Sullenger, *An Ecological Study of Omaha* (Omaha, NB: Municipal University of Omaha, 1938); William Louis Jolly Dee, "An Ecological Study of Mental Disorders in Metropolitan St. Louis" (MA thesis, Washington University, 1939); Raymod Bowers, "Ecological Patterning of Rochester, New York." *ASR* 4, no. 2 (1939): 180–189; Raymond Franklin Sletto, *A Catalogue of Maps, Charts, Projection Slides, and Tables in the Sociological Research Laboratory* (Minneapolis: University of Minnesota and WPA, 1939); Percy A. Robert, "Probationers in Essex County, New Jersey: A Study of the Social Backgrounds and Ecology of the Probationers of Essex County" (PhD diss., New York University, 1938); R. Clyde White, "The Relation of Felonies to Environmental Factors in Indianapolis," *SF* 10, no. 4 (1932): 498–509; Mary P. Smith and E. A. Taylor, "An Ecological Study of Juvenile Delinquency and Dependency in Athens County, Ohio," *PASS* 26 (1932): 144–149; Stuart A. Queen, "The Ecological Study of Mental Disorders," *ASR* 5: 2 (1940): 201–209; Edward Jackson Baur, "Delinquency among Mexican Boys in South Chicago" (MA thesis, University of Chicago, 1938). Such relationships with nonacademic organizations were not an entirely new development. At the University of Chicago, for example, faculty and students cultivated relationships with city institutions as a means of gathering data. Nels Anderson's dissertation, for example, was supported by the Committee on Homeless Men of the Chicago Council on Social Agencies.

95. Cited in Bulmer, *The Chicago School of Sociology*, 76.

96. Howard Harlan, "Zion Town: A Study in Human Ecology," *Phelps-Stokes Fellowship Papers* 13 (1935): 64, 63.

97. Roy Burroughs, "Urban Real Estate Mortgage Delinquency," *JLPUE* 11, no. 4 (1935): 364. See also Lewis F. Thomas, "The Sequence of Areal Occupance in a Section of St. Louis, Missouri," *AAAG* 21, no. 2. (1931): 75–90.

98. Sullenger, *An Ecological Study of Omaha*, no page.

99. Andrew Lind, "Some Ecological Patterns of Community Disorganization in Honolulu," *AJS* 36, no. 2 (1930): 220; Frazier, "Negro Harlem: An Ecological Study"; Bowers, "Ecological Patterning of Rochester, New York," *ASR* 4 (1939): 182–183; and Stuart Queen and Lewis Thomas, *The City* (New York: McGraw-Hill, 1939), 253.

100. Longmoor and Young, "Ecological Interrelationships of Juvenile Delinquency, Dependency, and Population Mobility," 602. In a testament to its broad and enduring influences, Burgess's model of the city was widely reprinted, not only in sociology texts such as Kimball Young's *Sourcebook for Sociology* (New York: American Book Company, 1939) but also in geographers' and economists' work, such as Stuart Queen and Lewis Thomas's *The City* (1939). When not reprinted, it was often described in textual form, as in Richard Ely and George Wehrwein's *Land Economics* (New York: Macmillan, 1940), 446–447.

101. George Renner, "Human Ecology—a New Social Science," online version, no page.

102. Queen and Thomas, *The City*, 307.

103. Tansley, *Practical Plant Ecology*.

104. Frederick Babcock, *The Valuation of Real Estate* (New York: McGraw-Hill, 1932), 61–62. This idea was not Babcock's alone. Since 1928, building on Hurd's work, several colleagues earlier had suggested how "the town and then the city became roughly star-shaped as a normal result of what has been called axial and central growth . . . so far as we can speak of cities in general they tend to be star-shaped." Although the urban pattern more closely resembled "the shape of concentric stars or irregular polygons than that of concentric circles," then, these observers saw Burgess's theory as valuable nonetheless. Dorau and Hinman, *Urban Land Economics*, 58–59, 65.

105. Charles Colby, "Centrifugal and Centripetal Forces in Urban Geography," *AAAG* 23, no. 1 (1933): 1.

106. Eleanor Gluck, "An Ecological Study of the Japanese in New York City" (MA thesis, Columbia University, 1940), 26, 34–35.

107. Hughes, *The Chicago Real Estate Board*, 2. On standards development, see also Dodge, "The Interpretation of Sequent Occupance," and Robert E. Park, "Succession, an Ecological Concept," *ASR* 1, no. 2. (1936): 171–179.

108. Whittlesey, "Sequent Occupance," 164.

109. Louis Wirth, Review of *The Social Insects* by William Morton Wheeler, *AJS* 36, no. 1 (1930): 140–142.

110. W. Elmer Ekblaw, review of *Geography: An Introduction to Human Ecology* by George Renner and C. Langdon White, *EG* 12, no. 4 (1936): 436.

111. James Quinn, "The Nature of Human Ecology: Reexamination and Redefinition," *SF* 18, no. 2 (1939): 161–168; idem, "The Burgess Zonal Hypothesis and Its Critics," *ASR* 5, no. 2 (1940): 210–218; Milla Alihan, *Social Ecology* (New York: Columbia University Press, 1938). Maurice Davie was representative of those critics suggesting Burgess's model had been too enthusiastically applied, for example, to St. Louis. Burgess, citing a student's paper on the city, had explicitly identified St. Louis as conforming to the general patterns he observed. See Burgess, "Residential Segregation in American Cities." Analyses from a multidisciplinary team of sociologists and geographers at Washington University, including Stuart Queen and Lewis Thomas, followed with similar conclusions. Yet, in a review of this work, Davie noted the disconnect between data gathered and interpretations made. As he praised the researchers' "numerous maps based on census tracts which would be excellent for the determination of ecological areas," he simultaneously complained that "full use of the St. Louis material is not made, however, because, although none of their data support it, the authors accept the Burgess hypothesis . . . There is nothing about the distributional maps which suggests concentric zones." Maurie R. Davie, review of *The City*, by Stuart Queen and Lewis Thomas, *AAAAS* (1939): 224.

112. Ecologists had implicitly validated some of human ecologists' observations, for example, in preparing a review of Noel P. Gist and Leroy A. Halbert, *Urban Society* (New York: Crowell, 1933), in *QRB* 9, no. 2 (1934): 232, or in Frederic Clements's reflections on the five-stage "sere" in the human community — "first trapper, then hunter, pioneer, homesteader, and finally urbanite." Cited in Donald Worster, *Nature's Economy* (San Francisco: Sierra Club Books, 1977), 218. These commentaries would be more explicit in their assessments.

113. Adams, *Guide to the Study of Animal Ecology*, 13.

114. Charles Adams, "The Relation of General Ecology to Human Ecology," *Ecology* 16, no. 3 (1935): 318.

115. Ibid., 329. In Adams's estimation, further refinements would be necessary to place the study of human ecology on par with the study of plants and animals, however. "It could not be expected that the older social surveys would escape all the errors that were made by the older natural history surveys. They did not." Yet his overarching conclusion was that much progress had been made in "technique and standardization, with the result that many significant comparisons are now possible."

116. Edward Haskell, Review of *Human Ecology*, by John Bews, *Ecology* 18, no. 3 (1937): 443.

117. The two were unable to attend, however, as described by Eugene Cittadino, "The Failed Promise of Human Ecology," in *Science and Nature: Essays in the History of the Environmental Sciences*, ed. Michael Shortland (London: British History of Science Society, 1993). Sears already had lectured on "Illustrative Material in Human Ecology" at Teachers College, *TCR* 39 (1937–1938): 440. Columbia University, in 1937, while on leave from the University of Oklahoma. Appearances by plant and animal ecologists at meetings of human ecologists were rare, although Charles Child delivered a paper at the American Sociological Association meeting in 1928.

118. Tansley, "British Ecology during the Past Quarter Century," 528–529.

119. Frederic Clements and Victor Shelford, *Bio-Ecology* (New York: Wiley, 1939), 1.

120. Exceptions include Charles M. Child, "Biological Foundations of Social Integration," *PASS* 22 (1928): 26–42; Warder Allee, "Co-operation among Animals," *AJS* 37, no. 3 (1931): 386–398; Eduard Lindeman, "Ecology: An Instrument for the Integration of Science and Philosophy," *EM* 10, no. 3 (1940): 367–372; Griffith Taylor, "The Ecological Basis of Anthropology," *Ecology* 15, no. 3 (1934): 223–242; and A. B. Hollingshead, "Human Ecology and Human Society," *EM* 10, no. 3 (1940): 354–366.

121. Roderick McKenzie, *Human Ecology (Selected Readings—Sociology 55)* (n.l.: n.p., 1929); Roderick D. McKenzie, ed., *Readings in Human Ecology* (Ann Arbor, MI: George Wahr, 1934); Steiner's collection, *Readings in Human Ecology*, appears to be an update to McKenzie's earlier work.

122. Robert E. Dickinson, "The Scope and Status of Urban Geography: An Assessment," *LE* 24, no. 3 (1948): 237; Robert Dickinson, Review of *Human Ecology*, by Amos Hawley, *EG* 26: 4 (1950): 328; William Koelsch, "The Historical Geography of Harland Barrows," *AAAG* 59 (1969): 632–651; Floyd N. House, *The Range of Social Theory* (New York: H. Holt, 1929), 65; Louis Wirth, "The Social Sciences," in *American Scholarship in the Twentieth Century*, ed. Merle Curti (Cambridge, MA: Harvard University Press, 1953); Roderick McKenzie, "Demography, Human Geography, and Human Ecology," in *The Fields and Methods of Sociology*, ed. L. L. Bernard (New York: R. Long & R. R. Smith, 1934). Apparent in research citations and reading assignments, this interdisciplinarity was visible in the circulation of maps as well. For example, sociologists, geographers, and economists, including Calvin Schmid, Robert Park, Roderick McKenzie, Lewis Thomas, and Homer Hoyt, all used maps from outside their own disciplines. See Park, "Sociology," in *Research in the Social Sciences*, ed. Wilson Gee (New York: Macmillan, 1929), 24; McKenzie, "The Ecological Approach to

the Study of the Human Community"; Schmid, "Land Values as an Ecological Index," *Research Studies of the State College of Washington* 9, no. 3 (1941): 16–36; Queen and Thomas, *The City*; Homer Hoyt, *One Hundred Years of Land Values in Chicago* (New York: Macmillan, 1933).

123. Roderick McKenzie, "The Scope of Human Ecology," *ASR* 20 (1926): 142.

124. E. Gordon Ericksen, *Introduction to Human Ecology* (Los Angeles: University of California, Los Angeles, 1949), 4.

125. Louis Wirth, "Human Ecology," *AJS* 50, no. 6 (1945): 488.

126. Robert McFall, "Urban Centralization and Decentralization," in *Economic Essays in Honor of Wesley Clair Mitchell* (New York: Columbia University Press, 1935), 298.

127. Ely and Morehouse, *Elements of Land Economics*, 86.

128. Charles T. Male, *Real Estates Fundamentals* (New York: D. Van Nostrand, 1932), 201.

Chapter Two: The City Is a National Resource

1. "Research on Neighborhood Conservation," *RER* (December 28, 1940): 6.

2. FHLBB, *Waverly: A Study in Neighborhood Conservation* (Washington, DC: FHLBB), 65.

3. Ibid., 53, 58, viii.

4. Scholars such as Henricka Kuklick and Amy Hillier have noted the influences of ecological models in these professions; see the "Essay on Sources" for further discussion.

5. As Charles Male explained in one real estate manual, "Fortunately, the colored people have a natural tendency to live in sections by themselves," suggesting Realtors' preference for homogenous city neighborhoods simply echoed the natural behavior of population groups. Male, *Real Estates Fundamentals* (New York: D. Van Nostrand, 1932).

6. Ecologists of this era were not all professionally identified by the title "ecologist" but rather worked as biologists, botanists, and zoologists. Thus, figures such as Walter Taylor, who called for agriculture to "become less an art, more a science . . . to place the emphasis on conservation rather than exploitation" worked at the Bureau of Biological Survey at the USDA. Taylor, "Man and Nature: A Contemporary View," *SM* 41 (October 1935): 355.

7. Wade DeVries, "The Michigan Land Economic Survey," *JFE* 10, no. 4 (1928): 516–524; Joint Committee on Bases of Sound Land Policy, *What about the Year 2000?* (Harrisburg, PA: Mount Pleasant Press, 1929); Charles C. Adams, *The Importance of Preserving Wilderness Conditions* (Albany: University of the State of New York, 1929); and "Henry Chandler Cowles, Physiographic Plant

Ecologist," *AAAG* 30, no. 1 (1940): 39–43; Sharon Kingsland, *Modeling Nature* (Chicago: University of Chicago Press, 1985).

8. *Proceedings of the National Conference on Land Utilization* (Washington, DC: Government Printing Office, 1932); Paul Sears, *Deserts on the March* (Norman: University of Oklahoma Press, 1935); U.S. Great Plains Committee, *The Future of the Great Plains* (Washington, DC: Government Printing Office, 1936); Victor Shelford and Herbert Hanson, "The Problem of Our Grasslands," in *Our Natural Resources and Their Conservation*, ed. A. E. Parkins and J. R. Whitaker (New York: Wiley, 1936); and Henry L. Henderson and David B. Woolner, eds., *FDR and the Environment* (New York: Palgrave Macmillan, 2005).

9. I refer to several scholars' work on the importance of conservation to presidents Hoover and Roosevelt; on the historiography of conservation policy, see the "Essay on Sources."

10. This early history of conservation is documented in the work of Samuel P. Hays and others; see the "Essay on Sources."

11. L. C. Gray, "The Social and Economic Implications of the National Land Program," *JFE* 18:2 (1936): 258. Whereas the environmental movement of recent decades has frequently framed its agenda in opposition to economic development, during the New Deal, the values of ecological science and economic development were linked. Many conservation promoters, recognizing the two fields shared roots ("oikos"), suggested that ecological principles applied to resource management would offer increased productivity for the long term in the form of "renewable" resources that might be conserved as they were used.

12. Criticized from its debut, Clements's work nevertheless was well received in range management, forest management, and soil conservation circles. Clements and his student and collaborator John Weaver became consultants on a number of ecological engineering projects. Hanson, a student of both men, also worked on rural rehabilitation. Christopher Masutti, "Frederic Clements, Climatology and Conservation in the 1930s," *Historical Studies in the Physical and Biological Sciences* 31, no. 1 (2006): 27–48, describes how Clements pragmatically revised his theories in light of criticism to remain relevant to national needs.

13. Frederic Clements, *Plant Succession and Human Problems* (Washington, DC: Carnegie Institution of Washington, 1935), 241.

14. Frederic Clements and Ralph Chaney, *Environment and Life in the Great Plains*. (Washington, DC: Carnegie Institution of Washington, 1937), 52.

15. Herbert Hanson, "Ecology in Agriculture," *Ecology* 20, no. 2 (1939): 117; C. W. Thornthwaite, "The Relation of Geography to Human Ecology," *EM* 10, no. 3 (1940): 343–348, expresses similar ideas. The public perception of poor farming practices as a root cause of the Dust Bowl, the explanatory framework that has dominated scholarly and popular histories of this era, was erroneous in

the view of historian Geoff Cunfer. Nonetheless, the strength of this belief during the period prompted great interest in studying the effects of humans on the environment. Among American scientists, the federal agriculture department in particular was recognized as a model for the marriage of scientific research and planning in the interwar period. See Cunfer, *On the Great Plains: Agriculture and Environment* (College Station: Texas A&M University Press, 2005); Edward Banfield, "Organization for Policy Planning in the U.S. Department of Agriculture," *JFE* 34, no. 1 (1952): 14–34; and Brian Balogh, *Chain Reaction: Expert Debate and Public Participation in American Commercial Nuclear Power, 1945–1975* (New York: Cambridge University Press, 1991).

16. The eight-part scheme, created by Hugh Bennett, is described in Roy Hockensmith and J. G. Steele, *Classifying Land for Conservation Farming* (Washington, DC: USDA, 1943), 9.

17. George Renner and William Hartley, *Conservation and Citizenship* (Boston: Heath, 1940), 128. According to Christopher Masutti, land use planners' scientific determinations drew heavily on ecological theories in the 1930s. This was not a totally new role for ecologists; they had participated in early land classification efforts such as the Michigan Land Economic Survey in the 1920s.

18. AAAS, *Conserving Our Natural Resources: A Selected List of Material Useful to Students and Discussion Clubs* (Washington, DC: AAAS, 1937); SCS, *Soil Conservation Districts for Erosion Control* (Washington, DC: SCS, 1937); Hugh Bennett, "Developing Enlightened Public Opinion in Conservation, 1940" (address by Hugh H. Bennett, Chief, Soil Conservation Service, USDA, before the Assembly on Use of Human and Natural Resources in Education, Seventy-eighth Annual Meeting, National Education Association, Milwaukee, Wisconsin, July 2, 1940), VF NAC 7560; idem, *Land Use and Soil Conservation Informational Material Available to the Public* (Washington, DC: SCS, 1940); idem, *Publications and Visual Information on Soil Conservation* (Washington, DC: USDA, 1941); John Babcock, "The Role of Public Discourse in the Soil Conservation Movement, 1865–1935" (PhD diss., University of Michigan, 1985); Henderson and Woolner, *FDR and the Environment*; Carroll Pursell, ed., *From Conservation to Ecology* (New York: Crowell, 1973). Such efforts were complemented by public relations campaigns from state and local authorities. Notably, what Kendrick Clements has called "managed voluntarism and controlled decentralization," the notion that expert-designed programs should be "administered as much as possible by the people most affected by them," had been a hallmark of Herbert Hoover's approach to conservation during his time in federal service; as the conservation effort expanded to private lands this strategy held great appeal to his presidential successor. Kendrick Clements, *Hoover, Conservation, and Consumerism: Engineering the Good Life* (Lawrence: University Press of Kansas, 2000).

19. W. Robert Parks, *Soil Conservation Districts in Action* (Ames: Iowa State College Press, 1952); Wayne Rasmussen, "The New Deal Farm Programs: What They Were and Why They Survived," *American Journal of Agricultural Economics* 65, no. 5 (1983): 1158–1162; Linda Gordon, "Dorothea Lange: The Photographer as Agricultural Sociologist," *JAH* 93, no. 3 (2006): 698–727; Theresa May, "Earth Matters: Ecology and American Theatre" (PhD diss., University of Washington, School of Drama, 2000); Randal Beeman and James Pritchard, *A Green and Permanent Land* (Lawrence: University Press of Kansas, 2001); Finis Dunaway, *Natural Visions: The Power of Images in American Environmental Reform* (Chicago: University of Chicago Press, 2005). These public relations efforts complemented the content of government reports for policy planning such as the U.S. Great Plains Committee's *The Future of the Great Plains* and *Supplementary Report of the Land Planning Committee to the National Resources Board* (Washington, DC: Government Printing Office, 1935–1938), which cited the work of ecologists, including Adams, Shelford, and Clements. Frederic Clements praised Bennett's work in particular. See Frederic E. Clements, "Climatic Cycles and Human Populations in the Great Plains," *SM* 47, no. 3 (1938): 193–210.

20. George S. Wehrwein, "An Appraisal of Resettlement," *JFE* 19, no. 1 (1937): 190; Gray, "The Social and Economic Implications of the National Land Program," 268.

21. Rexford Tugwell, "The Reason for Resettlement," address on the *National Radio Forum*, Washington, DC, 1935; USRA, *Helping the Farmer Help Himself* (Washington, DC: Government Printing Office, 1936); Cara Finnegan, *Picturing Poverty* (Washington, DC: Smithsonian Books, 2003); Sidney Baldwin, *Poverty and Politics* (Chapel Hill: University of North Carolina Press, 1968).

22. Daniel M Schuyler, "Constitutional Problems Confronting the Resettlement Administration," *JLPUE* 12, no. 3 (1936): 304–306.

23. Richard Kirkendall, *Social Scientists and Farm Politics in the Age of Roosevelt* (Columbia: University of Missouri Press, 1966); Harry McDean, "Social Scientists and Farm Poverty on the North American Plains, 1933–1940," *Great Plains Quarterly* (1983): 17–29; Mary Bosworth Treudley, "Community Structure and Organization," *JES* 19, no. 9 (1946): 576–585.

24. Building on his experience as a New York governor, President Franklin D. Roosevelt joined the conservation of "human resources" with the conservation of natural resources in approving programs, including the Civilian Conservation Corps and the Resettlement Administration. Textbooks of the period such as *Conservation and Citizenship* from Renner and Hartley featured human resource planning prominently in their discussions of the conservation movement. There was some precedent for these national discussions of human conservation in the

earlier administration of Theodore Roosevelt. At that time, concerns about future resource availability extended to anxieties about the caliber of the U.S. population, linking calls for human conservation to the burgeoning eugenics movement. The movement to conserve human resources in the later Roosevelt administration, by contrast, turned away from biological determinism to focus on skills development. On eugenics and conservation, see Gray Brechin, "Conserving the Race: Natural Aristocracies, Eugenics, and the U.S. Conservation Movement," *Antipode* 28, no. 3 (1996): 229–245.

25. Others such as Eduard Lindeman, whose chief interest was social welfare, also expressed interest in human conservation, organizing a conference on the "conservation and development of human resources." See "Current Items," *ASR* 3, no. 2 (1938), 244.

26. Hanson, "Ecology in Agriculture," 116. He was speaking of Carl Taylor and Conrad Tauber. This was consistent with Clements's view of the successional processes at the center of ecology as critical to national recovery. "As an instrument for the control of the entire range of human uses of vegetation and the land," Clements explained, "succession is wholly unrivalled." Frederic Clements, "Experimental Ecology in the Public Service." *Ecology* 16, no. 3 (1935): 345.

27. John D. Guthrie, "The CCC and American Conservation," *SM* 57, no. 5 (1943): 412. Scholars have noted how, beyond the public relations campaign, the sheer size of some programs—especially the human resource conservation work relief and job training programs such as the Civilian Conservation Corps and Tennessee Valley Authority and the public nature of many physical planning projects—helped to acquaint the nation with conservation during this period.

28. Paul Landis, "The New Deal and Rural Life," *ASR* 1, no. 4 (1936): 599.

29. Rasmussen, "The New Deal Farm Programs"; Henderson and Woolner, *FDR and the Environment*; Donald Worster, *Dust Bowl: The Southern Plains in the 1930s* (Oxford: Oxford University Press, 2004).

30. Parkins and Whitaker, *Our Natural Resources and Their Conservation*, 18.

31. Richard Hofstadter, *The Age of Reform* (New York: Knopf, 1955); William E. Leuchtenburg, *Franklin D. Roosevelt and the New Deal* (New York: Harper, 1963).

32. Their interest in management on scientific grounds was a general scientific management to be distinguished from the Taylorist approach popular earlier in the twentieth century.

33. Harvey Zorbaugh, "The Natural Areas of the City," *PASS* 20 (1926): 196.

34. Nels Anderson and Eduard Lindeman, *Urban Sociology* (New York: Al-

fred Knopf, 1928), 383–384; Frederic Thrasher, "The Study of the Total Situation," *JES* 1, no. 8 (1928): 599–612; Nels Anderson, "Zoning and the Mobility of Urban Population," *City Planning* (October 1925): 159.

35. George B. Ford, "The City Scientific," *Engineering Record* 67 (May 17, 1913): 551–552; George Ford, "Fundamental Data for City Planning Work," in *City Planning*, ed. John Nolen (New York: D. Appleton, 1929); Technical Advisory Corporation, *Planning the Comprehensive City Plan* (New York: The Corporation, November 1921), VF NAC 815; Advisory Committee of City Planning Experts of the Conference of Mayors and Other City Officials of New York State, *Questionnaire for City Planning Survey of the Cities of New York State* (New York: The Committee, 1924), VF NAC 815. Clementsian ecology viewed communities as organisms, which fit well with an older view of cities as bodies popular in the United States and in Europe.

36. Ernest Fisher, "Expansion of the Urban Land Area," in *The Metropolitan Community*, ed. Roderick McKenzie (New York: McGraw-Hill, 1933), 241–242; Louis Wirth, "The Urban Mode of Life," in *APCA* (1937); Nels Anderson, Letter to the Editor, *Millar's Housing Letter* 2, no. 26 (April 9, 1934): 6; Chauncy D. Harris, "Charles C. Colby, 1884–1965," *AAAG* 56, no. 2 (1966): 378–382; Karl Lohmann, *Principles of City Planning* (New York: McGraw-Hill, 1931); Harland Bartholomew, *Urban Land Uses* (Cambridge, MA: Harvard University Press, 1932); Mabel Walker, *Urban Blight and Slums* (Cambridge, MA: Harvard University Press, 1938).

37. In notable contrast to social scientists, these professionals would rarely cite the work of plant and animal ecologists directly. One exception was the decision by *Planners' Journal* to reprint an essay by ecologist Charles Adams. See Adams, "A Note for Social-minded Ecologists and Geographers," *Ecology* 19, no. 3 (1938): 500–502 in *Planners' Journal* 5, no. 1 (1939): 6–7.

38. Architects Club of Chicago, *Rehabilitating Blighted Areas* (Chicago: Architects Club, 1932), 9.

39. See, for example, Herbert Swan, "Land Values and City Growth," *JLPUE* 10, no. 2 (1934): 188–189.

40. Architects Club of Chicago, *Rehabilitating Blighted Areas*, 34.

41. George Renner, "National Regional Planning in Resource Use," in *Our Natural Resources and their Conservation*, 597. On charges of socialism made against planners, see Otis Graham, *Toward a Planned Society* (New York: Oxford University Press, 1976); and Marion Clawson, *New Deal Planning: The National Resources Planning Board* (Baltimore: Johns Hopkins University Press, 1981).

42. Harold S. Buttenheim and Associates, "Urban Land Policies: Staff Paper for Item 11 of the General Program of the Research Committee on Urbanism of the National Resources Committee" (Confidential—Not for Public Release. Prelimi-

nary and Incomplete, November 1936), Vol. 2 discusses the history of eminent domain in cities.

43. Such meetings included the American Planning and Civic Association (where presenters such as A. R. Mann analyzed the urban and rural land use surveys together in a single presentation); journals such as the *JLPUE* (where articles on urban and rural real estate appeared) also provided a space for cross-disciplinary conversation. See A. R. Mann, "The Urban and Rural Land-Use Survey," in *APCA* (1935): 241–248; Charles Kellogg, "A Method for the Classification of Rural Lands for Assessment in Western North Dakota," *JLPUE* 9, no. 1 (1933): 10–15; and Lewis A. Maverick, "Cycles in Real Estate Activity: Los Angeles County," *JLPUE* 9, no. 1 (1933): 52–56.

44. Banfield, "Organization for Policy Planning in the U.S. Department of Agriculture"; E. A. Foster and H. A. Vogel, "Cooperative Land Use Planning," in *Farmers in a Changing World* (Washington, DC: USDA, 1940); *Soil Conservation Districts for Erosion Control* (Washington, DC: SCS, 1937); Babcock, "The Role of Public Discourse in the Soil Conservation Movement, 1865–1935." This is not to ignore the controversies such programs generated. On the conflicts for land use planners surrounding acquisitions of forests and farms, see E. H. Wiecking, "Application of Research to Action Programs," *JFE* 19, no. 2 (1937): 594–602.

45. Harold L. Ickes, "The Place of Housing in National Rehabilitation," *JLPUE* 11, no. 2 (May, 1935): 111.

46. Ibid., 112.

47. Delano compared cities and orchards, and the need to "prune" both toward effective development; he did not use the term *resource*. See Joint Committee on Bases of Sound Land Policy, *What about the Year 2000?* 137–138.

48. Loyal Durand, "State and Local Planning," in *Our Natural Resources and Their Conservation*; Ben Minteer, "Regional Planning as Pragmatic Conservation," in *Reconstructing Conservation*, ed. Ben Minteer and Robert Manning (Washington, DC: Island Press, 2003). Planning education reflected these trends, combining social science instruction in human ecology with natural resource planning as students were trained to conceptualize planning for the needs of cities within the context of planning at the regional and national levels. "Conservation of Natural Resources" became a staple of the planning curriculum at numerous schools, as described by Hugh S. Morrison, *Education for Planners* (Boston: NRPB, Region I, 1942), VF NAC 960 M, and ASPO, *Planning Education: A Survey of the Curricula in 23 Colleges and Universities* (memorandum, Chicago, ASPO, May 1942). By the mid-1930s, when plant ecologist Charles Adams remarked on the similarities of "the methods of the regional and ecological surveys," he thus confirmed the enthusiastic convergence of traditions in the social science and planning fields. Adams, "The Relation of General Ecology to Human Ecol-

ogy," 329. Notably, regional planning was another field to which human ecologists, and "ecological" thinking more generally, made important contributions. On ties to regional planning see J. Paul Goode, *The Geographic Background of Chicago* (Chicago: University of Chicago Press, 1926); Fritiof M. Fryxell, *The Physiography of the Region of Chicago* (Chicago: University of Chicago Press, 1927); Robert Murray Haig, *Major Economic Factors in Metropolitan Growth and Arrangement* (New York: RPA, 1927); Selden Cowles Menefee, "A Plan for Regional Administrative Districts in the State of Washington: An Ecological Study," *University of Washington Publications in the Social Sciences* 8, no. 2 (1935): 29–80; Lorrie Nelson Douglas, "Human Ecology and Regional Planning: An Abstract of a Paper Read to the Halsted Street Philosophical Society" (Halsted Street Philosophical Society, Chicago, 1934), VF NAC 6100; Rupert Vance, "Implications of the Concepts 'Region' and 'Regional Planning'" in *PASS* 29, no. 3 (1935); Howard Odum, "The Role of Regionalism and the Regional Council in National Planning," *Proceedings of the National Conference of Planning* (Chicago: ASPO, 1941); Louis Wirth, "The Prospects of Regional Research in Relation to Social Planning," *PASS* 29 (1935): 107–114; and Adams, "A Note for Social-Minded Ecologists and Geographers." The importance of ecology at the Regional Planning Association of America, the most prominent regional planning organization in this period, is detailed in Marc Luccarelli, *Lewis Mumford and the Ecological Region: The Politics of Planning* (New York: Guilford Press, 1997).

49. National Resources Committee, *Our Cities: Their Role in the National Economy* (Washington: NRC, 1937), 73; NRC, *Our Cities* (brochure) (Washington: NRC, September 1937), 25. Comey, with a degree in landscape architecture from Harvard, had been a student of Frederick Law Olmsted and had worked on park development before turning to city planning. At this time, instruction in city planning often took place in programs of landscape architecture, for example, at Harvard and University of Illinois. Other NRPB affiliates not on the urbanism committee, including geographers Charles Colby and George Renner, made similar arguments in their work. As Colby observed at a meeting of the American Planning and Civic Association, "Town and country are so intimately related in modern times that rural and urban planning have much in common." Renner's textual and photographic juxtaposition of ruined landscapes, rural and urban, dramatized how the problems of nature and the problems of cities were urgent matters for the conservation movement to resolve. See Colby in John Black, "The Soil and the Sidewalk," in *APCA* (1939): 79; and Renner and Hartley, *Conservation and Citizenship*.

50. According to Marion Clawson, the primary reason this group had as its name "national" rather than "natural" resources was the presence of University of

Chicago political scientist Charles Merriam (a specialist on urban politics among other topics) on its executive board. When executive board members Charles Eliot, Wesley Clair Mitchell, and Charles Merriam met with Roosevelt in June 1934 to discuss the mission of the new organization following its first year in operation, the president offered up the phrase "land and water planning." Eliot (who, it should be noted, had extensive city planning experience) countered with "natural resources" and finally Merriam with "national resources." "The President then repeated the phrase several times, liked its sounds and remarked, 'That's right, friend Eliot, get that down, because that's settled.'" Clawson, *New Deal Planning*, 45, citing Arthur Schlesinger. From that time forward, for example, in his message to Congress in January 1935, the president — and, in turn, participants on the NRPB — would use the terms "national resources" and "natural resources" essentially interchangeably. Richard Andrews, "FDR's Environmental Legacy," in *FDR and the Environment*, 226.

51. National Planning Board, *Report* (Washington, DC: The Board, December 1, 1934).

52. U.S. Conference of Mayors, "Resolution on Federal Research into Urban Problems Passed at the Annual Conference of the United States Conference of Mayors Held in November 1934." In Box 886 (no folder), Central Office Records, Central Office Classified Correspondence and Related Records, 1931–1943, 415 Urbanism Reports, Transmittal — Urbanism Report, NRPBR.

53. NRC, *Our Cities*, vi.

54. Ibid., vii.

55. NRC, *Our Cities*, 64.

56. Ibid., v; NRC, *Our Cities* (brochure), 1–2.

57. NRC, *Our Cities*, xiii.

58. Phoebe Cutler, *The Public Landscape of the New Deal* (New Haven, CT: Yale University Press, 1985). There is some evidence that, several decades earlier, at the height of excitement about scientific management, a few writers had suggested the value of conservation for cities, including W. D. Foulke, "Conservation in Municipalities," *Proceedings of the Buffalo Conference for Good City Government* (Philadelphia: National Municipal League, 1910), 12–21.

59. President's Committee on Administrative Management, *Report* (Washington, DC: President's Committee, 1937). However, rather than explicitly link conservation to cities, the Brownlow report proposed city planning become the charge of a permanent National Planning Agency.

60. NRC, *Our Cities* (brochure), 4. "Asks Federal Hand in Decadent Cities; National Resources Committee Advises Reorganization, or Resettlement of Dwellers as in Poor Farming Areas," *NYT*, September 20, 1937, 7.

61. Wirth, "The Urban Mode of Life," 375.

62. NRC, *Our Cities*, 73; "Democratic Planning," *NYT*, September 24, 1937, 20.

63. NRC, *Our Cities*, x. They repeated this again later in the report with more specificity, calling for "a section for urban research which should perform for urban communities functions comparable to those now performed for rural communities by the Bureau of Agricultural Economics and Agricultural Engineering." Ibid., 79. A subsequent report would expand on this idea, calling for "study and research" undertaken "at universities in the different regions," an approach whose "wisdom . . . has long been recognized in the study of agricultural problems; a federal agency maintains a national research staff, and aids experiments at stations connected with agricultural colleges." Charles S. Ascher, *Our Cities, Building America* (Washington, DC: NRPB, 1942), 17. In fact, the agriculture department had begun to sponsor its own urban research at this time, led by Oliver E. Baker — who would serve on the urbanism committee as well. Baker had publicly equated human ecology and economic geography and published several articles on urban growth in *Real Estate* around this time. See O. E. Baker, "Will Cities Continue to Grow?" *RE* 50 (December 12, 1936): 7–8, 12; idem, "Population Affects Land Values," *RE* 51 (December 19, 1936): 9–10; idem, "Population Is a Value Factor," *RE* 52 (December 26, 1936): 9–13; O. E. Baker et al., "Round Table Conferences, Economics and Geography," *American Economic Review* 16, no. 1 (1926): 113.

64. Letter from M. L. Wilson to Charles Merriam, July 10, 1937, 1–2. In Box 886 (no folder), Central Office Records, Central Office Classified Correspondence and Related Records, 1931–1943, 415 Urbanism Reports, Transmittal—Urbanism Report, NRPBR.

65. *Supplementary Report of the Land Planning Committee to the National Resources Board* (Washington, DC: Government Printing Office, 1935–1938). The board released numerous reports on conservation during its tenure. Notably, one figure not cited in this work was eminent U.K. ecologist-planner Patrick Geddes, who had sought "conservation" for cities, although in using this term meant something closer to historical preservation.

66. Richard Ratcliff, Review of *Our Cities*, *JLPUE* 14, no. 2 (1938): 231.

67. Walker, *Urban Blight and Slums*, 31; *Supplementary Report of the Urbanism Committee to the U.S. National Resources Committee* (Washington, DC: Government Printing Office, 1939).

68. To Members of the American Society of Planning Officials, from Walter Blucher, September 20, 1937, Re: *Our Cities—Their Role in the National Economy* (report of the Urbanism Committee to the National Resources Committee), 2, VF NAC 1110.

69. Anderson and Lindeman, *Urban Sociology*.

70. Untitled document, Box 13, Records of the Federal Home Loan Bank System, Federal Home Building Service Plan, Program Subject Files, 1936–1942, Folder Pamphlets and Literature on Housing (Folder No. 1), 13, FHLBBR.

71. Marc A. Weiss, *The Rise of the Community Builders: The American Real Estate Industry and Urban Land Planning* (New York: Columbia University Press, 1987); and Jeffrey Hornstein, *A Nation of Realtors* (Durham, NC: Duke University Press, 2005).

72. J. C. Nichols, "Responsibilities and Opportunities of a Real Estate Board," in *Proceedings of the General Sessions of the National Association of Real Estate Boards* (Chicago: NAREB, 1924), 21. See also Richard Ely, "Research in Land and Public Utility Economics," *JLPUE* 1, no. 1 (1925), 1.

73. Richard Ely and Edward Morehouse, *Elements of Land Economics* (New York: Macmillan, 1924), 71; Frederic Ogg, *Research in the Humanistic and Social Sciences* (New York: Century, 1928), 188. Distinguished figures in agricultural economics, including L. C. Gray and Oliver E. Baker, were former students and reflected Ely's earlier influences on studies of agricultural land.

74. Like his University of Chicago colleagues, Ely stressed the importance of practical experience for advancing scholarly knowledge, a perspective reflected in the background of Institute faculty and student affiliates. From Arthur Mertzke (who arrived from a post as executive secretary of the Madison Real Estate Board) to Frederick Babcock (who worked simultaneously as a realtor and appraiser at William H. Babcock and Sons), at Northwestern as at Chicago, integrating theory and practice was an institutional priority. Coleman Woodbury, "Richard T. Ely and the Beginnings of Research in Urban Land and Housing Economics," *LE* 25, no. 1 (1949): 65.

75. Roderick McKenzie, "The Ecological Approach to the Study of the Human Community," *AJS* 30 (1924): 287–301; and Herbert Nelson, *The Administration of Real Estate Boards* (New York: Macmillan, 1925), 191.

76. Nelson, *The Administration of Real Estate Boards*, 191. On the use of social scientists' maps in real estate practice, see Nelson's chapter on "The Board as a Center of Real Estate Information."

77. Thomas Philpott, *The Slum and the Ghetto* (New York: Oxford University Press, 1978), 216. The scientific rationale for segregation by race, nationality, and economic status is widely discussed; for example, in Herbert Dorau and Albert Hinman, *Urban Land Economics* (New York: McGraw-Hill, 1928); Stanley McMichael, *McMichael's Appraising Manual* (New York: Prentice-Hall, 1931); AIREA, *Real Estate Appraisal* (Chicago: AIREA, 1935); and Frederick Babcock, *The Valuation of Real Estate* (New York: McGraw-Hill, 1932).

78. In 1920, the Farm Mortgage Bankers Association proposed to the National Association of Real Estate Boards that they join together as cosponsors of

this work; only the partnership between the Realtors' association and Ely's group was ultimately pursued.

79. While compared to matters of interest to city dwellers natural resource policy concerns continued to dominate the national political landscape, motivated by the belief that stabilizing the economic base for housing and addressing the issue of unemployment in the construction and building trades would contribute to larger economic recovery, the federal government took some actions on urban housing matters in the immediate wake of the nation's economic depression. See Leuchtenburg, *Franklin D. Roosevelt and the New Deal*.

80. Earlier and competing appraisal practices are discussed in Willis Clark and J. Harold Williams, "A Guide to the Grading of Neighborhoods," *Whittier State School Bulletin* 8 (July 1919); William Bailey, "Appraising Your City: A Scientific Method of Comparing the Advantages of Different Communities," *NREJ* 26 (May 18, 1925): 37–39; Stanley McMichael, "New Problems for the Appraiser," *JAIREA* 1, no. 4 (1933): 356–361; and Guy Stuart, *Discriminating Risk: The U.S. Mortgage Lending Industry in the Twentieth Century* (Ithaca, NY: Cornell University Press, 2003). Like their human ecologist colleagues who sought a generalizable model of the city, real estate appraisers were eager for scientific measures so that appraisers' assessments in one city would have meaning in another.

81. Marc Weiss, "Richard T. Ely and the Contribution of Economic Research to National Housing Policy, 1920–1940," *Urban Studies* 26, no. 1 (1989): 115–126, describes how the same figures from Northwestern University's Institute for Land Economics and Public Utilities who created the national real estate curriculum also influenced federal housing policies (especially at the Federal Housing Administration). Early issues of *Insured Mortgage Portfolio*, the agency journal, are filled with examples of ecological interpretations of urban patterns and processes. In "Residential Security Maps and Neighborhood Appraisals: The Home Owners' Loan Corporation and the Case of Philadelphia," *SSH* 29, no. 2 (2005): 207–233, Amy Hillier discusses the ecological theory as applied at HOLC. While the agencies adopted different approaches to assessing risk, both settled on rating systems that graded locations on a four-class scale from A to D. The difference was that while HOLC security maps made locational risk their central concern; the Federal Housing Administration's risk rating system (which also created location rating maps) integrated this information with quantitative ratings of the applicant property, borrower, and mortgage pattern to compute an overall index of risk. Kenneth Jackson, "Race, Ethnicity, and Real Estate Appraisal: The Home Owners' Loan Corporation and the Federal Housing Administration," *JUH* 6, no. 4 (1980): 434; Federal Housing Administration, *Underwriting Manual* (Washington, DC:

Government Printing Office, 1935); "New Manual Gives FHA Method of Rating Mortgage Risk," *NREJ* 36 (April 1935): 25–26; NAREB, "Babcock Outlines FHA Plans for Risk-Rating of Mortgages Before Huge National Conference of Real Estate Appraisers" (NAREB, Chicago, 1934), VF NAC 1434.9.

82. Collaborations between local appraisers and agency officials are documented in Homer Hoyt, "Exhibit J: The Preparation of Maps Showing the Dynamic Factors of City Growth: Prepared for the Research Division for the Underwriting Staff," January 8, 1935, 2, appended to "Program for the Study of Sixty-Two Cities." In Records Relating to the Economic Data System, 1936, 1941, 1943, Box 1, Records of the Housing and Home Finance Agency, Records of the Division of Research and Statistics of the FHA, "Economic Data System" Records Containing Data on Cities, 1937–1945, HUDR, and in HOLC files for individual cities at the U.S. National Archives. Both the FHA journal *Insured Mortgage Portfolio* and the FHLBB journal *Federal Home Loan Bank Review* printed numerous articles publicizing their methods. The *Underwriting Manual* was widely disseminated, but both agencies were secretive about their location rating maps. FHA maps were available for consultation in local insuring offices while HOLC maps were distributed "to the executive branch of the Federal Home Loan Bank Board and its affiliates" (NRPB, *Federal Aids to Local Planning, June 30, 1940* [Washington, DC: Government Printing Office, 1941], 95), together with citywide summary reports, such as *Metropolitan Chicago: Summary of Economic, Real Estate and Mortgage Finance Survey* (Washington, DC: Division of Research and Statistics, FHLBB, 1940). Certainly, just as all social scientists had not agreed with Burgess's four-stage model of invasion and succession, nor with his five-ring model of urban concentric zones, many individuals and institutions in the real estate community remained uncommitted to the four-stage standards promoted by the federal government. Such was the case at one Los Angeles bank, which employed a five-stage classification system of subdivision, growth, maturity, decline, or decadence in its appraisals. L. Elden Smith, "Measuring the Neighborhood Risk," *IMP* 2 (1938): 9–11, 22–23.

83. Although historians have debated the extent to which the Home Owners Loan Corporation's standards of risk estimation affected banks' lending decisions, that this agency, like the Federal Housing Administration, transformed ecological theory into practically oriented advice is not in doubt.

84. Herbert Dorau and Albert Hinman recognized as early as the 1920s how "eminent domain, like the police power, is limited only by the opinion of courts . . . in the final analysis, public purpose is defined by the courts . . . hence, like the police power, it is capable of indefinite expansion as the public, or social, point of view becomes more universal." Dorau and Hinman, *Urban Land Economics*, 273.

85. NRC, *Our Cities*, 62.

86. "NAREB Tackles Problems of Re-planning Blighted Areas," *NREJ* (April 1935), 44.

87. NAREB, *Neighborhood Protective and Improvement Districts* (Chicago: NAREB, 1935), 2.

88. Chicago sociology graduate student Everett Hughes's 1931 observation was typical: "The change from white to colored occupancy is said to be as irretrievable as from residence to business." Everett Hughes, *The Chicago Real Estate Board* (Chicago: University of Chicago Press), 94.

89. Babcock, *The Valuation of Real Estate*, 88–92. Not every observer concurred with the consensus interpreting cycles in urban real estate as tending toward decline. An example of the older more optimistic view is Ernest Fisher's *Advanced Principles of Real Estate Practice* (New York: McGraw-Hill, 1930), although his later writings in *Insured Mortgage Portfolio* suggest he moved to consensus with Babcock.

90. The "neighborhood unit" was developed and refined by Clarence Perry of the Regional Planning Association of America during the late 1920s. Clarence Arthur Perry, *The Neighborhood Unit* (New York: RPA, 1929); Clarence Perry, *The Rebuilding of Blighted Areas* (New York: RPA, 1933); and James Dahir, *The Neighborhood Unit Plan* (New York: Russell Sage, 1947).

91. ACPI Committee on the Proposed Statute of the National Association of Real Estate Boards, "Confidential to Members of the American City Planning Institute" (report of the ACPI, October 1, 1936), 1, VF NAC 1676.

92. Babcock, *The Valuation of Real Estate*, 58.

93. Harold S. Buttenheim, "Urban Land Policies," in *Supplementary Report of the Urbanism Committee*, ed. NRC (Washington, DC: NRC, 1939), 2:217.

94. Ibid., 151.

95. ACPI Committee on the Proposed Statute of the National Association of Real Estate Boards, "Confidential to Members," 3, 7.

96. Philpott, *The Slum and the Ghetto*, 190. The 1924 Code of Ethics is printed in Nelson, *The Administration of Real Estate Boards*. Fisher also had some role in writing this. Letter from Frederick Babcock to Albert E. Landvoight, October 7, 1936. In Folder Babcock, Frederick, Box 2, Records of the Central Housing Committee, Records of the Committee on Appraisal and Mortgage Analysis, 1935–1942, HUDR.

97. As Thomas Philpott has described, these efforts were paid for by individual property owners as well as by neighborhood institutions such as churches and businesses. Philpott, *The Slum and the Ghetto*, 191. A range of parties in the real estate community had promoted the idea of giving community organizations legal status in the 1930s toward similar ends. See Arthur Holden, "Eliminating

Block Depreciation by Group Management," *RER and Builders Guide* (March 24, 1934): 6–8; idem, "Removing the Obstacles to Group Rehabilitation of Real Properties," *JLPUE* 12, no. 2 (1936): 161–168; and Robert B. Mitchell, "Prospects for Neighborhood Rehabilitation," in *Housing Yearbook*, ed. NAHO (Chicago: NAHO, 1938), 143.

98. Herbert Nelson, "Rehabilitation of the Blighted District: The Share of the Realtor," in *Planning for the Future of Cities*, ed. ASPO (Chicago: ASPO, 1935).

99. NAREB, *The Neighborhood Improvement Act* (Chicago: NAREB, 1937), 1. Nelson, "Rehabilitation of the Blighted District," 90–94, makes this clear.

100. Ibid. See also "The Next Step: Neighborhood Management," *RER* (May 15, 1937): 41. Passed by Michigan's legislature, the governor vetoed the act. "Adopt Act to Help City Neighborhoods: Action by Michigan to Improve Poor Areas Called first of Type in Country," *NYT*, July 25, 1937, 154.

101. Efforts to encourage individual property owners to maintain housing quality through "repair," "rehabilitation," "modernization," "reclamation," and "reconditioning" date to the earliest years of the FHA and HOLC. At the FHA, where Title 1 of the 1934 Housing Act gave access to housing loans, a better housing initiative organized publicity efforts to convince homeowners to take advantage of the program, publicity that included a series of films, brochures, and radio spots directed at homeowners' associations, men's and women's clubs, banks, labor organizations, and other groups to popularize modernization and repair. At the HOLC, a mortgage rehabilitation division, a staff of regional reconditioning supervisors, and a Federal Home Building Service Plan that encouraged high standards for new construction worked toward complementary ends. None of these tactics was especially new. From the appearance of new hobbyist products and home magazines in the late nineteenth century to the Hoover administration's Better Homes campaign and domestic engineering movement of the early twentieth century, middle-class home improvement has had a long history. The popularity of such measures lay as much in their associations with patriotism, individualism, and productive leisure as they did with actual changes to living environments. The difference between these earlier efforts and their federally sponsored counterparts under the two new housing agencies was the level of systematic investment and analysis involved. By 1937, for example, nearly 1.5 million modernization loans had been made through Title I. At both agencies, such improvement programs were subjected to intensive study. Early results appeared highly promising, as internal reports and external publicity shared how, in most cases, the value added to a property far outweighed the costs of home repair. See "Loans Made for Modernization," *NYT*, May 23, 1937, 196; and the collected news clippings in Box 6 Records of the Federal Home Loan Bank System, Federal Home Building Service Plan, Program Subject Files 1936–42, Folder Exhibits, FHLBBR.

102. In contrast to city planners, who interacted with colleagues in natural resources planning at a variety of conferences on regional and national planning, encounters between real estate economists working on rural and urban issues were rare. For this reason, it appears that interactions with planners were the inspiration for the adoption of conservation by members of the real estate profession as a strategy for promoting city improvement. More specifically, as preparations for the experiments in Baltimore and Chicago went on, Miles Colean, Arthur Goodwillie, Jacob Crane, and John Ihlder were recruited by planners from the National Resources Planning Board to participate in a study of rural and urban land policies, bringing them into direct contact with the colleagues for whom conservation had become a dominant frame for discussions of comprehensive planning, land assembly, and scientific management for cities. The final document they collaboratively produced offered a range of proposals for expanding the public purpose of eminent domain to cities, framed by a discussion of how in the national enthusiasm for natural resources conservation, Americans now largely had accepted the idea that privately as well as publicly held farms and forests were within the purview of the "public purpose" that defined the legal requirements for usage of eminent domain, yet by contrast with the exception of clearing land for public housing, "public acquisition of urban land is neither as well established nor as well accepted as the acquisition of rural land." NRPB, *Public Land Acquisition in a National Land-Use Program* (Washington, DC: Government Printing Office, 1940), vol. 2, cover letter to Frederick Delano. One of the few points of contact that might have familiarized the real estate community with activities in the conservation movement was the Joint Committee on Appraisal and Mortgage Analysis, part of the U.S. Central Housing Committee, which assembled representatives from federal agencies, including the Federal Housing Administration, Home Owners Loan Corporation, and the Farm Credit Administration with heads of the American Institute of Real Estate Appraisers and Society of Residential Appraisers.

103. Untitled Document, Box 13, Records of the Federal Home Loan Bank System, Federal Home Building Service Plan, Program Subject Files, 1936–1942, Folder Pamphlets and Literature on Housing (Folder 1), 10, FHLBBR. This document appears to be a draft of portions of FHLBB, *Waverly* (Washington, DC: FHLBB, 1940).

104. NRC, *Our Cities*, 73; Untitled Document, Box 13, Records of the Federal Home Loan Bank System, Federal Home Building Service Plan, Program Subject Files, 1936–1942, Folder Pamphlets and Literature on Housing (Folder 1), 10, FHLBBR.

105. Untitled document, Box 13, Records of the Federal Home Loan Bank System, Federal Home Building Service Plan, Program Subject Files, 1936–1942, Folder Pamphlets and Literature on Housing (Folder 1), 8–9, FHLBBR.

106. Ibid.

107. Glenn Frank, "The Realtor as the New Pioneer," *AREP* (Chicago: NAREB, 1928): 3–4; J. C. Nichols, "Why I Am in the Real Estate Business," St. Louis, Missouri, January 29, 1937. In Planning for Permanence: the Speeches of J. C. Nichols, Western Historical Manuscript Collection — Kansas City uses the comparison between agricultural and urban development to call for more scientific forms of city management.

108. Richard Ely, "Research in Land and Public Utility Economics," *JLPUE* 1, no. 1 (1925): 4. See also Dorau and Hinman, *Urban Land Economics*, 297.

109. *Conservation* was a common economic term meaning the preservation of economic value. Economics "might well be called the science of conservation, for the terms economy and conservation, when properly understood, convey one and the same meaning," wrote Herbert Dorau and Albert Hinman in 1928, noting how "Zoning laws are usually accepted as valid since they accomplish an object which has been called the 'conservation of economic values,' in the sense that land values are stabilized and preserved by a rational, orderly development of a city." Dorau and Hinman, *Urban Land Economics*, 199.

110. Clifford J. Hynning, *State Conservation of Resources* (Washington, DC: NRC, 1939). Particularly relevant was the work of the Soil Conservation Service, which had organized demonstration programs as a public relations tool to convince states to adopt laws to enable the creation of soil conservation districts. See USDA Land Policy Committee, *A Standard State Soil Conservation Districts Law* (Washington, DC: Government Printing Office, 1936); R. Neal Sampson, *For Love of the Land* (League City, TX: National Association of Conservation Districts, 1985).

111. FHLBB, *Waverly*, viii. With one of the fundamental tenets of these federal agencies' approach to real estate appraisal grounded in the belief that area character influenced property valuation, it was increasingly recognized that although do-it-yourself protection efforts "considerably benefited surrounding properties by increasing the value of the latter and by inspiring neighbors to recondition their homes . . . once disintegration and decay have really begun their march in any area, the individual effort of a single property owner, even a very considerable one, cannot alone preserve a district from ultimate destruction." Untitled Document, Box 13, Records of the Federal Home Loan Bank System, Federal Home Building Service Plan, Program Subject Files, 1936–1942, Folder Pamphlets and Literature on Housing (Folder 1), 15, FHLBB. The theory behind the soil conservation effort, which called for establishing soil conservation districts based on the recognition that land use concerns did not stop at an individual owner's property line, had relevance here as well.

112. "Waverly: A Demonstration of Neighborhood Conservation," *FHLBR*

6 (1940): 330. Some of the participants occasionally referred to Waverly and Woodlawn as projects in neighborhood or municipal "housekeeping," a vocabulary choice linking these undertakings to the City Beautiful Movement. HOLC also used the term *stabilization*, suggesting how like efforts for agricultural stabilization so too urban "neighborhood stabilization" would need to be pursued.

113. Donald H. McNeal, "Waverly: A Study in Neighborhood Conservation," in *Proceedings of the National Conference on Planning*, ed. ASPO (Chicago: ASPO, 1941): 222. The Home Owners Loan Corporation had surveyed more than two hundred cities to depict graphically the state of individual neighborhoods in the urban life cycle. Rating neighborhoods A, B, C, or D by coloring in green, blue, yellow, or red on maps, this choice had enduring significance as the use of red for the areas characterized as least desirable is said to have laid the foundation for the real estate practice of "redlining." Corporation records explain how these maps translated ecological concepts into assessments of the character of specific urban locations: Rated A were " 'hot spots' . . . not yet fully built up . . . homogenous" communities. Rated B were "completely developed" areas "like a 1935 automobile—still good, but not what the people are buying today." Rated C were communities that "have reached the transition period," places "lacking homogeneity" or "characterized by age, obsolescence, and change of style; expiring restrictions or lack of them; infiltration of a lower grade population; the presence of influences which increase sale resistance." Rated D were "those neighborhoods in which the things that are now taking place in the C neighborhoods, have already happened." According to agency instructions, the implication was that lenders "with available funds" should "make their maximum loans" in grade A neighborhoods. By contrast, they might "refuse to make loans" or "lend only on a conservative basis" in communities rated D. [HOLC] Division of Research and Statistics with Cooperation of the Appraisal Department, Baltimore, Maryland, Untitled Document, May 29, 1937, Box 106, Records of the Federal Home Loan Bank Board, Home Owners Loan Corporation, Records Relating to the City Survey File, 1935–40. Ohio, Maine, Maryland Box, Folder Baltimore, Maryland Area Descriptions and Map No.1, no page, FHLBBR. However, as Amy Hillier has observed in several studies, much lending, including to areas outlined in red, preceded the maps' construction. The conservation project suggested a new line of interpretation regarding the significance of these classifications.

114. On the Waverly experiment, see also Adolf Waterval, *The Neighborhood Conservation Program, Waverly Area* (Baltimore: Waverly Conservation League, 1940); Fred Catlett, "Rehabilitation of Substandard Housing Areas" (unpublished paper), VF NAC 1613; Harvey Pinney, "A Blueprint for Urban Conservation," *National Municipal Review* 30 (March 1941): 157–159; and "Research on Neighborhood Conservation," *RER* (January 18, 1941): 7–8.

115. "Waverly," *FHLBR* 6 (1940): 333, 331.

116. Untitled Document, Box 13, Records of the Federal Home Loan Bank System, Federal Home Building Service Plan, Program Subject Files, 1936–1942, Folder Pamphlets and Literature on Housing (Folder 1), 18–19, FHLBBR.

117. Federal Housing Administration, *Underwriting Manual* (1935 edition); "Urges Renovation of Old Dwellings," *NYT*, July 6, 1941, RE4; "Research on Neighborhood Conservation," *RER* (December 28, 1940): 7. Of Waverly's 1,610 homes, the corporation owned 20 homes and held 122 mortgages.

118. Untitled Document, Box 13, Records of the Federal Home Loan Bank System, Federal Home Building Service Plan, Program Subject Files, 1936–1942, Folder Pamphlets and Literature on Housing (Folder 1), 20, FHLBBR.

119. Ibid., 16–17.

120. "Home Sales Spurred by Clean-Up Drive," *NYT*, September 1, 1940; Memorandum To: General Manager, Assistant General Managers Mr. Blouke, Mr. Charters, Mr. Downie, Mr. Stalling, Mr. Follin, Mr. Ferrens, From: DH McNeal, September 14, 1937, Box 2, Records of the Federal Home Loan Bank System, Federal Home Building Service Plan, Program Subject Files 1936–42, Folder Charts, Graphs and Maps, FHLBBR.

121. "Waverly," *FHLBR* 6 (1940): 330.

122. The Waverly Conservation League mobilized heads of several existing area groups — the Waverly Improvement Association, Chestnut Hill Improvement Association, and Greenmount Improvement Association.

123. On Woodlawn, see Paul Healy, "Launch Woodlawn Home Survey," *CDT*, April 28, 1940, sw1; "WPA to Make Realty Survey in Woodlawn," *CDT*, January 14, 1940; Robert Mitchell, "Neighborhood Conservation in Chicago: Wood-lawn Conservation Project," *AC* 55 (May 1940): 64–65; idem, "Here's Vital Conservation," *RE* (April 27, 1940): 19; idem, "Here's Vital Conservation," *RE* (May 4, 1940): 13, 22–23.

124. Philpott, *The Slum and the Ghetto*, 147.

125. Area Description of C-216 from "Security Map of Metropolitan Chicago, Ill.," October 1939, Box 84, Records of the Federal Home Loan Bank Board, Home Owners Loan Corporation, Records Relating to the City Survey File, 1935–40, Illinois, Folder Chicago Sec II No. 2, FHLBBR.

126. Ibid. Area Description of D-78.

127. Ibid. Area Description of D-79. Such assessments by HOLC appraisers often were at odds with the economic picture of the area. In the case of Woodlawn, as a 1940 study showed, area nonwhites were better off than area whites, with higher rates of home ownership and occupancy in nicer units. WPA, *Housing in Chicago Communities, Community Area Number 42, Preliminary Release* (Chicago: CPC, 1940).

128. "Notes on the Organization and History of Certain Neighborhood Improvement Associations," Appendix C in NHA, *Preliminary Report on Conservation of Middle-Aged Neighborhoods and Properties. For Internal Use Only* (Washington, DC: NHA Office of the Administrator, Urban Development Division, October 16, 1944), 37, VF NAC 1613. Although the "general decline and deterioration *within* the neighborhoods" subsequently became "of some concern."

129. Even after the closing of Ely's institute, Chicago maintained its status as headquarters to the nation's real estate organizations, as well as planning and housing groups.

130. Healy, "Launch Woodlawn Home Survey," sw1; Mitchell, "Neighborhood Conservation in Chicago," 65.

131. However, while many portrayals of the university's role in the city's urban redevelopment have emphasized how the institution steeled itself against blighting influences from the surrounding area, in the eyes of appraisers, the University of Chicago was a major cause of area decline. They attributed this to the pressure to convert many homes around Woodlawn into apartment rentals for students. Discussing parcel C-218 in preparing the Chicago security map, one appraiser explained, "The University of Chicago has definitely changed the whole picture of Woodlawn with so many students taking post graduate courses, many of them have to seek employment in the vicinity, and it is changing more and more into a rooming house area." Area Description of C-218 from "Security Map of Metropolitan Chicago, Ill.," October 1939, Box 84, Records of the Federal Home Loan Bank Board, Home Owners Loan Corporation, Records Relating to the City Survey File, 1935–40, Illinois, Folder Chicago Sec. II, No. 2, FHLBBR.

132. Robert Mitchell, "Woodlawn: A Program of Neighborhood Conservation." Part 1. "Objectives — Preliminary Draft," undated. In Folder Mitchell, Bob, Box 12, Interoffice Correspondence, Regional Correspondence, Correspondence by Person, Records of the Office of the Director, Records of the Urban Section, November 1941–June 1943, NRPBR. Mitchell's initial proposal for conservation in Woodlawn envisioned the elimination of some structures together with the short-term and long-term maintenance of others. If successful, from that time forward, "Rebuilding would be a continuous process . . . and would be in accordance with an evolutionary plan for gradually improved development of the community." Ibid., 32.

133. Mitchell, "Prospects for Neighborhood Rehabilitation," 141; idem, "Neighborhood Conservation in Chicago," 64–65; Robert Mitchell, "Here's Vital Conservation," *RE* (April 27, 1940): 19; idem, "Here's Vital Conservation," *RE* (May 4, 1940): 13, 22–23. Notably the letter Mitchell wrote to Frederick Babcock first proposing this program did not use the term *conservation*. See letter from Robert Mitchell to Frederick Babcock, November 30, 1937, Folder The University

of Chicago, Chicago, Illinois, Box 5, Correspondence with Universities, Records of the Central Housing Committee, Records of the Committee on Appraisal and Mortgage Analysis, 1935–1942, HUDR.

134. Scholars describe both unstable meanings of the term during a given period, and transformations to the meaning of the term over time. As Sarah Phillips has suggested in an account focused on farm policy, "the ideological characteristics that made the New Conservation such a potent political force . . . also reduced its potential for meaningful reform." Sarah T. Phillips, *This Land, This Nation* (New York: Cambridge University Press, 2007), 74.

135. On these and other interest groups, see Beeman and Pritchard, *A Green and Permanent Land*; Worster, *Dust Bowl*; and Phillips, *This Land, This Nation*. Sarah Phillips has suggested how, as the implementation of programs went on, scientific managers' vision was eclipsed by that of the farm lobby and other business interests such that large landowners benefited rather than sharecroppers or tenants. "The end result would be a form of government intervention that ultimately served to increase the value of private property without requiring much else." As a result, many of the scientific planners, including Bennett and Tugwell, together with ecologists such as Paul Sears and Aldo Leopold, sought alternative means to promote their ecological agenda, for example, becoming involved with the organization Friends of the Land. Yet, as Phillips notes (*This Land, This Nation,* 82), "The fallout from these compromises would not be immediately apparent," suggesting why for urban conservation promoters the concept likely held continued appeal.

136. R. I. Nowell, "Experience of Resettlement Administration Program in Lake States," *JFE* 19, no. 1 (1937): 216.

137. Nelson, "Rehabilitation of the Blighted District"; FHLBB, *Waverly*; Ruth Berman, "Neighborhood Conservation," *FHLBR* 12 (March 1946): 171–175.

138. In many middle-class neighborhoods, local residents already had some record of participation in nature conservation through garden clubs and other associations. Stephen Fox, *The American Conservation Movement* (Madison: University of Wisconsin Press, 1986); Conservation Committee of the Garden Club of America, *Conservation Guide* (New York: Garden Club of America, 1939).

139. FHLBB, *Waverly*, 58.

140. "Home Sales Spurred by Clean-Up Drive," *NYT*, September 1, 1940. As Marc Weiss and John Metzger note, even as foreclosures in Maryland were up, Waverly property values rose—and homes on the market sold. Marc Weiss and John Metzger, "The American Real Estate Industry and the Origins of Neighborhood Conservation," in *Proceedings of the Fifth National Conference on American Planning History*, ed. Laurence C. Gerckens (Hilliard: SACRPH, 1994).

141. "Factors Are Cited Harmful to Cities," *NYT*, June 23, 1940, RE8.

142. "Urges Renovation of Old Dwellings," *NYT*, July 6, 1941, RE4.

143. "Waverly," *FHLBR* 6 (1940): 330; "Civic Efforts Aid Suburban Center," *NYT*, June 23, 1940.

144. Mitchell, "Neighborhood Conservation in Chicago," 65.

145. Untitled Document, Box 13, Records of the Federal Home Loan Bank System, Federal Home Building Service Plan, Program Subject Files, 1936–1942, Folder Pamphlets and Literature on Housing (Folder 1), 13, FHLBBR; "Research on Neighborhood Conservation," *RER* (December 28, 1940): 6; Untitled Document, Box 13, Records of the Federal Home Loan Bank System, Federal Home Building Service Plan, Program Subject Files, 1936–1942, Folder Pamphlets and Literature on Housing (Folder 1), 8–9, FHLBBR.

146. McNeal, "Waverly," 217.

147. Ibid., 217. These results did not suggest that cities' life cycles would forever be subject to human control. With the agency acknowledging that "integration and disintegration is a never ending cycle that is common alike to inanimate and to animate matter," neighborhood conservation's chief spokesperson made clear there were limits to what intervention might achieve. Untitled Document, Box 13, Records of the Federal Home Loan Bank System, Federal Home Building Service Plan, Program Subject Files, 1936–1942, Folder Pamphlets and Literature on Housing (Folder 1), 8–9, FHLBBR. "I do not mean to argue that all residential areas can be made safe forever," Donald McNeal cautioned, "or that we can entirely eliminate slum tendencies." McNeal, "Waverly," 221.

148. This is also referred to as the Neighborhood Conservation Program in NRPB, *Federal Aids to Local Planning* (Washington, DC: NRPB, June 1940), 94.

149. Untitled Document, Box 13, Records of the Federal Home Loan Bank System, Federal Home Building Service Plan, Program Subject Files, 1936–1942, Folder Pamphlets and Literature on Housing (Folder 1), 17, FHLBBR.

150. Ibid., 16.

151. This report also noted related legislation introduced but not yet passed in New York: the Urban Redevelopment Corporations Act.

152. McNeal, "Waverly," 220.

153. This was "'State Laws for Better Land Use,' dealing not only with zoning, soil conservation districts, etc. but with more effective tax delinquency foreclosure, reorganization of rural local government, etc." Charles Ascher, "Proposed Program for Urban Conservation and Development, 1942–1943" (confidential, Executive Office of the President, National Resources Planning Board Washington, DC, n.d.), 1, 5, 6, VF NAC 110 As.

154. Charles Ascher, "For the Transition." February 3, 1942. In Folder Federal Urban Organizations, Box 19, Alphabetical Correspondence A-F, Records of

the Office of the Director, Records of the Urban Section, November 1941–June 1943, NRPBR.

155. See NRPB, *Public Land Acquisition in a National Land-Use Program*.

156. Charles Abrams, *The City Is the Frontier* (New York: Harper and Row, 1965), 242.

Chapter Three: A Life Cycle Plan for Chicago

1. Richard Nelson, "Neighborhood Conservation," Appendix A, in NHA, *Preliminary Report on Conservation of Middle-Aged Neighborhoods and Properties*. For Internal Use Only (Washington, DC: NHA Office of the Administrator, Urban Development Division, October 16, 1944), 1–2. Nelson, who worked at the Downs-Mohl firm, was son of National Association of Real Estate Boards secretary Herbert Nelson.

2. CPC, *Master Plan of Residential Land Use of Chicago* (Chicago: CPC, 1943), 10.

3. Martin Millspaugh suggested that it was only with the 1953 presidential advisory commission on housing issues that the two groups were brought together. Martin Millspaugh, "Objectives and Criteria of Urban Renewal," *Proceedings of the Academy of Political Science*, 27, no. 1 (1960): 49–56. Certainly, up to 1940, many planners were advocating demolition while others focused on repair. Arthur Holden, "The Menace of Urban Blight," *AJ* 8, no. 3 (1940), lays out the dispute between the two groups. Yet, exceptions such as Herbert Swan, "Land Values and City Growth," *JLPUE* 10, no. 2 (1934): 188–189, suggest the story is more complex.

4. Letter from Charles Abrams to Albert Cole, August 5, 1953, Box 762, Program Files, 1940–65, President's Advisory Committee on Housing Policies and Programs, 1953–54, Folder Material from Citizens' Housing Groups, 9, HUDR.

5. On Homer Hoyt's career, see Robert Beauregard, "More Than Sector Theory: Homer Hoyt's Contributions to Planning Knowledge," *Journal of Planning History* 6 (2007): 248–271.

6. Homer Hoyt, *One Hundred Years of Land Values in Chicago* (Chicago: University of Chicago Press, 1933).

7. Wesley Clair Mitchell, *Business Cycles* (Berkeley: University of California Press, 1913); Roderick McKenzie, "The Scope of Human Ecology," *PASS* (1920): 141–154.

8. Frederick Babcock, *The Valuation of Real Estate* (New York: McGraw Hill, 1932). Ernest Fisher, *Advanced Principles of Real Estate Practice* (New York: McGraw-Hill, 1930).

9. Homer Hoyt, *According to Hoyt* (Washington, DC: Hoyt, 1966), 526. He

also cited the earlier work of economist Richard Hurd, characterized by social scientists as a proto-ecological thinker for his contributions to urban theory in *Principles of City Land Values* (New York: Record and Guide, 1903).

10. On Hoyt's work at FHA, see Henrika Kuklick, "Chicago Sociology and Urban Planning Policy: Sociological Theory as Occupational Ideology," *Theory and Society* 9 (1980): 821–845.

11. The maps were to be used in the preliminary screening of applications; those passing would be subjected to more scrutiny. On the field organization of agency underwriting, see "The Field Organization of the FHA," *IMP* (May 1940): 17.

12. Homer Hoyt, "Program for the Study of Sixty-Two Cities." In Records Relating to the Economic Data System, 1936, 1941, 1943, Box 1, Records of the Housing and Home Finance Agency, Records of the Division of Research and Statistics of the FHA, "Economic Data System" Records Containing Data on Cities, 1937–1945, HUDR, and in HOLC files for individual cities at the U.S. National Archives.

13. Peyton Stapp, *Urban Housing: A Summary of Real Property Inventories Conducted as Work Projects, 1934–1936* (Washington, DC: WPA, 1938). The survey technique was revised several times during the two-year period, and *Technique for a Real Property Survey* (Washington, DC: WPA, 1935) reports some of the details. Launched by the Department of Commerce and Civil Works Administration, the Real Property Inventories were completed under the supervision of the Work Progress Administration (later Works Projects Administration).

14. "17 Big Cities' Growth Shown in FHA Maps: Series will be valuable to Real Estate and Mortgage Firms," *WP*, March 28, 1937, R9.

15. These time-interval studies, called Dynamic Factor Maps, depicted past urban change over a period of approximately thirty years. They were used to forecast the likely rate and direction of future transformations in specific neighborhoods. The preparation of these maps relied on participation from the prominent figures in each city's real estate industry, who also served as the chief source of area data for cities lacking Real Property Inventories.

16. Hoyt, *According to Hoyt*, no page.

17. Examples of Hoyt's interest in forest and farm cartography include Homer Hoyt, Review of Raphael Zon and William N. Sparhawk, *Forest Resources of the World*, and Review of John D. Black, *Research in Farm Real Estate Values: Scope and Method*, in *JPE* 33, no. 1 (1925): 123–124, and *JPE* 43, no. 6 (1935): 850; he praised the former as "a monumental contribution" furnishing "the technical knowledge necessary for the understanding of forest conservation as one phase of social control." Hoyt's interest in map work at the Home Owners Loan Corporation is apparent in the files he created in connection with the Federal

Housing Administration's Economic Data System, one component of the revisions to location rating. See Volume 27, Cross Reference to Security Area Maps; Maps Prepared 1936–1940, "Economic Data System" Records Containing Data on Cities, 1937–1945, Records of the Division of Research and Statistics of the Federal Housing Administration and Volume 19, Binder Titled Confidential Reports on Economic Conditions in Various Cities; Source: Home Loan Bank Board, "Economic Data System" Records Containing Data on Cities, 1937–1945. Records of the Division of Research and Statistics of the Federal Housing Administration, HUDR.

18. Articles on the history and growth of cities, including Detroit, Miami, New York, Chicago, and Washington, DC, appeared between December 1936 and April 1937. FHA, *The Structure and Growth of Residential Neighborhoods in American Cities* (Washington, DC: Government Printing Office, 1939).

19. Robert M. Haig, "Toward an Understanding of the Metropolis," *QJE* 40 (May 1926): 407.

20. FHA, *The Structure and Growth of Residential Neighborhoods in American Cities*, 44.

21. Ibid., 3.

22. Ibid., 115–116. Hoyt further develops his ideas of filtration in "Rebuilding American Cities after the War," *JLPUE* 19, no. 3 (1943): 364–368. This model spread quickly to the appraisal community. In its 1940 survey of Chicago, for example, Home Owners Loan Corporation appraisers described how "the degree of detrimental influence in many cases depends upon the areas into which the races in question are filtering." HOLC Division of Research and Statistics, "Confidential Report of a Re-survey of Metropolitan Chicago, IL," June 1940, Box 85, Records of the Federal Home Loan Bank Board, Home Owners Loan Corporation, Records Relating to the City Survey File, 1935–40, Illinois, Folder Chicago Re-survey Report 2, Vol. 1 (5), 34, FHLBBR.

23. Hoyt, *One Hundred Years of Land Values in Chicago*, 316. Chicago Realtor John Usher Smyth provided Hoyt with this list.

24. Nelson, "Neighborhood Conservation," 14.

25. Indeed, this is how many social scientists of Hoyt's own time understood his work, citing him alongside the human ecologists for their similar perspectives on urban structures and processes, as in James Quinn, "Topical Summary of Current Literature on Human Ecology," *AJS* 46 (1940): 191–226.

26. Fisher, *Advanced Principles of Real Estate Practice*, 126–127. Fisher, notably, was an economist Roderick McKenzie praised as among those doing excellent ecological work. See McKenzie, "The Scope of Human Ecology," 142.

27. An alternative seen repeatedly was what geographers Chauncy Harris and Edward Ullman eventually formalized as the "multiple nuclei" model in 1945.

Their work had many predecessors. See Norman Scott Brien Gras, *An Introduction to Economic History* (New York: Harper, 1922); Shideler, "The Chain Store" (PhD diss., University of Chicago, 1927); Fisher, *Advanced Principles of Real Estate Practice*, 127; Robert McFall, "Urban Centralization and Decentralization," in *Economic Essays in Honor of Wesley Clair Mitchell* (New York: Columbia University Press, 1935), 309; and special issue, *Urban Geography* 18, no. 1 (1997) on Harris's and Ullman's work.

28. Arthur Weimer and Homer Hoyt, *Principles of Urban Real Estate* (New York: Ronald Press, 1939).

29. FHA, *The Structure and Growth of Residential Neighborhoods in American Cities*, 3.

30. Ibid., 17.

31. Kuklick, "Chicago Sociology and Urban Planning Policy."

32. The technical advisory group later expanded significantly to include representatives from Chicago's newspapers, city departments, real estate firms, academic institutions, and other high-profile agencies, such as Ferd Kramer of the Chicago Real Estate Board, Samuel Stouffer of University of Chicago, Herbert Simpson of Northwestern, Robert Mitchell of the Woodlawn Conservation Project, and D. W. Mackelmann of the Metropolitan Housing Council.

33. William E. Leuchtenburg, *Franklin D. Roosevelt and the New Deal* (New York: Harper, 1963).

34. CHA, *Information in Regard to the Proposed South Park Gardens Project* (Chicago: CHA, February 17, 1938), 27.

35. Hugh Young, "The Story of Chicago's Physical Development: Arrested Growth and Blight Mark the Close of the Pioneer Era," *RE* (April 13, 1940): 9–21.

36. For example, H. Evert Kincaid, "Quarterly Report on Master Plan Progress," *Quarterly Review of the City Planning Advisory Board of Chicago* (May 1944): 9, uses the same text as in CPC, *Summary of the Residential Land Use Plan for Chicago* (Chicago: CPC, June 1943), 1, VFNAC 6827 Chi 1943m, a document Hoyt prepared based on the land use survey data. The plan commission's prior engineering focus is noted in Thomas Philpott, *The Slum and the Ghetto* (New York: Oxford University Press, 1978), 246.

37. See University of Chicago Population Research and Training Center, *Population Research and Training Center and the Chicago Community Inventory: Statement of Program and Summary Report, 1947–58* (Chicago: University of Chicago, Chicago, 1958). By the late 1930s, research and teaching in human ecology and city planning had grown close, with urban studies scholars increasingly preoccupied with planning questions and requirements for professional certification in planning demanding social science knowledge. See Jesse Steiner, *Read-*

ings in Human Ecology (Seattle: University of Washington Bookstore, 1939); Stuart Queen and Lewis Thomas, *The City* (New York: McGraw-Hill, 1939); Dorothy King, "City Planning: Selected Bibliography for Use in Preparation for Civil Service Examinations in City Planning," Municipal Reference Library, New York, 1939, VF Z NAC 250; "Suggested Bibliography for Course on City and Regional Planning," University of Chicago, University College, Political Science 305, 1932, VF Z NAC 250.4; Hugh Morrison, *Education for Planners* (Boston: NRPB Region I, 1942); National Conference on City Planning, *A Suggestion for a College Course in City Planning* (National Conference on City Planning, 1936), VF NAC 935; Edwin Burdell, "Sociology for City Planners," *JAIP* 2, no. 2 (1936): 163–168. Hoyt's hiring of social scientists to his planning research staff, however, was more rare.

38. He also hired a former affiliate of Ely's Institute, Helen Monchow. Among the research projects making use of plan commission work were Robert Klove, "The Park Ridge-Barrington Area" (PhD diss., University of Chicago, 1942); Donald Foley, "An Index of the Physical Quality of Dwellings in Chicago Residential Areas" (Master of Science thesis, University of Chicago, 1942). Because of the opportunities that Hoyt created for interdisciplinary exchange, and his uncanny ability to mobilize disparate communities of urban professionals toward common ends, it was the Chicago Plan Commission, even more than the much-studied University of Chicago Local Community Research Committee, that offered an opportunity to translate social science theory into city planning practice. That committee's work is detailed in Sudhir Venkatesh, "Chicago's Pragmatic Planners: American Sociology and the Myth of Community," *SSH* 25, 2 (2001): 275–317, and Martin Bulmer, "The Early Institutional Establishment of Social Science Research: The Local Community Research Committee at the University of Chicago, 1923–30," *Minerva* 18, no. 1 (1980): 51–110.

39. Examples of this literature include Robert O'Brien, "Beale Street: A Study in Ecological Succession," *Sociology and Social Research* 26, no. 5 (1942): 430–436; Earl Lomon Koos, ed., *Rochester, New York III: An Atlas of Ecological Patterns of the City's Social Problems* (Rochester, NY: Council of Social Agencies, 1944); J. Wreford Watson, "Urban Developments in the Niagara Peninsula," *CJEPS* 9, no. 4 (1943): 463–486; Edward Ackerman, "Sequent Occupance of a Boston Suburban Community," *EG* 17, no. 1 (1941): 61–74; George Zipf, *National Unity and Disunity: The Nation as a Bio-social Organism* (Bloomington, IN: Principia Press, 1941); Gladys Engel-Frisch, "Some Neglected Temporal Aspects of Human Ecology," *SF* 22, no. 1 (1943): 43–47; Harlan W. Gilmore, "The Old New Orleans and the New: A Case for Ecology," *ASR* 4 (1944): 385–394; Chauncy Harris and Edward Ullman, "A Theory of Location for Cities," *AJS* 46 (1941): 853–864. Cleveland and New York real property surveys are discussed in Howard

Green, "Cultural Areas in the City of Cleveland," *AJS* 38 (1932): 356–367; Roy Burroughs, "Urban Real Estate Mortgage Delinquency," *JLPUE* 11, no. 4 (1935): 357–367; and E. B. Olds, "The Use of NYA Workers in Ecological Studies," *SF* 20, no. 2 (1941): 218–223. Also relevant was Hugh Young's predilection for an ecological explanation of the city's development, expressed even before Hoyt's arrival at the plan commission. See Hugh Young, "The Story of Chicago's Physical Development," *RE* (April 13, 1940): 9–21.

40. CPC, *Chicago Land Use Survey* (Chicago: CPC, 1942), Vol. 2, Accompanying cover letter.

41. Homer Hoyt, "Instructions for Dividing the City into Neighborhoods" undated, Carton 18, Folder 0–109, "Neighborhood Rating Technique: Miscellaneous," Research and Statistics, FHAR.

42. The survey schedule is reprinted in the land use survey. A copy of the FHA's map of mortgage risk districts, redrawn by the CHA in 1938, is available in the University of Chicago Library's Special Collections.

43. CPC, *Chicago Land Survey*, 1:xxvii. In addition, a two-week mini-course was organized to standardize enumeration practices for the many hired hands, and procedures were put in place to check and recheck figures. WPA, *Housing in Chicago Communities, Community Area Number 7, Preliminary Release* (Chicago: CPC, 1940), 1–2.

44. Donald Foley, personal communication with author, January 2003.

45. CPC, *Chicago Land Use Survey*, 1:xxi.

46. Ibid., xxii.

47. WPA, *Housing in Chicago Communities, Community Area Number 7, Preliminary Release*, 2.

48. For example, the work of Ernest Burgess, Louis Wirth, Margaret Furez, and Charles Newcomb.

49. Only occasionally was nationality mentioned in the supplementary volumes to the report, for example, in the description of Community Area 7, Lincoln Park, which observed, "44 per cent of the families in the area in 1934 were foreign-born" and "an additional 25 per cent were of foreign born-or mixed parentage. Predominant among the foreign-born heads of households were German." WPA, *Housing in Chicago Communities, Community Area Number 7, Preliminary Release*, 23.

50. Data provided at the block level included average monthly rental, number of dwelling units, percent of structures built before 1919, percent owner occupied, percent unfit or needing repairs, number of business or commercial structures, percent of households of a race other than white, and percent with no toilet or bath.

51. WPA, *Housing in Chicago Communities, Community Area Number 29, Preliminary Release* (Chicago: CPC, 1940), 7, 11.

52. CPC, *Chicago Land Use Survey*, 1:206–210.

53. McFall, "Urban Centralization and Decentralization," 298; NRC, *Our Cities* (Washington, DC: NRC, 1937).

54. To maximize the use of the survey, Hoyt and his colleagues formatted these data in several ways: by city blocks (of special interest to planners), by census tracts (of special interest to social scientists and social service agencies), and by half miles (of special interest to the real estate community). This latter unit of measurement had been popularized by George Olcott's *Land Values Blue Book of Chicago*, the annual compendium of real estate information on which the local industry relied — and which Hoyt had used a central data source in *One Hundred Years of Land Values in Chicago*.

55. HOLC Division of Research and Statistics, "Confidential Report of a Re-Survey of Metropolitan Chicago, IL," June 1940, Box 85, Records of the Federal Home Loan Bank Board, Home Owners Loan Corporation, Records Relating to the City Survey File, 1935–40, Illinois; Folder Chicago Re-Survey Report No. 2 Vol. 1 (5), 29, FHLBBR.

56. Charles S. Ascher, *Our Cities, Building America* (Washington, DC: NRPB, 1942), 20.

57. Thomas Hanchett, "Federal Incentives and the Growth of Local Planning, 1941–1948," *JAPA* 60 (1994): 197–208; Jennifer Light, *From Warfare to Welfare: Defense Intellectuals and Urban Problems in Cold War America* (Baltimore: Johns Hopkins University Press, 2003).

58. The racial implications of this study are especially interesting given Goodwillie's assistance from several Howard University researchers in its preparation.

59. Arthur Goodwillie, "The Rehabilitation of Southwest Washington as a War Housing Measure, a Revised Memorandum to the Federal Home Loan Bank Board" (Washington, DC: n.p., 1942), 4, 1. An earlier document, *The Rehabilitation of Southwest Washington as a War Housing Measure* (Washington, DC: Conservation Service, Home Loan Corporation, January 2, 1942), is cited in Arthur Holden, "Urban Redevelopment Corporations," *JLPUE* 18, no. 4 (1942): 412–422. See also Arthur Goodwillie, *Neighborhood Conservation in War-time and Peace-time* (Washington, DC: HOLC, 1942).

60. Arthur Goodwillie, "Rehabilitation of Blighted Areas as a War Housing Measure," *AC* 57, no. 3 (1942): 37.

61. Hal Rothman, *The Greening of a Nation? Environmentalism in the United States since 1945* (Fort Worth: Harcourt Brace College, 1998). Although some scholars emphasize the new wartime orientation as an extension of scientific

and technocratic approaches to conservation, others emphasize the business development opportunities. Regardless of their interpretive emphases, they agree that the war aligned conservation with security concerns. See, for example, Henry Henderson and David Woolner, eds., *FDR and the Environment* (New York: Palgrave Macmillan, 2005), and Sarah Phillips, *This Land, This Nation*.

62. Historians suggest that, compared with other scientific fields, ecology played a limited role in the war effort, despite federal investments in other areas of environmental science. On broader links between the history of warfare, environmentalism, and environmental science in the United States, see Edmund Russell, *War and Nature: Fighting Humans and Insects with Chemicals from World War I to Silent Spring* (New York: Cambridge University Press, 2001); and Ronald Doel, "Constructing the Postwar Earth Sciences: The Military's Influences on the Environmental Sciences in the USA after 1945," *Social Studies of Science* 35, no. 5 (2003): 635–666.

63. "Police Ordered to Get on Toes for Halloween," *CDT*, October 24, 1942; and "Gay Halloween Goblins Do Bit to Win the War," *CDT*, October 31, 1943, 3.

64. Hugh Hammond Bennett, *Soils and Security* (Washington, DC: Government Printing Office, 1941), 25; Roy Hockensmith and and J. G. Steele, *Classifying Land for Conservation Farming* (Washington, DC: USDA, 1943), 1, back cover; H. P. Rusk, "The Soil Conservation District as a Democratic Institution for Conservation of Agricultural Resources," *APCA* (1944); Arthur Bunce, "War and Soil Conservation," *JLPUE* 18, no. 2 (1942): 121–133; Sherman Johnson, "Adapting Agricultural Programs for War Needs," *JFE* 24, no. 1 (1942), 1–16; Raymond C. Smith, "Social Effects of the War and the Defense Program on American Agriculture," *JFE* 23, no. 1 (1941): 15–27; James Burdett, *Victory Garden Manual* (Chicago: Ziff-Davis, 1943); War Food Administration, *Your Victory Garden Counts More Than Ever!* (Washington, DC: War Food Administration, 1945); Terrence H. Witkowski, "World War II Poster Campaigns — Preaching Frugality to American Consumers," *Journal of Advertising* 32, no. 1 (2003): 69; Day Monroe, "Using Family Resources Wisely in Wartime," *Marriage and Family Living* 5, no. 3 (1943): 52–54.

65. Another aspect of conservation's changing definition was Goodwillie's proposal for relocation.

66. Arthur Goodwillie, "The Rehabilitation of Southwest Washington as a War Housing Measure," 39; Charles Mercer, "20 Millions More for D.C. Housing Asked," *WP*, February 27, 1942, 1; Merlo Pusey, "Wartime Washington; Mr. Goodwillie's Captivating Plan," *WP*, March 10, 1942, 11.

67. Carolyn Goldstein, *Do It Yourself: Home Improvement in 20th-Century America* (New York: Princeton Architectural Press, 1998); "Bibliography on Ur-

ban Reconstruction," *JLPUE* 19, no. 3 (1943): 368–369; "Urban Redevelopment," *Yale Law Journal* 54, no. 1 (1944): 116–140.

68. CPC, *Master Plan of Residential Land Use of Chicago*, 11. See also Homer Hoyt, "Rebuilding American Cities after the War," *JLPUE* 19, no. 3 (1943): 364–368.

69. CPC, *Master Plan of Residential Land Use of Chicago*, 10.

70. Untitled Document, Box 13, Records of the Federal Home Loan Bank System, Federal Home Building Service Plan, Program Subject Files, 1936–1942, Folder Pamphlets and Literature on Housing (Folder 1), 16, FHLBBR.

71. "Waverly," *FHLBR* 6 (1940): 330.

72. Robert B. Mitchell, "Prospects for Neighborhood Rehabilitation," in *Housing Yearbook*, ed. NAHO (Chicago: NAHO, 1938), 134–135.

73. Edward Graham, *Natural Principles of Land Use* (New York: Oxford University Press, 1944), 228–229.

74. Bennett, *Soils and Security*, 19.

75. CPC, *Master Plan of Residential Land Use of Chicago*, 75. The near-complete omission of race and nationality among these factors (the report contains only spare mentions of links between racial change and community decline) is striking given its general importance in Hoyt's prior work and the more specific revision of Chicago's survey schedules to incorporate more detailed demographic information. Additionally, in a similar assessment of Richmond, Virginia for the Federal Housing Administration, Hoyt had layered "Negro areas" on age, condition, and rent maps to depict that city's worst housing areas. See FHA, *The Structure and Growth of Residential Neighborhoods in American Cities*, 47.

76. Ibid., 67. The definition of the planning areas follows: Blighted Areas (50% or more built before 1895 and 50% or more of units substandard and 20% or more of units in need or repair or unfit for use); Near-Blighted Areas (either 50% or more built before 1895, or 50% or more of units substandard and 50% or more of units renting for less than $25 per month); Conservation Areas (50% or more built 1895–1914 and 50% or more of units renting for more than $25 per month); Stable Areas (50% or more built 1915–1929), Arrested Development Areas (less than 10% built since 1929 and only 10%-50% of land in residential use); Progressive Development areas (10–50% of units built since 1929 and 10%-50% of land in residential use); New Growth Areas (50% or more units built since 1929 and more than 10% in residential use); and Vacant Areas (less than 10% in residential use). CPC, *Housing Goals for Chicago* (Chicago: CPC, 1946), 219. As with Hoyt's earlier *The Structure and Growth of Residential Neighborhoods in American Cities*, here a set of transparent plastic overlay maps—an age map, a condition map, and a rent map—when juxtaposed, revealed new patterns signaling an area's status as blighted, near blighted, conservation or stable.

77. Clarence Wiley, "Settlement and Unsettlement in the Resettlement Administration Program," *LCP* 4, no. 4 (1937): 456–471; Homer Hoyt, "A Practical Plan for Rebuilding the Existing Homes of a City," *NREJ* (September 1943), reprinted in Hoyt, *According to Hoyt*, 599. Such statements contrasted with those of earlier land surveyors, for example, staff on the land economic survey organized by the Michigan Department of Conservation, who observed how "the lines between super-marginal, marginal, and sub-marginal lands seem to be indicated with even greater precision than was to be hoped for and without involving any form of arbitrary or debatable assertion. In this, perhaps, lies one of the most important discoveries of the Michigan survey." P. S. Lovejoy, "Theory and Practice in Land Classification," *JLPUE* 1, no. 2 (1925): 170.

78. Robert Klove, "A Technique for Delimiting Chicago's Blighted Areas," *JLPUE* 17 (November 1941): 483–484; Donald L. Foley, "An Index of Housing in Chicago," *JLPUE* 18, no. 2 (1942): 209–213. Donald Foley discovered that developing a scientific index for delimiting housing quality could serve both as a master's thesis topic and as a useful rating system for the plan commission. Foley was motivated by the belief that housing programs to date had been unscientific—they did not come "anywhere near to what the physical scientist" does, and that a lack of adequate measures for housing quality was one obstacle. Foley, "An Index of the Physical Quality of Dwellings in Chicago Residential Areas," iii. Notably, while Klove wrote up the details of the Chicago Plan Commission's method for delimiting Chicago's blighted areas for publication, his own PhD thesis under Charles Colby, which also contained maps delineating a "time sequence pattern of neighborhood development," included eight categories distinct from those of the *Master Plan of Residential Land Use of Chicago*.

79. That Hoyt, like University of Chicago geography professor Charles Colby, and numerous scholars in sociology, came to view urban patterns in a similar light, as composed of generally concentric zones, makes a great deal of sense in light of the fact that the three departments shared a map room and statistical laboratory—and the assistance of social science research staff Charles Newcomb and Mae Schiffman (later Maizlish), who helped these researchers in the preparation of statistical charts and maps.

80. CPC, *Summary of the Residential Land Use Plan for Chicago*, 2–3. Planners preferred "belts," like climate belts, over zones; this is likely because in the early 1920s "zoning" had taken on a precise legal meaning in the context of city planning.

81. The concentric model had been an even better fit earlier in the city's history: "The central core of blight extends around the loop within a radius of two to three miles on the north and west sides and two to four miles on the south. This

concentric belt of early housing once formed an almost continuous ring around the central business district, but the expansion of factories and railroads into areas originally occupied by homes has split the central nucleus of blight into five main parts. These five segments are in effect the remnants of Chicago's residential development before 1890. CPC, *Master Plan of Residential Land Use of Chicago*, 76.

82. Homer Hoyt, "Recent Distortions of the Classical Models of Urban Structure," *LE* 40, no. 2 (1964): 199–212.

83. Kincaid, "Quarterly Report on Master Plan Progress," 9.

84. FHA, *The Structure and Growth of Residential Neighborhoods in American Cities*, 3.

85. CPC, *Master Plan of Residential Land Use of Chicago*, 69.

86. Homer Hoyt and Leonard Smith, "Valuation of Land in Urban Blighted Areas," *AJ* (July 1943): 199–209.

87. Chicago Committee on Sub-standard Housing, *The Chicago Program for Demolition and Rehabilitation of Sub-standard Housing* (Chicago: The Committee, 1935). Coleman Woodbury had been involved in this project, as had the University of Chicago's Social Science Research Committee.

88. Herbert Dorau and Albert Hinman, *Urban Land Economics* (New York: McGraw-Hill, 1928), 223; Slum Clearance Committee of New York, *Maps and Charts Prepared by the Slum Clearance Committee of New York, 1933–34* (New York, The Committee, 1934). Indeed, pictures in the *Master Plan of Residential Land Use of Chicago* praised these public housing efforts and urged the usage of present and future legislation to "achieve the same or similar results." CPC, *Master Plan of Residential Land Use of Chicago*, 77.

89. WPA, *Housing in Chicago Communities, Community Area Number 29, Preliminary Release*.

90. Graham, *Natural Principles of Land Use*.

91. Frederic Clements and Ralph Chaney, *Environment and Life in the Great Plains*. (Washington, DC: Carnegie Institute, 1937), 51.

92. One wonders why Hoyt alternated between these two classification schemes. Eight classes of lands might have called to mind the work of the Soil Conservation Service. Certainly, four classes of land invoked the rating systems of the Federal Housing Administration and Home Owners Loan Corporation—as well as Clements' four-stage model of succession to climax.

93. Homer Hoyt, "The Need for Master Plans for American Cities," *NREJ* 44, no. 5 (1943): 32–34.

94. Hoyt and Smith, "Valuation of Land in Urban Blighted Areas," reprinted in Hoyt, *According to Hoyt*, 457; and Hoyt, "A Practical Plan for Rebuilding the Existing Homes of a City," 602. This viewpoint was apparent in his analysis of the

Waverly experiment, which he praised but then critiqued for not being citywide. See Homer Hoyt, Review of *Waverly: A Study in Neighborhood Conservation*, by Arthur Goodwillie, *AJS* 47, no. 5 (1942): 788.

95. Neighborhood unit planning is discussed in CPC, *Rebuilding Old Chicago* (Chicago: CPC, 1941); idem, *Building New Neighborhoods* (Chicago: CPC, 1943); *Preliminary Comprehensive Plan* (Chicago: CPC, 1946), and H. Evert Kincaid, "Organization of Neighborhood Units," *APCA* (1947).

96. CPC, *Master Plan of Residential Land Use of Chicago*, 10.

97. Ibid., 134.

98. Anderson cited in Ernest Burgess, "Urban Areas in Chicago," in *Chicago: An Experiment in Social Science Research*, ed. T. V. Smith and L. D. White (Chicago: University of Chicago Press, 1929), 130. Indeed, Hoyt's views on slum treatment echoed Anderson's ideas more generally that "except for moving the people, slum clearance has never contributed much to the solution of the housing problem . . . but, on the contrary, has promoted new slums." Nels Anderson, "Letter to the Editor," *Millar's Housing Letter* 2, no. 26 (1934): 6.

99. K. Loenberg-Holm, "Time Zoning as a Preventative of Blighted Areas," *Architectural Record* (November 1933).

100. CPC, *Chicago Land Use Survey*, 1:xxxiv.

101. CPC, *Master Plan of Residential Land Use of Chicago*, 67. "The five areas comprising most of the built-up residential sections of the city, in the order in which they will require treatment are: 1. Blighted areas 2. Near-Blighted areas 3. Conservation areas 4. Stable areas 5. New Growth areas. By the end of the second generation from now the Conservation areas of today should have been rebuilt, and the present Stable areas will then be first on the list for clearance as blighted" (ibid.).

102. CPC, *Master Plan of Residential Land Use of Chicago*, 98. Twenty years was considered the average length of real estate cycles in U.S. cities.

103. "Waverly," *FHLBR* 6 (1940): 330.

104. CPC, *Master Plan of Residential Land Use of Chicago*, 97.

105. See, for example, U.S. Office of Civilian Defense, *The Block Plan of Organization for Civilian War Services* (Washington, DC: Office of Civilian Defense, 1942); and Charles R. Hoffer, "Impact of War on American Communities," *Review of Educational Research* 13, no. 1 (1943): 5–12.

106. Paul Healy, "Launch Woodlawn Home Survey," *CDT*, April 28, 1940, sw1; CPC, *Woodlawn Community: Proposed Improvement Plan* (Chicago: CPC, 1945); idem, *Woodlawn: A Study in Community Conversation* (Chicago: CPC, 1946).

107. Donald Foley, personal communication with author, January 2003.

108. Al Chase, "Open Woodlawn 'Conservation' Fight on Blight," *CDT*, January 28, 1945, A6; CPC, *A Program for Community Conservation in Chicago and an Example: "The Woodlawn Plan"* (Chicago: CPC, 1946), 4. Hoyt's publication record alone—spanning economics, sociology, and geography, as well as the trade literature of banking, real estate, and planning—marked his success in sharing ideas and setting precedents across professions.

109. CPC, *A Program for Community Conservation in Chicago and an Example*, 73. The original conservation programs in Waverly and Woodlawn had invested in some public improvements such as street closures, but this plan envisioned building new schools, parks, and other facilities for neighborhood use.

110. Chase, "Open Woodlawn 'Conservation' Fight on Blight," A6; Hoyt's praise for the potentials of Woodlawn are apparent in CPC, *Master Plan of Residential Land Use of Chicago*, 97.

111. Thomas Furlong, "Chicago Declares War on Blight," *CDT*, November 24, 1946, B8; "Woodlawn Plan to Rebuild Area Put in Motion: 8 Point Program Called Model for City," *CDT*, January 19, 1947, 24. CPC, *A Program for Community Conservation in Chicago and an Example*, 74. Notably, while it was not a resource conservation project, among "the many rehabilitation projects being carried out" as plans for the neighborhood unfolded under plan commission guidance was "a front-lawn chrysanthemum display contest sponsored jointly by the University of Chicago, the Woodlawn Business Men's Association, and the Woodlawn Property Owners League." It was University of Chicago botanist Ezra Kraus whose "extensive botanical research for the perennial flower best suited to Chicago's climate" had stimulated the undertaking. CPC, *A Program for Community Conservation in Chicago and an Example*, 24.

112. CPC, *Chicago Looks Ahead* (Chicago: CPC, 1945); idem, *Preliminary Comprehensive City Plan*; idem, *Housing Goals for Chicago* (Chicago: CPC 1946), cover letter; Robert Klove, "City Planning in Chicago: A Review," *GR* 38, no. 1 (1948): 127–131; Planning Committee of the Metropolitan Housing Council, *The Necessary Elements in a Comprehensive Plan for Chicago* (Chicago: Planning Committee of the Metropolitan Housing Council, October 9, 1946), VF NAC 6827 Chi.

113. CPC, *Housing Goals for Chicago*, cover letter.

114. As contributors to the National Resources Planning Board's report on public land acquisition would explain, in the national enthusiasm for natural resources conservation, Americans now largely had accepted the idea that privately as well as publicly held farms and forests were within the purview of the "public purpose" that defined the legal requirements for usage of eminent domain. By contrast "public acquisition of urban land is neither as well established nor as

well accepted as the acquisition of rural land" — limited under the 1937 Housing Act to clearing areas on which public housing would be built. NRPB, *Public Land Acquisition in a National Land-Use Program* (Washington, DC: NRBB, 1940), Vol. 2, cover letter to Frederick Delano. Other efforts to expand legislation are detailed in Alexander von Hoffman, "A Study in Contradictions: The Origins and Legacy of the Housing Act of 1949," *HPD* 11, no. 2 (2000): 299–326, as well as CPC, *Master Plan of Residential Land Use of Chicago*, 83–85; CPC, *Rebuilding Old Chicago* (Chicago: CPC, 1941), 21; Guy Greer and Alvin Hansen, *Urban Redevelopment and Housing* (Washington, DC: National Planning Association, 1941); ULI, *Outline for a Legislative Program to Rebuild Our Cities* (Washington, DC: ULI, 1942); and FHA, *A Handbook for Urban Redevelopment for Cities in the United States* (Washington, DC: FHA, 1942). In these proposals, assembling large urban land parcels was the central issue, with related economic and political matters considered as well. For example, given the comparatively low costs of building homes outside the city limits (which additionally did not already house occupants who would need to be relocated), there was much discussion about incentives for developers and real estate investors to build attractive urban alternatives. Notably, this report also praises the neighborhood unit concept.

115. CPC, *Master Plan of Residential Land Use of Chicago*, 83. First proposed in 1938, the bill had passed in the state Senate but not in the House; however, this first attempt became the model for similar legislation adopted in New York. Thomas Furlong, "Green Praises Slum Bill in Original Form," *CDT*, May 7, 1941, 33. With the plan commission's role in such work merely advisory, this legal development appealed more to real estate than to planning professionals. CPC, *Rebuilding Old Chicago*. Miles Colean, *Renewing Our Cities* (New York: Twentieth Century Fund, 1953), suggests that as a piece of legislation, it came close to achieving the goals of the model statute the National Association of Real Estate Boards had proposed a decade earlier.

116. MHPC, *Conservation* (Chicago: MHPC, 1953), Vol. 1.

117. Robert Mitchell, "Memorandum: Establishment of a Federal Urban Conservation Service." Appended to letter from Robert Mitchell to Charles Ascher, June 6, 1941, 2. Folder Mitchell, Bob, Box 12, Interoffice Correspondence, Regional Correspondence, Correspondence by Person, Records of the Office of the Director, Records of the Urban Section, November 1941–June 1943, NRPBR.

118. "Memo for Mr. Geo. Duggar, January 7, 1941, Subject: Analogy for Urban Development Based on Dept. of Ag. Org." In Folder Federal Urban Organizations, Box 19, Alphabetical Correspondence A–F, Records of the Office of the Director, Records of the Urban Section, November 1941–June 1943, NRPBR.

119. Marion Clawson, *New Deal Planning: The National Resources Planning Board* (Baltimore: Johns Hopkins University Press, 1981).

120. When the city's Redevelopment Land Agency turned to postwar rebuilding, it chose a site in Southwest Washington, DC, but "rejected as too timid" the Goodwillie plan. Harold Gillette, "A National Workshop for Urban Policy: The Metropolitanization of Washington, 1946–1968," *Public Historian* 7, no. 1 (1985): 14. The exact date of the Conservation Service's closing is unknown. There are public references to it as late as 1944 but none after Goodwillie's death in 1946.

121. "Waverly," *FHLBR* 6 (1940): 359; "Notes on the Organization and History of Certain Neighborhood Improvement Associations," Appendix C in *Preliminary Report on Conservation of Middle-Aged Neighborhoods and Properties*. For Internal Use Only. (NHA Office of the Administrator, Urban Development Division, October 16, 1944) VF NAC 16313, 36; Ruth Berman, "Neighborhood Conservation," FHLBR (March 1946), 173. This even though to appease planning interests in the city Waverly's organizers insisted the conservation work fit into Baltimore's general plan.

122. Cleveland Regional Association, *Neighborhood Conservation: A Handbook for Citizen Groups* (Cleveland, OH: Cleveland Regional Association, 1943), VF NAC 6827 Cle-K18, no page. Cleveland had been poised to undertake conservation in the early 1940s, and even after a wartime interruption revived the program, but a conservation effort was not broadly implemented until urban renewal.

123. Berman, "Neighborhood Conservation," 171. Berman blamed the overreliance on citizen action and called for more government investment in the cause.

124. CPC, *Building New Neighborhoods* (Chicago: CPC, 1943); Carl Condit, *Chicago, 1930–70: Building, Planning, and Urban Technology* (Chicago: University of Chicago Press, 1974); and Homer Hoyt and Morton Bodfish, "Growth of Negro Population in Chicago," *Savings and Homeownership* (March 1951): 1.

125. Earlier figures for the African American population in Chicago stand at 30,150 in 1900; 44,103 in 1910; 109,458 in 1920; and 233,903 in 1930. *Distribution of Population of Chicago by Square Mile Sections* (Chicago: n.p., 1920); CHA, *Information in Regard to the Proposed South Park Gardens Project*, 4.

126. Arvarh E. Strickland, *History of the Chicago Urban League* (Urbana: University of Illinois Press, 1966), 158.

127. Charles Abrams, *Forbidden Neighbors: A Study of Prejudice in Housing* (New York: Harper, 1955); and Matthew Frye Jacobson, *Whiteness of a Different Color: European Immigrants and the Alchemy of Race* (Cambridge, MA: Harvard University Press, 1998).

128. Walter MacCornack, "America's New Frontier: Concerted Effort Is Es-

sential for the Successful Solution of Problems of Replanning and Reconstruction," *Technology Review* 46, no. 4 (1944): 200. MacCornack headed the American Institute of Architects Committee on Postwar Reconstruction.

129. Ibid., 224.

130. Miriam Kligman, "Human Ecology and the City Planning Movement," *SF* 24, no. 1 (1945): 95. Despite ecologists' influences in planning in the 1920s and 1930s, that planners had not yet fully taken their ideas to heart was expressed by a number of observers. See also Richard Ratcliff, who in 1944 called for the city planning profession to take better notice of sociologists', geographers', and economists' work. Richard Ratcliff, "A Land Economist Looks at City Planning," *JLPUE* 20, no. 2 (1944): 106–8.

131. Charles Ascher, "Proposed Program for Urban Conservation and Development, 1942–1943" (confidential, Executive Office of the President, National Resources Planning Board, NRPB, Washington, DC, n.d.), VF NAC 110 As, 5; Ralph R. Temple, "Bibliography RE Outline of Report on 'Social Control of the Use of Urban Land.'" In Folder Land Policy, Box 4, General Reference File I-L, Records of the Office of the Director, Records of the Urban Section, November 1941–June 1943, NRPBR. In connection with this work, the National Resources Planning Board gathered a list of state statutes on redevelopment. See the three folders titled "Legislation" (ibid.).

Chapter Four: From Natural Law to State Law

1. MHPC, *Conservation* (Chicago: MHPC, 1953), 3:258.

2. Ibid., 260.

3. Ibid., 259–260.

4. One exception is Joe Bailey, *Social Theory for Planning* (London: Routledge and Kegan Paul, 1975).

5. John T. Metzger, "Planned Abandonment: The Neighborhood Life-Cycle Theory and National Urban Policy," *HPD* 11, no. 1 (2000): 7–40; and Rick Cohen, "Neighborhood Planning and Political Capacity," *UAQ* 14, no. 3 (1979): 337–362.

6. Henry L. Henderson and David B. Woolner, eds., *FDR and the Environment* (New York: Palgrave Macmillan, 2005).

7. Sarah T. Phillips, *This Land, This Nation* (New York: Cambridge University Press, 2007), 74.

8. M. B. Schnapper, ed., *The Truman Program: Addresses and Messages by President Harry S. Truman* (Washington, DC: Public Affairs Press, 1949), 219; and Karl Brooks, "A Legacy in Concrete: The Truman Presidency Transforms America's Environment," in *Harry's Farewell: Interpreting and Teaching the Tru-*

man Presidency, ed. Richard Kirkendall (Columbia: University of Missouri Press, 2004).

9. U.S. President's Materials Policy Commission, *Resources for Freedom* (Washington, DC: Government Printing Office, 1952).

10. On civil rights and race relations under President Truman, see Michael Gardner, *Harry Truman and Civil Rights* (Carbondale: Southern Illinois University Press, 2002); Arnold R. Hirsch, "Choosing Segregation: Federal Housing Policy between Shelley and Brown," in *From Tenements to the Taylor Homes*, ed. John F. Bauman, Roger Biles, and Kristin M. Szylvian (University Park: Pennsylvania State University Press, 2000), 206–225; Carol Anderson, "Clutching at Civil Right Straws: A Reappraisal of the Truman Years and the Struggle for African American Citizenship," in *Harry's Farewell*; Mary Dudziak, "Desegregation as a Cold War Imperative," *Stanford Law Review* 41, no. 1 (1988): 61–120.

11. As Richard Watt put it, "As our society is now organized, winning a war is apparently the only objective that compels us to evolve a comprehensive plan of action and to rush through all obstacles, the legal ones included." Watt, "Urban Redevelopment—'Law-as-Usual' a Barrier to Progress," *JH* 3, no. 12 (1946): 286–287.

12. NAREB, "A Program for Redevelopment of American Cities by Private Capital and the Use of Incentive Taxation" (Confidential, not for publication, NAREB, 1944), VF NAC 1613; Schnapper, *The Truman Program*, 190–192; Walker Trohan, "Truman Levels Blast at Real Estate," *CDT*, March 22, 1949, 4; Peter Dreier, "Labor's Love Lost? Rebuilding Unions' Involvement in Federal Housing Policy," *HPD* 11, no. 2 (2000): 327–391. Notably, Hoyt's life cycle plan was invoked in these discussions about the need for a federal housing policy. See, for example, *Housing Charts Accompanying the Testimony of John B. Blandford, Jr., Administrator of the National Housing Agency, before the Senate Banking and Currency Committee, November 27, 1945* (Washington, DC: NHA, 1945).

13. Winning the hearts and minds of the international community through social policies and programs had nearly equal standing to generating the resources for securing military victory. As Mary Dudziak has described, Truman's long-standing interest in civil rights protection was reframed as a matter of America's defense. Mary Dudziak, *Cold War Civil Rights* (Princeton, NJ: Princeton University Press, 2002).

14. Reginald Isaacs, "Criteria for Selection of Initial Redevelopment Areas," in *Planning 1948*, ed. ASPO (Chicago: ASPO, 1948), 49–50. According to Watt, "Urban Redevelopment—'Law-as-Usual' a Barrier to Progress," a precursor to the 1947 Illinois law was introduced and failed in 1945.

15. Arnold Hirsch, "Searching for a 'Sound Negro Policy:' A Racial Agenda

for the Housing Acts of 1949 and 1954," *HPD* 11, no. 2 (2000): 393. Many states passed similar enabling legislation; for example, California's 1945 Community Redevelopment Act encouraged the formation of local redevelopment agencies to combat blight. On the evolution of state redevelopment legislation, see Miles Colean, *Renewing Our Cities* (New York: Twentieth Century Fund, 1953).

16. Julian Levi, "The Politics of Urban Renewal," in *Proceedings of Renewing Chicago in the Sixties* (Chicago: University of Chicago University College, 1961); "All-Developing Body for Blight Areas Is Urged," *CDT*, December 1, 1946, 27; Gladys Priddy, "Land Clearance Body Will Mark Its 5th Birthday," *CDT*, September 25, 1952, A1.

17. Arnold Hirsch, for example, who documents the circulation of prominent businessmen from city planning advisory boards to local planning boards as part of a multipronged attack of urban problems, has singled out Milton Mumford from Marshall Field and Company (who later became the city's housing and redevelopment coordinator) and Holman Pettibone from Chicago Title and Trust Company as two key participants in the Chicago story.

18. One hundred sixty acres was the standard measure used in earlier homesteading and reclamation policies in the United States; in the gridded city of Chicago, it also constituted one "section" of land.

19. Technicians across the city's varied housing and planning agencies contributed to the report. CPC, *Ten Square Miles of Chicago* (Chicago, CPC, 1948). The rating system is explained in Appendix C. As a comparison point, note that five years earlier the *Master Plan of Residential Land Use of Chicago* (Chicago: CPC, 1943) had classified 9.3 square miles as blighted and 13.3 as near blighted.

20. The score was on a scale of 0–100 penalty points. Earlier housing rating systems had employed the "penalty point" strategy, for example the American Public Health Association. "A New Method for Measuring the Quality of Urban Housing," *American Journal of Public Health* 33, no. 6 (1943): 729–740. The land clearance commission's rating instrument was a modification of Hoyt's original system, but like Hoyt's earlier revisions to the classic theories of Chicago sociology, these changes remained consistent with a naturalistic framework in which urban neighborhoods experienced a predictable life cycle, and could be quantitatively mapped and compared with one another along a spectrum of development and decay.

21. For example, they are used in CLCC, "Site Designation Report for Slum and Blighted Area: Redevelopment Project 7" (CLCC, Chicago, August 1953), 19; and CLCC, "Site Designation Report for Slum and Blighted Area: Redevelopment Project 6" (CLCC, Chicago, January 1953), published a decade after the *Master Plan of Residential Land Use*. Outdated aerial photographs were used as well. Thus, earlier plans to update the land use inventory were not carried out. Other

cities did keep their land use information more current, however, most notably Cleveland.

22. "Warns Chicago to Clear Land for Factories," *CDT*, April 9, 1951, B9; CLCC, "Redevelopment Project 2" (CLCC, Chicago, June 1951); idem, "Report for 1950" (CLCC, Chicago, 1951); Edward Lally and Phil Doyle, "Chicago Land Clearance Commission Final Relocation Report" (Project No. UR ILL. 6–3, West Central Industrial District, CLCC, Chicago, April 1958); CPC, "A Study of Blighted Vacant Land" (CPC, Chicago, 1950); CLCC, "Redevelopment Project No. 3" (CLCC, Chicago, June 1951). The frequent use of naturalistic terminology in other reports, including the city's cyclical real estate patterns, the wavelike mechanisms that propelled settlement outward from the Loop, or the areas of invasion and succession, also suggests his influences.

23. The newsletter delivered executive summaries of his sector and filtering theories and life cycle–based planning to readers. For example, one of the 1949 issues, "The Structure and Pattern of Cities," combined a brief on Hoyt's 1939 report for the Federal Housing Administration with his ideas about plans for Chicago based on the *Chicago Land Use Survey* and the *Master Plan of Residential Land Use*. Like many of the city's eminent real estate professionals, Bodfish had been affiliated with Ely's Institute during its tenure in Chicago.

24. There was a farm provision in the bill, Title V, to be administered by the agriculture department; some suggested that, given the powerful farm lobby, this provision was included to get the bill passed in the first place. See Testimony of Mr. Deane, Testimony on 1949 Housing Act. 81-H1247-1, 188. On the history of the 1949 Housing Act, see the special issue of *HPD* 11, no. 2 (2000): entire issue.

25. CPC, *Master Plan of Residential Land Use of Chicago*, 67; SSPB, *Your Investment in the South Side Planning Board Will Pay Dividends* (Chicago: SSPB, 1951), no page, VF NAC 1613.g27 Chi. Foreword by Walter Gropius.

26. 1947 also marked the appearance of two classic sources on the "ecology" of public administration. John Gaus, *Reflections on Public Administration* (Birmingham: University of Alabama Press, 1947), and Robert A. Dahl, "The Science of Public Administration: Three Problems," *Public Administration Review* 7, no. 1 (1947): 1–11.

27. Gary Fine, ed., *A Second Chicago School? The Development of a Postwar American Sociology* (Chicago: University of Chicago Press, 1995), xvi.

28. Philip Hauser, Preface to Otis Dudley Duncan and Beverly Davis Duncan, *The Negro Population of Chicago: A Study of Residential Succession* (Chicago: University of Chicago Press, 1957), vii.

29. Sudhir Venkatesh, "Chicago's Pragmatic Planners," *SSH* 25, no. 2 (2001): 275–317; Martin Bulmer, *The Chicago School of Sociology* (Chicago: University of Chicago Press, 1984). The Chicago Community Inventory is now the university's Population Research Center.

30. Wiebolt had sponsored an earlier generation of human ecologists, including Ernest Mowrer and Robert Faris, to conduct studies such as Ernest Russell Mowrer and Harriet Rosenthal Mowrer, *Domestic Discord: Its Analysis and Treatment* (Chicago: University of Chicago Press, 1928); and Ellsworth Faris, Ferris Finley Laune, and Arthur James Todd, *Intelligent Philanthropy* (Chicago: University of Chicago Press, 1930).

31. Philip Hauser, "Ecological and Population Factors in Urban Planning," in *Report of the Urban Planning Conferences under the Auspices of the Johns Hopkins University* (Baltimore: Johns Hopkins University Press, 1944), 17–26. Albert Reiss also briefly served as acting director of the Chicago Community Inventory. A parade of associate and assistant directors included Eleanor Bernert, Donald Bogue, Otis Dudley Duncan, and Evelyn Kitagawa—all distinguished social scientists, early in their careers.

32. They worked alongside other social scientists such as Selma Monsky, a National Opinion Research Center–affiliated economist trained in field study and survey methods.

33. E. Gordon Ericksen, *Introduction to Human Ecology* (Los Angeles: University of California, Los Angeles, 1949), 94. Indeed, despite the recognition that, as University of Wisconsin economic geographer George Wehrwein (formerly of Ely's institute) put it, the idealized scheme "has so many exceptions that it is sometimes difficult to prove the rule!" in these myriad studies Burgess's concentric model would have substantial staying power. George S. Wehrwein, "The Rural-Urban Fringe," *EG* 18, no. 3 (1942): 219. Thus, scholars at Laval University's sociology department sent their draft map of Quebec to Burgess for his assessment and comments. Jean-Charles Falardeau, "Problems and First Experiments of Social Research in Quebec," *CJEPS* 10, no. 3 (1944): 365–371. In his PhD dissertation at the University of Toronto, geographer Roy Wolfe documented the presence of the zones in Toronto. Roy I. Wolfe, "Summer Cottagers in Ontario," *EG* 27, no. 1 (1951): 10–32. The map would continue to be reprinted in urban studies across the social sciences, from Ericksen, *Introduction to Human Ecology*, to Griffith Taylor, *Urban Geography* (New York: Dutton, 1949), to Raleigh Barlowe, *Land Resource Economics* (Englewood Cliffs, NJ: Prentice Hall, 1958). Even critics continued to use the concentric model as an essential reference point for characterizing the internal structure of cities. Geographer Robert Dickinson, who in earlier work had followed Norman Scott Brien Gras and Ernest Burgess in characterizing the "Radial-Concentric (or Spider-Web) System" of urban growth, suggested that the historical development of "the physical structure of almost every west European city" could be seen as "a series of more or less concentric zones," but like Charles Colby found there were only three. Robert Dickinson, "The Morphology of the Medieval German Town," *GR* 35, no. 1 (1945): 75;

Robert E. Dickinson, "The Scope and Status of Urban Geography," *LE* 24, no. 3 (1948): 226–227. K. N. Venkatarayappa found seven rather than five zones in Bangalore: the business zone, the factory zone, the agriculture zone, the culture zone, the middle-class residential zone, the retired people's residential zone, and the military zone (also a residential district). Venkatarayappa, *Bangalore: A Socio-ecological Study* (Bombay: University of Bombay, 1957), 123. Presenting their "multiple nuclei model" as a template for cities in the postwar period, geographers Chauncy Harris and Edward Ullman made sure to explain how for some urban areas Burgess's concentric description continued to apply. Harris and Ullman, "The Nature of Cities," *AAAPSS* 242 (1945): 7–17. As late as 1953, Burgess would still be telling audiences how "the critics of this theory have been rather obtuse in not realizing that this theory is an ideal construction, and that in actual observation many factors other than radial expansion influence growth . . . Whereas all of the critics have been looking for exceptions, I have looked for generic aspects." Ernest Burgess, "The Ecology and Social Psychology of the City" (mimeographed abstract of a talk given in a seminar at the University of Chicago, 1953). Cited in Richard Redick, "A Study of Differential Rates of Population Growth and Patterns of Population Distribution in Central Cities in the United States 1940–1950" (PhD diss., University of Chicago, 1954), 11. Ironically, one of the most fervent critics of the ecological paradigm, Walter Firey, was himself the subject of critical attention when reviewers revealed how his own data validated rather than undermined Burgess's and Hoyt's urban models. John Hames, "A Critique of Firey's *Land Use in Central Boston*," *AJS* 54, no. 3 (1948): 228–234.

34. There were a few direct replications, for example, with two studies in the 1950s and one in the 1960s updating Nels Anderson's prior work on the city's skid row Howard Bain, "A Sociological Analysis of the Skid-Row Lifeway" (MA thesis, University of Chicago, 1950); NORC and CCI, *Who Lives on Skid Row and Why?* (Chicago: NORC and CCI, December 1957); Gerald Newman, ed., *The Homeless Man on Skid Row* (Chicago: Tenants Relocation Bureau, 1961). Proposals for new directions for research are discussed in Donald Bogue ed., *Needed Urban and Metropolitan Research* (Chicago: Scripps Foundation for Research in Population Problems, Miami University; and Population Research and Training Center, University of Chicago, 1953), which summarized the conclusions of a 1953 interdisciplinary seminar on Population, Urbanism, and Ecology.

35. The seminar was organized in the fall of 1954, and an initial report was published two years later; they brought these ideas to an even wider audience with the publication of *The Negro Population of Chicago: A Case Study of Residential Succession*.

36. Historians of ecology, including Sharon Kingsland and Gregg Mitman, have pointed out that mathematical explorations of population (called mathemati-

cal ecology or population ecology) came to the fields of plant and animal ecology from statistics, biology, and other disciplines only in the 1920s—around the same time that the first generation of human ecologists were developing their ideas. In his 1923 guide to ecological research, for example, Arthur Tansley had observed statistics to be of limited value—chiefly in preparing maps—and as late as 1935 Charles Adams complained that "many of the problems of the ecologist have not yet reached the stage that permits precise and mathematical formulation. Charles Adams, "The Relation of General Ecology to Human Ecology," *Ecology* 16, no. 3 (1935): 321. By mid-century, however, plant and animal ecologists regularly worked with a variety of quantitative techniques—cluster analysis, sampling, population pyramids, point pattern analysis, and nearest neighbor indices. A similar shift was under way in ecological studies in the social sciences. See, for example, W. S. Robinson, "Ecological Correlations and the Behavior of Individuals," *ASR*, 15, no. 3 (1950): 351–357, and Otis Dudley Duncan, Ray P. Cuzzort, and Beverly Davis Duncan, *Statistical Geography: Problems in Analyzing Areal Data* (Glencoe, IL: Free Press, 1961). The growth of "population studies" as an academic field appears to have offered new opportunities for plant, animal, and human ecologists to share ideas, building on a long history of exchanges between the biological and social sciences that included Thomas Malthus's influences on Charles Darwin.

37. Paul Hatt, "The Concept of Natural Area," *ASR* 11, no. 4 (1946): 423–427; idem, "Spatial Patterns in a Polyethnic Area," *ASR* 10, no. 3 (1945): 352–356; idem, "The Relation of Ecological Location to Status Position and Housing of Ethnic Minorities," *ASR* 10, no. 4 (1945): 481–485; E. Gordon Ericksen, "The Superhighway and City Planning: Some Ecological Considerations with Reference to Los Angeles," *SF* 28, no. 4 (1950): 429–434. That an ecological orientation to understanding cities prospered through the 1940s and 1950s was not unique to the University of Chicago. At the departments of Sociology at University of Michigan (under Amos Hawley, Roderick McKenzie, and Otis Dudley Duncan) and University of Washington (under Jesse Steiner, Clarence Schrag, Norman Hayner, and Calvin Schmid) and the Department of Geography at University of California, Berkeley (under Carl Sauer), among others, the tradition of fieldwork, surveys, and statistical mappings of urban areas also endured. New programs were established where the field was central and where University of Chicago remained the academic model, for example, at Laval University in Quebec, which invited Everett Hughes to help set up the program in social research. Examples of postwar publications in this cross-disciplinary field include Robert Park, ed., *Human Communities: The City and Human Ecology* (Glencoe, IL: Free Press, 1952); Nels Anderson, *The Urban Community: A World Perspective* (New York: Holt, Rinehart and Winston, 1959); Calvin Schmid, "Generalizations Concerning the Ecol-

ogy of the American City," *ASR* 15, no. 2 (1950): 264–281; Lewis F. Thomas, "Decline of St. Louis as Midwest Metropolis," *EG* 25, no. 2 (1949): 118–127; P. J. Smith, "Calgary: A Study in Urban Pattern," *EG* 38, no. 4 (1962): 315–329; Nicholas Demerath and Harlan Gilmore, "The Ecology of the Southern City," in *The Urban South*, ed. Rupert B. Vance and Nicholas J. Demerath (Freeport, NY: Books for Libraries Press, 1954); James Quinn, *Human Ecology* (New York: Prentice Hall, 1950); C. Langdon White and George T. Renner, *Human Geography: An Ecological Study of Society* (New York: Appleton-Century-Crofts, 1948); E. Gordon Ericksen, *The Territorial Experience: Human Ecology as Symbolic Interaction* (Austin: University of Texas Press, 1980); T. Earl Sullenger, *Ecological Study of the Woodson Center Neighborhood as a Basis for Program Planning* (Omaha, NB: Woodson Center Board of Trustees, 1960); Leo Kuper, *Durban: A study in Racial Ecology* (New York: Columbia University Press, 1958); Robert A. Murdie, *Factorial Ecology of Metropolitan Toronto, 1951–1961* (Chicago: University of Chicago Department of Geography, 1969); Roy I. Wolfe, "Summer Cottagers in Ontario," *EG* 27 no. 1 (1951): 10–32; Rose Hum Lee, "Occupational Invasion, Succession, and Accommodation of the Chinese of Butte, Montana," *AJS* 55, no. 1 (1949): 50–58; Noel P. Gist, "The Ecology of Bangalore, India: An East-West Comparison," *SF* 35, no. 4 (1957): 356–365; and Hiroshi Kawabe and Abdul Aziz Muhammad Farah, "An Ecological Study of Greater Khartoum," in *Urbanization and Migration in Some Arab and African Countries* (Cairo: Cairo Demographic Centre, 1973).

38. Robert Park, "The Concept of Position in Sociology," *PASS* 20 (1925): 1–14; CCI, *Chicago's Negro Population, Characteristics and Trends* (Chicago: CCI, 1956), 13. Not all human ecologists in this later generation held such patterns of segregation to be unnatural. See, for example, Ericksen, *Introduction to Human Ecology*, 36. Figures on restrictive covenants are drawn from Harold Kahen, "Validity of Anti-Negro Restrictive Covenants: A Reconsideration of the Problem," *UCLR* 12 (1945); Gardner, *Harry Truman and Civil Rights*, 57. Thomas Philpott has described how the University of Chicago had been a key sponsor of restrictive covenants around its campus. Thomas Philpott, *The Slum and the Ghetto* (New York: Oxford University Press, 1978), 252–254.

39. Ericksen, *Introduction to Human Ecology*, 8. Similar comments on the field's multidisciplinary character are offered by contributors to Bogue, *Needed Urban and Metropolitan Research*, and Louis Wirth, "The Social Sciences," in *American Scholarship in the Twentieth Century*, ed. Merle Curti (Cambridge, MA: Harvard University Press, 1953). The cross-disciplinary publication patterns that characterized first-generation human ecology were sustained through the postwar period. See Paul K. Hatt and Albert J. Reiss, eds., *Reader in Urban Sociol-*

ogy (Glencoe, IL: Free Press, 1951), and Harold Mayer and Clyde Kohn, eds., *Readings in Urban Geography* (Chicago: University of Chicago Press, 1959); George A. Theodorson, ed., *Studies in Human Ecology* (Evanston, IL: Row, Peterson and Company, 1961); Dickinson, "The Scope and Status of Urban Geography"; Harold Mayer, "Patterns and Recent Trends of Chicago's Outlying Business Centers," *JLPUE* 18, no. 1 (1942): 4–16; Amos H. Hawley and Don J. Bogue, "Recent Shifts in Population: The Drift toward the Metropolitan District, 1930–40," *RES* 24, no. 3 (1942): 143–148; Leo Schnore, "Geography and Human Ecology," *EG* 37, no. 3 (1961): 207–217. Venkatarayappa, *Bangalore (A Socio-ecological Study)*, articulates some of the differences in emphasis between the different disciplinary approaches to similar materials. Thus, University of Toronto sociologist Aileen Ross captured the social psychological dimensions of invasion in her study of English to French neighborhood change in Canada. Aileen B. Ross, "The Cultural Effects of Population Changes in the Eastern Townships," *CJEPS* 9, no. 4 (1943): 447–462. University of Wisconsin land economist Richard Ratcliff observed ecological shifts with respect to industries in a review of work from the New York Regional Plan Association. Richard U. Ratcliff, Review of *The Economic Status of the New York Metropolitan Region in 1944*, *JLPUE* 21, no. 1 (1945): 90–91. And, approaching the topic of industrial location "ecologically rather than economically — that is, by regarding industries not as economic entities, but rather as functional forms in a large geographical pattern," geographer George Renner saw symbiosis, life cycles and climax — "the law of industrial ecology" — at work in organizing the locations of industries and businesses in the United States. George T. Renner, "Geography of Industrial Localization," *EG* 23, no. 3 (1947): 167, 183. (Indeed, looking back on developments in the early decades of the twentieth century, McGill's F. Kenneth Hare suggested in 1947 how "the whole tendency of geography in the past generation," had been to move "towards an ecological approach." F. Kenneth Hare, "Regionalism and Administration: North American Experiments," *CJEPS* 13, no. 4 (1947): 565. Geographer Robert Dickinson seconded this point of view, finding in ecological approaches to urban study "common ground to geography, sociology, economics, and anthropology," yet echoing Carl Sauer's earlier lament that geographers' work had been overshadowed by colleagues in sociology. Dickinson, "The Scope and Status of Urban Geography," 237; Carl Sauer, "Forward to Historical Geography," *AAAG* 31, no. 1 (1941): 1–24; Robert Dickinson, Review of Amos Hawley, *Human Ecology*, *EG* 26, no. 4 (1950): 328.

40. Reflections on analogy were more often found among the American plant and animal specialists such as Lee Dice, Marston Bates, and Eugene Odum, who continued to express interest, both complementary and critical, in the human ecology field. See William Gordon, "Can Traditional Ecology Embrace Human

Ecology?" *Ecology* 31, no. 3 (1950): 489; Frederick Smith, "Ecology and the Social Sciences," *Ecology* 32, no. 4 (1951): 763–764; Edward Deevey, "Recent Textbooks on Human Ecology," *Ecology* 32, no. 2 (1951): 347–350; idem, "General and Human Ecology," in *The Urban Condition*, ed. Leonard Duhl (New York: Simon and Schuster, 1963).

41. Sharon E. Kingsland, *The Evolution of American Ecology, 1890–2000* (Baltimore: Johns Hopkins University Press, 2005). However, ecology texts continued to be reviewed in social science journals as human ecology texts were reviewed in science journals, and observations from plant and animal ecologists such as William McDougall still appeared in urban studies textbooks. Review of Rupert Vance, *All These People: The Nation's Human Resources in the South*, *QRB* 22, no. 2 (1947): 177; Rupert B. Vance, Review of Eugene P. Odum, *Fundamentals of Ecology*, *SF* 32, no. 4 (1954): 375–376; Svend Riemer, *The Modern City* (New York: Prentice Hall, 1952), 119.

42. Allee made other efforts to speak to social science audiences, for example, in W. C. Allee, "Social Biology of Subhuman Groups," *Sociometry* 8, no. 1 (1945): 21–29.

43. "News and Notes," *AJS* 47, no. 4 (1942): 625; Robert Redfield, ed., *Levels of Integration in Biological and Social Systems* (Lancaster, PA: Jacques Cattell Press, 1942). Robert Redfield, the collection's editor, was professor of anthropology and dean of the university's Social Science Division.

44. "Symposium on Viewpoints, Problems, and Methods of Research in Urban Areas," *SM* 73, no. 1 (1951): 37–50.

45. "Population Studies: Animal Ecology and Demography," *Cold Spring Harbor Symposia on Quantitative Biology* 22 (1957).

46. University of Chicago, *Announcements of the Division of the Social Sciences: For Sessions of 1949, 1950* (Chicago: University of Chicago, July 1, 1949), VF NAC 980 C; University of Chicago Population Research and Training Center, *Population Research and Training Center and the Chicago Community Inventory* (Chicago: University of Chicago, 1958).

47. Philpott, *The Slum and the Ghetto*, 252–254; *PASS* 29, no. 3 (1935). Other affiliations in the city included the recreation and prison inquiry commissions, although these were less directly related to planning at the citywide scale.

48. Later, he collaborated with architect Ernest Grunsfeld to design an alternative scientific plan for Chicago, published after his death. Ernest Grunsfeld and Louis Wirth, "A Plan for Metropolitan Chicago," *Town Planning Review* 25, no. 1 (1954): 5–32. His comments on the mutual benefits of relationships between human ecology and planning are presented in Louis Wirth, "Human Ecology," *AJS* 50, no. 6 (1945): 483–488.

49. Letter from Louis Wirth to Charles Merriam, August 26, 1941. In Folder

Wirth, Louis, Box 12, Interoffice Correspondence, Regional Correspondence, Correspondence by Person, Records of the Office of the Director, Records of the Urban Section, November 1941–June 1943, NRPBR. Wirth's ruminations on the department of agriculture as a model for a federal urban agency were one of many from board researchers in the two years before it was shut down. Correspondence to and from Charles Ascher, Robert Mitchell, George Duggar, and Robert Hartley in files of the National Resources Planning Board took up this subject as well.

50. University of Chicago Population Research and Training Center, *Population Research and Training Center and the Chicago Community Inventory*, iv. Between 1954 and 1962, for example, CCI wrote seventeen reports for Chicago's city planning agency alone, not to mention its many studies, maps, and statistical analyses for other agencies, committees, boards, and commissions. See CPC, "Inventory of Publications, Chicago Plan Commission and Department of City Planning from 1909 through 1962" (CPC, Chicago, 1962).

51. Beverly Davis Duncan and Philip Hauser, *Housing A Metropolis — Chicago* (Glencoe, IL: Free Press, 1960), xviii.

52. CCB, *Reading List for Urban Renewal* (Chicago: CCB, 1961), VF Z-NAC 1613.5 C; CCHR, *Selling and Buying Real Estate in a Racially Changing Neighborhood in 1962* (Chicago: CCHR, 1962), VF NAC 1434.4g27 Chi; MHPC, Press Release, March 26, 1953, VF NAC 1434.4g27; CCHR, *Excerpts for Study in Connection with Proposed Open Occupancy Legislation Gathered by Staff* (Chicago: CCHR, January 1958). This would persist into the 1960s in documents such as City of Chicago, *Housing and Urban Renewal Program Report* (Chicago: CDUR, 1963); City of Chicago, *Housing and Urban Renewal Program Report* (Chicago: CDUR, 1964).

53. CCI, *Chicago 1965, Report to the Housing and Redevelopment Coordinator* (Chicago: CCI, 1956), 6.

54. For example, Charles Zueblin had taught a course in the Sociology Department on American "municipal progress" in 1905. Political Science 305, City and Regional Planning was on the books in the early 1930s, with class readings spanning political science, economics, sociology, and geography, including the key human ecologists, in addition to city planning. In the Geography Department alone, Harland Barrows's Conservation of Natural Resources; Charles Colby's Urban Land Use and Land Use Problems; Henry Leppard's Graphics and Cartography; and Wellington Jones's Land Classification all covered materials that the American Society of Planning Officials had characterized as relevant to planners' work; Charles Zueblin, *American Municipal Progress: Syllabus of a Course of Six Lecture-Studies* (Chicago: University of Chicago Press, 1905), VF NAC 980; "Suggested Bibliography for Course on City and Regional Planning University of Chicago University College — Political Science 305" (unpublished, 1932), VF-Z NAC

250; ASPO, *Planning Education — a Survey of the Curricula in 23 Colleges and Universities, Memorandum, May 1942* (Chicago: ASPO, 1942). See also Harold M. Mayer, "Urban Geography and Chicago in Retrospect," *AAAG* 69, no. 1 (1979): 114–118. Their continuing interest in national resources is apparent in "National Resources: Progress and Poverty. An NBC radio discussion by Edward Ackerman, William Vogt, and Gilbert White" (Chicago: University of Chicago, 1949), VF NAC 547 A. Although an Institute of Planning headed by Rexford Tugwell was active by 1946, it did not become a full-fledged degree program until 1947. Rexford Tugwell, "Planning Education," in *Planning 1946*, ed. ASPO (Chicago: ASPO, 1946).

55. Its mission statement, which characterized the program as one that built on the university's long legacy in the social sciences, elaborated how while planning education traditionally had been "a by-product of the curriculums of schools of architecture and design, with a focus upon physical city planning," there was now a recognition of the importance of "social, economic, and political affairs," both as independent aspects of planning and as "considerations in the physical approach itself." University of Chicago, *Announcements: The Program of Education and Research in Planning: For Sessions of 1948, 1949* (Chicago: University of Chicago, April 15, 1948), 1.

56. Blucher was head of the American Society of Planning Officials; Ludlow had consulted for the National Resources Planning Board. There were a few exceptions to this pattern such as chemist-turned-planner Richard Meier. Leland S. Burns and John Friedmann, eds., *The Art of Planning: Selected Essays of Harvey S. Perloff* (New York: Plenum Press, 1985).

57. Charles S. Ascher, *Our Cities, Building America* (Washington, DC: NRPB, 1942), 17. The American Institute of Planners had seconded this point of view, observing in 1946 that "so great is the difference between today's kind of planning and its direct ancestor, 'civic art' and the City Beautiful movement," that planning schools must look beyond the usual talents in architecture and engineering in their student recruitment, less they "uselessly deprive the field of planning . . . of many potentially valuable practitioners." AIP Committee on Personnel Education and Standards, *The Content of Professional Curricula in Planning* (Cambridge, MA: AIP, 1947).

58. At the University of Michigan in the early 1950s, city planning was taught at the Department of Conservation (where University of Chicago's Meier moved in the late 1950s). Students enrolled in courses across the disciplinary spectrum, including Planning Procedures, Integrated Land Management and Conservation Policy, and a Seminar in Land Utilization and Regional Planning that spanned rural and urban planning. Notably, human ecologist Amos Hawley was one of the instructors for the latter class. See University of Michigan School of Nat-

ural Resources *Announcement 1952–1953* (Ann Arbor: University of Michigan, 1952), VF NAC 980 Uni.

59. While in the planning program's first year all the allied courses were in the social sciences (geography, anthropology, economics, history, political science, psychology, business, and law), botany and zoology appeared as recommended subjects beginning the following year. See University of Chicago, *Announcements: The Program of Education and Research in Planning: For Sessions of 1948, 1949*, 3; University of Chicago, *Announcements of the Division of the Social Sciences: For Sessions of 1949, 1950*, 141.

60. University of Chicago, *Announcements of the Division of the Social Sciences: For Sessions of 1949, 1950*, 141.

61. William E. Leuchtenburg, *Franklin D. Roosevelt and the New Deal* (New York: Harper, 1963), 32; Rexford Tugwell, "The Resettlement Idea," *Agricultural History* 33 (1959): 159–164. Taylor's influences are apparent in Tugwell's belief in the "one best way" to plan. As he wrote in 1948, "Planning is no more than scientific appraisal of resources and energies and evolvement of a one-best-way which is embodied in a development plan and budget, together with an outline of ways to proceed." Rexford Tugwell, "The Utility of the Future in the Present," *PAR* 8, no. 1 (1948): 56.

62. Leuchtenburg, *Franklin D. Roosevelt and the New Deal*, 35.

63. Rexford Tugwell, "The Study of Planning as a Scientific Endeavor," in *Tugwell's Thoughts on Planning*, ed. Salvador Padilla (San Juan: University of Puerto Rico Press, 1975). In his collected works was an unpublished manuscript, *Notes on Human Ecology*, whose title suggests it fleshed out these themes. See also Tugwell, "Variations on a Theme by Cooley," *Ethics* 59, no. 1 (1949): 233–243.

64. Banfield's previous work also included stints at the U.S. Forest Service, the New Hampshire Farm Bureau Federation, and the U.S. Farm Security Administration (the successor to the Resettlement Administration in the agriculture department).

65. Edward Banfield, *Government Project*, foreword by Rexford G. Tugwell (Glencoe, IL: Free Press, 1951); idem, "Ten Years of the Farm Tenant Purchase Program," *JFE* 31, no. 3 (1949): 473; idem, "Organization for Policy Planning in the U.S. Department of Agriculture," *JFE* 34, no. 1 (1952): 14–34; idem, "Rural Rehabilitation in Washington County, Utah," *JLPUE* 23, no. 3 (1947): 261–270.

66. Banfield, "Organization for Policy Planning in the U.S. Department of Agriculture." Other faculty research interests supported the view of cities as ecological communities and national resources. Louis Wirth's continuing work in the area of applied ecology until his death in 1952 is well known. Martin Meyerson praised the ecological work of Philip Hauser as an especially valuable intellectual contribution to the empirical basis for urban planning. Donald Innis would later

depart for the University of California, Berkeley, to study for a PhD in geography under Carl Sauer with a thesis on the human ecology of Jamaica. Harvey Perloff (who served as the unit's second director) did not call himself a human ecologist, but he praised the field on several occasions; his "ecological thinking" was noted by contemporaries, and his later planning work as head of the Program of Urban and Regional Studies at Resources for the Future underscored a perspective on cities as national resources; Martin Meyerson, Review of *The Report of the 1943 Urban Planning Conferences under the Auspices of the Johns Hopkins University* in *AAAPSS* 241 (1945): 189–190. See also idem, "The Future Ecology of the City: The Case of the Downtown Core," in *Australian Cities: Chaos or Planned Growth?* ed. John Wilkes (Melbourne: Angus and Robertson, 1966), 34–59; Donald Innes, "Human Ecology in Jamaica with a Detailed Study of Peasant Agriculture in the Mollison District of Northern Manchester" (PhD diss., University of California, Berkeley, 1959); Harvey Perloff, "How Shall We Train the Planners We Need?" in *Planning 1951*, ed. ASPO (Chicago: ASPO, 1951), 19; idem, "Knowledge Needed for Comprehensive Planning," in *Needed Urban and Metropolitan Research*; Harvey Perloff, ed., *The Quality of the Urban Environment: Essays on "New Resources" in an Urban Age* (Baltimore: Johns Hopkins University Press, 1969); Eric Lampard, "American Historians and the Study of Urbanization," *American Historical Review* 67, no. 1 (1961): 60n25.

67. Richard Dewey, "The Neighborhood, Urban Ecology, and City Planners," *ASR* 15 (1950): 502–507, Richard Dewey, "Peripheral Expansion in Milwaukee County," *AJS* 54, no. 2 (1948): 118–125; Nicholas Demerath, "Ecology, Framework for City Planning," *SF* 26 (1947): 62–67; E. Gordon Ericksen, "The Superhighway and City Planning"; University of North Carolina, "Department of City and Regional Planning Announcements for the Session 1954–1955," *University of North Carolina Record*, April 26, 1954, 7; University of Michigan, School of Natural Resources *Announcement* 1952–1953, VF NAC 980 Uni; University of Oklahoma, *Urban Studies Curriculum* (Norman: University of Oklahoma, 1955), VF NAC 980 U. Anticipating continuing needs for trained planning experts, in the immediate postwar period many institutions created new programs for planning education—University of North Carolina in 1946, University of Oklahoma in 1955, University of Southern California in 1955, and University of Pittsburgh in 1959. Urban studies centers also were opening around this time, for example, at Oklahoma, where an Institute for Community Development was created in 1945, and at University of Pennsylvania, where an Institute for Urban Studies was established in 1951.

68. Thus in contrast to planners such as National Resources Planning Board conservation unit head Charles Ascher, who supported neighborhood unit planning at the same time he urged the avoidance of "a pattern of stratification, . . .

which will produce self-contained colonies either of manual laborers or intellectuals or enterprises, which will perpetuate area marked as the exclusive preserve of persons of one language group or national origin," this later generation saw the neighborhood unit as incompatible with their broader ideals. Ascher, *Our Cities, Building America*, 8.

69. Reginald Isaacs, "The Neighborhood Theory — a Basis for Social Disorganization" (unpublished manuscript, April 20, 1947), VF NAC 1676 I, 2.

70. Ibid., 1.

71. Reginald Isaacs, "The Neighborhood Theory: An Analysis of its Adequacy," *JAIP* 14 (1948): 15–23; idem, "The 'Neighborhood Unit' Is an Instrument for Segregation," *JH* 5, no. 8 (1948): 215–219. See also Frederick J. Adams, Svend Riemer, Reginald Isaacs, Robert B. Mitchell, and Gerald Breese, "The Neighborhood Concept in Theory and Application," *LE* 25, no. 1 (1949): 67–88.

72. Jean-Louis Sarbib, "The University of Chicago Program in Planning: A Retrospective Look," *JPER* 2, no. 2 (1983): 77–81.

73. See Herbert Emmerich, "Cooperation among Administrative Agencies," *American Journal of Economics and Sociology* 15, no. 3 (1956): 237–244. There would be other collaborations as well, for example, the five-year Urban Redevelopment Study (1948–1953), sponsored by the Spellman Fund and directed by Coleman Woodbury.

74. John Dyckman, *A Proposal for Industrial Development in Chicago's Central South Side* (Chicago: South Side Industrial Study Committee, 1950), 1:11.

75. University of Chicago, *Announcements: The Program of Education and Research in Planning. For Sessions of 1948, 1949*, 2.

76. "Kennelly Sings a Neighborhood Air in Right Key," *CDT*, March 12, 1947, 4. The article paraphrased the mayor's remarks.

77. "Woodlawn Plan on Realty Sale Pact Announced," *CDT*, July 3, 1948, A7; Richard Nelson, "Neighborhood Conservation," Appendix A, in NHA, *Preliminary Report on Conservation of Middle-Aged Neighborhoods and Properties* (Washington, DC: NHA Office of the Administrator, Urban Development Division, October 16, 1944), 30, VF NAC 1613; Jack M. Siegel and C. William Brooks, *Slum Prevention through Conservation and Rehabilitation* (Washington, DC: SURRC, 1953).

78. Arnold Hirsch, "Searching for a 'Sound Negro Policy,'" 405; Alva Maxey, "The Block Club Movement in Chicago," *PQ* 18, no. 2 (1957): 124–131; North Lawndale Citizens Council Newsletter December 14, 1953, Box 1, Folder 1953, GLCCR.

79. On the history of the HHFA's race relations service, see Hirsch, "Searching for a 'Sound Negro Policy.'" Oakland-Kenwood was located close to Hyde Park and Woodlawn.

80. To Raymond M. Foley from Frank S. Horne, "Subject: Bi-weekly report—Racial Relations Service," November 26, 1948, Box 7, Department of Housing and Urban Development, Subject Correspondence Files, Raymond M. Foley Administrator HHFA, 1947–53, Folder Report of Activities, Racial Relations Service (1 of 2), HUDR; CCHR, *Memorandum on Community Conservation Agreement* (Chicago, CCHR, 1948). As Peter H. Rossi and Robert A. Dentler, *The Politics of Urban Renewal: The Chicago Findings* (New York: Free Press of Glencoe, 1961) describe, such agreements were included in contracts for home purcheses in lieu of restrictive covenants.

81. Howard Oxley, "The Civilian Conservation Corps and the Education of the Negro," *Journal of Negro Education* 7, no. 3 (1938): 375–382; Olen Cole Jr., *The African-American Experience in the Civilian Conservation Corps* (Gainesville: University Press of Florida, 1999); John A. Salmond, "The Civilian Conservation Corps and the Negro," *JAH* 52, no. 1 (1965): 75–88; "CCC after 7 Years," *CD*, April 6, 1940, 13; Edgar Brown, "More Race Workers Added to U.S. CCC Camps," *CD*, December 22, 1934, 12; Edgar Brown, "Thousands of Race Boys Now Serving in National CCC Campus, Survey Reveals," *CD*, March 16, 1935, 12; "Race Is Well Represented on the CCC Corps," *CD*, November 2, 1935, 2.

82. Sherman Briscoe, "The Negro in Agriculture," *CD*, September 26, 1942, B16; "War Dept. Asks to Conserve on Wool," *CD*, February 21, 1942, 22; "To Stress Food Conservation," *CD*, November 11, 1944, 12; "Extension Leaders Confer," *CD*, June 23, 1945, 4; "Half Million School Children Join in Rationing Program," *CD*, February 27, 1943, 24; "FSA Supervisors Visit Model Farm of Ala. Negro," *CD*, August 5, 1944, 18; "Victory Gardens Get Advice from Rose Expert," *CDT*, February 13, 1943, 16.

83. Philpott, *The Slum and the Ghetto.*

84. To Raymond M. Foley from Frank S. Horne, "Subject: Bi-weekly report—Racial Relations Service," November 26, 1948, Box 7, Department of Housing and Urban Development, Subject Correspondence Files, Raymond M, Foley Administrator HHFA, 1947–53, Folder Report of Activities, Racial Relations Service (1 of 2), HUDR.

85. "Woodlawn Plan on Realty Sale Pact Announced," *CDT*, July 3, 1948; Siegel and Brooks, *Slum Prevention through Conservation and Rehabilitation*, 63; NCCC, *The Pastor and the Conservation of his Parish* (Washington, DC: NCCC, 1952), VF NAC 1676 N; "History of Lincoln Park Conservation Association," LPNDC. On the church's hostility to clearance, see Testimony of Right Reverend Monsignor John O'Grady, Secretary, National Conference of Catholic Charities on 83-H1442-7, Monday, March 15, 1954, 645; Barth Zurkhammer, "The Relationship between DePaul University and the LPCA," Accession No. lp.lpnc.dpu .0004 LPNDC.

86. NCCC, *The Pastor and the Conservation of his Parish*, 5; NCCC, *The Pastor and Neighborhood Conservation* (Washington, DC: NCCC, 1953), VF NAC 1676-N; Samuel Stritch, *Neighborhood Conservation: Housing as a Pastoral Problem* (Washington, DC: NCCC, 1953), VF NAC 1426g27 Chi.

87. NCCC, *The Pastor and Neighborhood Conservation*, 1.

88. NCCC, *The Pastor and the Conservation of His Parish*; NCCC, *The Pastor and Neighborhood Conservation*, 1. Later, Cardinal Samuel Stritch established the Cardinal's Committee on Conservation.

89. Reginald Isaacs, "Hospital Engages in Planning Survey," *JAIP* 11, no. 4 (1945): 39.

90. The city had released a preliminary comprehensive plan in 1946 detailing a vision of the city to 1965 but had made little progress on implementation. See CPC, *Chicago Tomorrow: An Interpretation of the Preliminary Comprehensive City Plan* (Chicago: CPC, 1946), VF NAC 544 and CPC, *Preliminary Comprehensive City Plan of Chicago* (Chicago: CPC, 1946).

91. John Dyckman, *Community Appraisal Study, A Summary of Current Proposals* (Chicago: SSPB, Spring 1952), 1, 36; Harvard University Graduate School of Design, "Community Appraisal (second draft, A Program for Redevelopment, Rehabilitation and Conservation, Central South Side, Kenwood and Hyde Park Communities, Chicago, IL)" (Cambridge, MA: Harvard University Graduate School of Design, 1950), 2, VF NAC 1613.5g27 H. To give a sense of the area's size, it spanned four-and-a-half square miles and contained a population of one hundred and eighty thousand—comparable to Salt Lake City at that date; in the *Master Plan of Residential Land Use*, the entire territory had been listed as a conservation area.

92. Lead participants from Illinois Institute of Technology included Mies van der Rohe and Ludwig Hilbersheimer. University of Chicago participants included planners Martin Meyerson and Richard Meier, geographer Harold Mayer, and sociologists Philip Hauser and Louis Wirth (who died as the project was wrapping up in 1952). Harvard design school participants included Lester Collins, Hidayo Sasaki, William Wheaton, Roger Creighton, and Walter Gropius. Also enlisted were Egbert Schietinger, who had prepared a master's thesis in sociology on racial succession and changing property values in Chicago; Herbert Thelen from the University of Chicago Department of Education; several affiliates of the National Opinion Research Center, including Donald Bogue, Clyde Hart, Fred Meyer, and Josephine Williams; and other faculty and students from IIT, including Albert Biderman, Frank Cliffe, and Don Smithberg. Many of these participants already knew one another from work in Chicago and beyond. For example, Wirth, van der Rohe, and Hilbersheimer had worked together at the South Side Planning Board's Planning Committee. Mayer and Meyerson had worked together in Philadelphia

at the Planning Commission. Wirth and Blucher had participated on race relations work for Chicago civic organizations.

93. Walter Gropius and Martin Wagner, "Epilogue: The New City Pattern for the People and by the People," in *The Problem of the Cities and Towns*, ed. Guy Greer (Cambridge: n.p., 1942), VF NAC 613, 112, 101.

94. In an ideal scenario, "the most rational application of planning method," as University of Chicago planning student John Dyckman explained, "the survey has come to be considered the key 'pre-planning' instrument, and tentative plans customarily await tabulation of the survey findings." In the Community Appraisal Study, however, surveys and planning took place concurrently, for practical reasons of course scheduling. Thus, while planners could make use of early survey returns to prepare area "profiles," after the "social accounting" was completed some modifications had to be made. Nonetheless, participants were proud of the opportunities for information-based planning their work had created. "The mere assembly of all pertinent available materials on the communities under study, the inventory of information and the addition of new syntheses, along with the tremendous volunteer interview undertaking and physical inventory," Dyckman explained, "provided a framework for informed planning decisions seldom available to the planner, governmental or independent." Dyckman, *Community Appraisal Study*, 2–3, 10. Extensive study questions and tasks for the varied participants are laid out in Harvard University Graduate School of Design, "Community Appraisal," 2.

95. The questionnaire was "prepared with the guidance of the Chicago Community Inventory and the Department of Sociology of the University of Chicago" but was turned over to be "carried out principally by the Hyde-Park Kenwood Community Conference, aided by students" at IIT, Roosevelt College, and Wright Junior College. Dyckman, *Community Appraisal Study*, 2.

96. Ibid. Meyerson and Meier also jointly led workshops on "university-community problems."

97. "Eighteen Photographs of Models Made by Students in Connection with Collaborative Program on the Redesign of Part of Chicago's South Side, Done under the Direction of Professor Reginald R. Isaacs" (Cambridge: Harvard Graduate School of Design, 1951); Dyckman, *Community Appraisal Study*, illustrations between pp. 13 and 14; HPKCC, *A Report to the Community: A Preliminary Review of Area Problems and Possibilities* (Chicago: HPKCC, 1951).

98. Dyckman, *Community Appraisal Study*, 2.

99. Ibid.

100. On the real estate community's continuing advocacy for neighborhood racial homogeneity, see Charles Abrams, *Forbidden Neighbors* (New York: Harper, 1955).

101. HPKCC, *A Report to the Community* included a "Discussion Draft of a Community Conservation Agreement."

102. Although scholars at the Chicago Community Inventory and Program for Education and Research in Planning recognized that racial minorities were not the cause of blight this had little effect on the behaviors of individual white families, an effect exacerbated by present valuation practices at federal and local levels. The FHA, for example, still was practicing discriminatory lending, and standard real estate texts continued to maintain that homogenous neighborhood were preferred. Anderson, "Clutching at Civil Right Straws: A Reappraisal of the Truman Years and the Struggle for African American Citizenship," in *Harry's Farewell*; AIREA, *Appraisal Terminology and Handbook* (Chicago: AIREA, 1950); William H. Brown Jr., "Access to Housing: The Role of the Real Estate Industry," *EG* 48, no. 1 (1972): 66–78. That the influx of minority residents did not cause blight had, however, by this time become more widely recognized in other quarters, for example, the Catholic Church and among some members of the real estate community, such as Downs-Mohl principal and Real Estate Research Corporation founder James Downs, who consistently stressed in public presentations how it was instead illegal conversions, property deterioration, and failures of law enforcement that encouraged blight to spread. NCCC, *The Pastor and Neighborhood Conservation*, 1. The study group's explicit rejection of the neighborhood unit idea is noted in Dyckman, *Community Appraisal Study*, 11–12.

103. Dyckman, *Community Appraisal Study*, 12; HPKCC, *A Report to the Community*, 33–34.

104. Rossi and Dentler, *The Politics of Urban Renewal* 2; Julia Abramson, *A Neighborhood Finds Itself* (New York: Harper, 1959).

105. Rossi and Dentler, *The Politics of Urban Renewal*, 127.

106. Untitled document, Box 13, Records of the Federal Home Loan Bank System, Federal Home Building Service Plan, Program Subject Files, 1936–1942, Folder Pamphlets and Literature on Housing (Folder 1), 11, FHLBBR; HPKCC, *A Report to the Community*, 10.

107. Rossi and Dentler, *The Politics of Urban Renewal*, 107–109.

108. "Memo for Mr. Geo. Duggar, January 7, 1941, Subject: Analogy for Urban Development Based on Dept. of Ag. Org." In Folder Federal Urban Organizations, Box 19, Alphabetical Correspondence A–F, Records of the Office of the Director, Records of the Urban Section, November 1941–June 1943, NRPBR.

109. FHLBB, *Waverly* (Washington, DC: FHLBB, 1940), 37.

110. HPKCC, *A Report to the Community*, 33.

111. Herbert A. Thelen and Bettie Belk Sarchet, *Neighbors in Action: A Manual for Community Leaders* (Chicago: University of Chicago Human Dynamics Laboratory, 1954); Herbert Thelen, *Dynamics of Groups at Work* (Chicago: University of Chicago Press, 1954), 1; idem, "Shall We Sit Idly By?" *ETC: A*

Review of General Semantics 9, no. 1 (1953); idem, "Social Process versus Community Deterioration," *Group Psychotherapy* 4 (1951): 206–212.

112. Thelen, *Dynamics of Groups at Work*, 29–30.

113. Ibid., 275–276.

114. Ibid., 9.

115. Dyckman, *Community Appraisal Study*, 29.

116. HPKCC, *A Report to the Community*, 32; Bettie Belk Sarchet and Herbert Thelen, *Block Groups and Community Change* (Chicago: Human Dynamics Laboratory, University of Chicago, 1955).

117. Sarchet and Thelen, *Block Groups and Community Change*; Audrey Probst, "The Community That Saved Itself," *Nation*, March 5, 1955; Siegel and Brooks, *Slum Prevention through Conservation and Rehabilitation*; Harvey Perloff, *Urban Renewal in a Chicago Neighborhood* (Chicago: Hyde Park Herald, 1955), VF NAC 1613.5g27 Chi; Herbert Thelen, Review of *A Neighborhood Finds Itself*, by Julia Abramson, *School Review* 67, no. 4 (1959): 469–473.

118. Dyckman, *Community Appraisal Study*, 9. Titles included *Church Congregation Migration, Community Leadership*, and *Insurance Rates for Negro Owned Property*.

119. Thelen, *Dynamics of Groups at Work*, 24. Other training courses are described in Thelen and Sarchet, *Neighbors in Action*, 68.

120. This was the result of their growing appreciation that, as planners long ago had articulated, pockets of slum removal or even slum clearance would be insufficient to stop blight from spreading when a predictable life cycle continued to drive growth, decay, and change across the diverse neighborhoods of any city. "From time to time," the South Side Planning Board had explained in 1947, using an earlier generation's medical vocabulary for city planning, "business leaders have advanced the theory that blight could be segregated, and its effect cut off from the rest of the community," yet history already had shown that "this theory has proved no more effective than a theory that placing a tourniquet on the cancerous arm of a human being would stop cancer" SSPB, *Prospectus* (Chicago: SSPB, 1947).

121. Harvard University Graduate School of Design, *An Approach to Redevelopment: A Panel Discussion* (Chicago: Michael Reese Hospital, 1951), p. 18, VF NAC 980 Har. This meeting was one of Isaacs's efforts to rally support to the conservation cause, a gathering of visiting critics to evaluate student work that included Walter Blucher (American Society of Planning Officials), Ira Bach (Chicago Land Clearance Commission), Otto Nelson (the New York Life Insurance Company), and Ferd Kramer (Chicago's Metropolitan Housing and Planning Council and Draper and Kramer real estate).

122. NAREB, *Recommendations on Rehabilitation by the Committee on Rehabilitation* (Washington, DC: NAREB, 1952), VF NAC 1613 NAREB, 2. The continued conviction that predictable life cycles defined the character of urban areas is expressed in one 1952 manual. "This, in a general way, is what happens to residential neighborhoods," offered the American Institute of Real Estate Appraisers: "They usually grow in desirability for a while after they are established, and are built up quite rapidly. They attain a peak of desirability, remain stable for a time, and then deteriorate in quality. Or another way of stating it is that there are three stages of neighborhood status—integration, equilibrium, and disintegration." AIREA, *The Appraisal of Real Estate* (Chicago: R. R. Donnelly and Sons, 1952), 105. In a nod to the continuing relevance of ecological analysis, many appraisers in this period called for the preparation of "sequent occupance" maps (another term for time-interval maps) as part of the evaluation process. See Eugene van Cleef, "Maps for Appraisals," *Appraisal Journal* 17 (1949): 219–231; Stanley McMichael, *Appraising Manual* (New York: Prentice Hall, 1951), 575. Some real estate texts even printed Burgess's concentric model, for example, Ernest Fisher and Robert Fisher, *Urban Real Estate* (New York: Holt, 1954).

123. Harvard University Graduate School of Design, *An Approach to Redevelopment*, 27.

124. Already having "made a study of conservation in West Kenwood as Housing [and Planning] coordinator in 1951," Downs soon affiliated himself with Chicago's other ongoing conservation efforts—from those sponsored by the Catholic Church to the multi-institution Community Appraisal Study. NCCC, *The Pastor and Neighborhood Conservation*, 1; Rossi and Dentler, *The Politics of Urban Renewal*, 77–78. Earlier, in 1931, Downs had created the Real Estate Research Corporation. The company's work pioneering forecasting techniques, and the fact that it had anticipated the real estate downturn of the depression had given Downs an aura of authority in the arena of real estate prediction.

125. Rossi and Dentler, *The Politics of Urban Renewal*, 77.

126. Rossi and Dentler, *The Politics of Urban Renewal*, 78.

127. MHPC, *Conservation*; William K Brussat, *Neighborhood Conservation* (Chicago: Housing and Redevelopment Coordinator, 1956). See also Jack Siegel, "Observations on Neighborhood Conservation in Chicago and Other Cities" (Chicago: Committee on Housing, May 15, 1952), VF NAC 1676g27 Chi, which discusses conservation programs in Baltimore and Philadelphia. The intellectual overlap between this private commission's work and the earlier inter-university collaboration is evident in the similarities of its text and maps; in the similarities of participating personnel (with the addition of Harvey Perloff, who had been on leave during the period of the appraisal study); and in the report's acknowledg-

ment (1:72) that "in large measure, the present Conservation Study . . . is a direct outgrowth of the Community Appraisal Study."

128. MHPC, *Conservation*, 1:5.

129. Juxtaposing its map with Hoyt's *Master Plan of Residential Land Use of Chicago* finds conservation designations nearly identical to those prepared in 1943.

130. MHPC, *Conservation*, 1:25.

131. Interim Commission on Neighborhood Conservation, *Preventing Tomorrow's Slums* (Chicago: Interim Commission on Neighborhood Conservation, October 1952), VF NAC 1613.1g27 Chi. A brochure of that same name would be distributed to residents across the city. See City of Chicago, *Preventing Tomorrow's Slums: A Citizen Action Program* (January 1955) in Box 2, Folder Jan–March 1955, GLCCR.

132. MHPC, *Conservation*, 3:223.

133. Ibid., 3:260–259.

134. "Urban Conservation Gets a Boost," *CDT*, September 27, 1954, 18; Rossi and Dentler, *The Politics of Urban Renewal*, 86.

135. Siegel and Brooks, *Slum Prevention through Conservation and Rehabilitation*. On the conservation conference, see MHPC, *Conservation*, Vol. 3, acknowledgments.

136. The initial proposal from Chicago's city council was for 160 acres. See "Aldermen Act to Keep Slums from Growing" *CDT*, July 11, 1953, A7.

137. There was some early debate about whether to administer the conservation program inside an existing agency or whether to create a new one. Homer Hoyt's 1943 plan had envisioned a life cycle–based program guided by the city's plan commission, but the commission's comparatively weak stature in the 1950s suggested to conservation study authors the need to place responsibility for the conservation program in a separate department. On the debate about how to administer the conservation program, see letter from E. O. Griffenhagen to Mayor Martin Kennelly, August 5, 1953, LPNDC; MHPC, *Conservation*; and Rossi and Dentler, *The Politics of Urban Renewal*.

138. Robert B. Mitchell, "Prospects for Neighborhood Rehabilitation," in *Housing Yearbook*, ed. NAHO (Chicago: NAHO, 1938), 141.

139. "Research on Neighborhood Conservation," *RER* (January 18, 1941): 8.

140. Nelson, "Neighborhood Conservation."

141. Rossi and Dentler, *The Politics of Urban Renewal*, 77–78.

142. Dyckman, *Community Appraisal Study*.

143. Ibid., 32.

144. Everett Hughes, *The Chicago Real Estate Board* (Chicago: University of Chicago Press, 1933), 106.

145. MHPC, *Conservation*, 3:286.

146. Helping citizens to help themselves was a major theme in New Deal conservation. See, for example USRA, *Helping the Farmer Help Himself*. Slayton, like Isaacs and Siegel, similarly was disappointed by conservation, noting how in Woodlawn, even after the Chicago Plan Commission had issued an official conservation plan for the area, little action had occurred. "The program (if that is the proper word) that followed the report has been a disappointment, to put it mildly." He too used this as an argument for strengthened enforcement. William Slayton, "Urban Redevelopment Short of Clearance: Rehabilitation, Reconditioning, Conservation, and Code Enforcement in Local Programs," in *Urban Redevelopment Problems and Practices*, ed. Coleman Woodbury (Chicago: University of Chicago Press, 1953), 379.

147. William Slayton and Richard Dewey, "Urban Development and the Urbanite," in *The Future of Cities and Urban Redevelopment*, ed. Coleman Woodbury (Chicago: University of Chicago Press, 1953), 469. The Tennessee Valley Authority was widely praised as the embodiment of democratic values. See David E. Lilienthal, *TVA: Democracy on the March* (New York: Harper & Brothers, 1944); Philip Selznick, *TVA and the Grassroots* (Berkeley: University of California Berkeley Press, 1984).

148. CCB, *The Chicago Conservation Program* (Chicago: CCB, 1956), VF NAC 1613.1g27 Chi. Notably, several of the community groups that would play a role in conservation, for example, the Near West Side Planning Board, had their origins in the Chicago Area Project, a program of the Illinois Institute for Juvenile Research, which began in the 1930s under the direction of University of Chicago faculty Ernest Burgess and Clifford Shaw.

149. Untitled and undated document, Box 756, Program Files, 1940–65, Presidents Advisory Committee on Housing Policies and Programs, 1953–54, Folder Presidential Advisory Committee—Administrator's Shirtsleeve Conference, Chicago, HUDR.

150. The Interim Commission's report echoed the Chicago Plan Commission's 1951 assessment that there were fifty-six square miles of conservation areas in the city, simply reprinting a 1951 plan commission map.

151. H. L. Dunton, "Organizing the District and Promoting Educational Activities within the District," in *Effective Operations of the Soil Conservation Districts* (Washington, DC: SCS, 1941), VF NAC 7560 US; Letter from James Downs to Maida Steinberg, November 12, 1953, Box 1, Folder 1953, GLCCR; D. E. Mackelmann, "Community Organization in Conservation; Community Organization in Chicago" (speech to the monthly meeting of the Associated Clubs of Woodlawn, Inc., October 19, 1960, at the Woodlawn Regional Library), 4; and CCB, *Checklist for Neighborhood Analysis* (Chicago: CCB, n.d.), 1, lp.lpccc.dur

.0020 and lp.lpccc .dur.0019 from LPNDC. Indeed, the soil conservation district appears to have been the model for such community organizations, a unit that while not "units of local government like cities or counties" nevertheless "possessed authority to conduct research, manage demonstration projects, acquire property, and enforce land use regulations mutually agreed on." Sarah T. Phillips, *This Land, This Nation* (New York: Cambridge University Press, 2007), 137.

152. "History of Lincoln Park Conservation Association," 16, LPNDC. This would be no small task: In Chicago's Lincoln Park, more than two years of work were required before a conservation designation was forthcoming from the city.

The enthusiastic response of another community's residents to the official designation as a "conservation area" shows the success of this strategy: "Dear Neighbor," explained a special edition of the North Lawndale Citizens Council Newsletter. "THIS IS A SPECIAL NEWSLETTER BECAUSE WE HAVE SOMETHING SPECIAL TO TELL YOU. BECAUSE OF THE LETTERS YOU HAVE SENT—AND OTHER EVIDENCE OF MAKING NORTH LAWNDALE A BETTER PLACE—OUR COMMUNITY HAS BEEN APPROVED AS A CONSERVATION AREA BY THE CHICAGO COMMISSION ON NEIGHBORHOOD CONSERVATION!" North Lawndale Citizens Council Newsletter December 14, 1953, Box 1, Folder 1953, 1, GLCCR.

153. HPKCC, *A Report to the Community*, 32.

154. City of Chicago, CCB and Department of Buildings, *Conservation Survey: Instructions for Use of Structure Schedule* (Chicago: CCB, October 1956), VF NAC 1613.1g27 Chi.

155. South East Chicago Commission executive director and assistant to University of Chicago Chancellor Lawrence Kimpton Julian Levi worked to amend the 1941 Neighborhood Redevelopment Corporation Act that same year to permit conservation; the Chicago Housing Authority also embarked on a program of conservation at this date; "Conservation, a New Area for Urban Redevelopment," *UCLR* 21 (1953–1954), 494; "Clinic: Urban Conservation," in ASPO, *Planning 1953* (Chicago: ASPO, 1953). According to Rossi and Dentler, the Metropolitan Housing and Planning Council's work on the Urban Community Conservation Act and the work of Julian Levi and the South East Chicago Commission on updating the neighborhood redevelopment act "were not coordinated." Rossi and Dentler, *The Politics of Urban Renewal*, 85.

Chapter Five: A Nation of Renewable Cities

1. H. Warren Dunham, "The City: A Problem in Equilibrium and Control," in *The City in Mid-Century*, ed. H. Warren Dunham (Detroit: Wayne State University Press, 1957), 167.

2. Maurice Parkins, *Neighborhood Conservation: A Pilot Study* (Detroit:

DCPC, 1958), 41. In Box 11, Folder Neighborhood Conservation, A Pilot Study 1958, DCPCR; Mel Ravitz, "The Sociologist as a Participant in an Urban Planning Program," *JAIP* 20, no. 2 (1954): 76–80; Mel Ravitz and Adelaide Dinwoodie, "Detroit Social Workers Mobilize Citizen Aid for Urban Renewal," *JH* (July 1956).

3. On the dominance of clearance in the historiography of urban renewal, see the "Essay on Sources."

4. Indeed, it was not until 1954, with the marriage of rehabilitation and conservation to clearance, that "urban renewal" officially got under way.

5. City of Toledo Housing and Urban Redevelopment Commission, *A Recommended Program for Slum Prevention and Neighborhood Rehabilitation, October 5, 1953*, 2, in Box 625, Program Files, 1940–65, "HHFA Archives," 4-Program Planning and Administration 4/21/Title I (URA Program, General Information) — 4-21-1/UR Directories (1962–present), Folder 4-21 Title I (URA Program General Information), HUDR.

6. Elmo R. Richardson, *Dams, Parks, and Politics: Resource Development and Preservation in the Truman-Eisenhower Era* (Lexington: University of Kentucky Press, 1973).

7. On the conflict between short- and long-term assessments of U.S. conservation policy, see the "Essay on Sources."

8. Stephen Rauschenbush, "Conservation in 1952," in *From Conservation to Ecology*, ed. Caroll Pursell (New York: Crowell, 1973), 61; *AAAPSS* 28 (1952): entire issue on "The Future of Our Natural Resources."

9. Sarah Phillips, *This Land, This Nation* (New York: Cambridge University Press, 2007).

10. As Randal Beeman and James Pritchard, *A Green and Permanent Land* (Lawrence: University of Kansas Press, 2001) note, this was one reason Bennett joined Friends of the Land.

11. Eli Ginzberg, *The Negro Potential* (New York: Columbia University Press, 1956).

12. Pursell, *From Conservation to Ecology*, 3.

13. Government policies did not back away from conservation entirely, however, as Phoebe O'Neall Faris Harrison, *Books, Booklets, Bulletins on Soil and Water Conservation* (Washington, DC: Government Printing Office, 1953) attests. Soil Conservation Service, *Our Productive Land* (Washington, DC: SCS, 1953), for example, shows the continued usage of land capability mapping.

14. A. Dan Tarlock, "Rediscovering the New Deal's Environmental Legacy," in Henry Henderson and David Woolner, eds., *FDR and the Environment* (New York: Palgrave Macmillan, 2005), 167. However, environmental historians have articulated how Clements's influences lingered in both ecological theory and resource planning practice, even as his ideas were publicly abandoned. In agreement

that Clements's powers of persuasion peaked during the interwar period before being supplemented by alternative schools of thought such as systems ecology, scholars simultaneously note that aspects of the organismic theory of succession to climax endured for several decades after Clements's death. Pointing to systems ecologists' assumptions about cycles and balance in natural systems, Joel Hagen and Ashley Schiff have described how, even as many of these scholars sparred with the deterministic and teleological character of Clements's contributions, they nevertheless depended on the intellectual groundwork he laid. Citing the persistence of specific practices in forest, farm, and range management, L. A. Joyce and Christopher Masutti have documented how the theory of succession to equilibrium climax continued to shape action on natural resources planning as well. Hagen, "Research Perspectives and the Anomalous Status of Modern Ecology," *Biology and Philosophy* 4(1989): 433–455; Schiff, "Innovation and Administrative Decision Making: The Conservation of Land Resources," *Administrative Science Quarterly* 11, no. 1 (1966): 1–30; Joyce, "The Life Cycle of the Range Condition Concept," *Journal of Range Management* 46, no. 2 (1993): 132–138; Christopher Masutti, "Frederic Clements, Climatology, and Conservation in the 1930s," *Historical Studies in the Physical and Biological Sciences* 31, no. 1 (2006): 27–48. Ashley Schiff makes the case that many users would not recognize the extent of their dependence on this earlier template.

15. Richardson, *Dams, Parks, and Politics,* 74.

16. Edward L. Schapsmeier and Frederick H. Schapsmeier, "Eisenhower and Agricultural Reform: Ike's Farm Policy Legacy Appraised," *American Journal of Economics and Sociology* 51, no. 2 (1992): 147–159. The Bureau of Agricultural Economics, home to many New Deal era scientific land use managers, was dismantled at this time.

17. Eisenhower had witnessed firsthand the problems of the "unfit" population of young men recruited for military service. As president of Columbia University, he had asked economist Eli Ginzberg to undertake a study of human resources in the United States; the university established a research center for the project in 1950 later called the Eisenhower Center for the Conservation of Human Resources. This interest in human conservation does not appear to have shaped his urban policies, however.

18. Nathaniel Keith, *Relocation Problems in Urban Redevelopment* (Washington, DC: HHFA, 1952), 1.

19. As Truman's Housing and Home Finance Agency administrator expressed only two years into the program, his staff already was concerned about "extending financial aid to a city in the name of slum clearance without assurance that its application will not spread blight to new and greater areas, merely piling up new future problems." Raymond Foley, *Address by Raymond M. Foley, Hous-*

ing and Home Finance Administrator, before the National Conference of Catholic Charities (Washington: HHFA, 1951), VF NAC 1430.4 F, 6.

20. Arguing that after years of Democratic control, Republicans needed time for research and analysis before program implementation, Eisenhower appointed several advisory groups to study matters of national interest, including but not limited to housing.

21. D. Bradford Hunt, "How Did Public Housing Survive the 1950s?" *Journal of Policy History* 17, no. 2 (2005): 193–216; Nell MacNeil, "Cole Announces Overhaul Of U.S. Housing Programs," *WP*, June 28, 1953, R3; Anthony Leviero, "Eisenhower Picks 21 Housing Aides," *NYT*, September 13, 1953, 1.

22. This subcommittee did not represent the first federally organized study group on slum prevention. In 1947, for example, in the lead-up to the national housing act, housing officials had commissioned analyses of conservation with an eye toward policy recommendations. Slum removal was perceived as a more urgent priority, however, and conservation set aside. See Memo, August 22, 1947, Earl von Storch to Raymond M. Foley, Box 9, Subject Correspondence Files, Raymond M. Foley Administrator HHFA, 1947–53, Folder Report of Activities Urban Studies Staff, 2, HUDR.

23. Howard Gillette Jr., "Assessing James Rouse's Role in American City Planning," *JAPA* 65 no. 2 (1999): 150–167; Marc Weiss and John Metzger, "The American Real Estate Industry and the Origins of Neighborhood Conservation," in *Proceedings of the Fifth National Conference on American Planning History*, ed. Laurence C. Gerckens (Hilliard: SACRPH, 1994); *Lead the Way to a "Baltimore Plan" for Your Community* (Wilmette, IL: Encyclopaedia Britannica Films, 1953), VF NAC 543 N; Ralph Johnson, Huntington Williams, and Roy O. McCaldin, "The Quality of Housing 'before' and 'after' Rehabilitation," *American Journal of Public Health Nations Health* 45, no. 2 (1955): 189–196. A generation earlier, "rehabilitation" had referred to total demolition and rebuilding; by the 1950s, its technical meaning in the urban professions had shifted closer to conservation.

24. As was "redevelopment" as opposed to "slum clearance." The controversies of clearance were such that many urban professionals preferred a shift of vocabulary in the hopes of minimizing negative citizen response. In Greenville, South Carolina, for example, the local real estate board proposed "that the words 'slum clearance' be dropped and 'Neighborhood Redevelopment' substituted when referring to the elimination of sub-standard homes and completely 'run down' neighborhoods." Greenville Real Estate Board, *Plan of Neighborhood Reclamation and Redevelopment* (Greenville, SC: The Board, 1952), VF NAC 1613.1g27 Gre.

25. "Appendix 2: Report of the Subcommittee on Urban Redevelopment, Rehabilitation, and Conservation," in President's Advisory Committee on Government Housing Policies and Programs, *Report to the President* (Washington,

DC: n.p., 1953). Although the committee documented ongoing slum prevention activities across the United States, Chicago served as its primary point of reference, an emphasis seconded in letters of support for members' policy proposals, for example, from John Searles Jr., NAHO Redevelopment Section chair, who praised Illinois' 1953 law and called for its possible expansion. Letter from John Searles Jr., to Subcommittee on Urban Redevelopment, Rehabilitation and Conservation, October 25, 1953, Box 769, Program Files, 1940–65, President's Advisory Committee on Housing Policies and Programs, 1953–54, Folder Presidential Advisory Commission — Subcommittee on Urban Redevelopment, Rehabilitation and Conservation, Material Prepared for Subcommittee (Folder 2 of 2), HUDR.

26. Jack M. Siegel and C. William Brooks, *Slum Prevention through Conservation and Rehabilitation* (Washington, DC: SURRC, 1953).

27. Ibid., 7.

28. President's Advisory Committee on Government Housing Policies and Programs, *Report to the President*. In this way, they echoed prior observers on the inadequate implementation of programs for soil conservation, charging "damage is undoubtedly spreading faster than control measures are being applied." Beeman and Pritchard, *A Green and Permanent Land*, 33.

29. Siegel and Brooks, *Slum Prevention through Conservation and Rehabilitation*, 53.

30. "Comments: Conversion Control and Neighborhood Conservation," *Northwestern University Law Review* 48 (1953): 599; Siegel and Brooks, *Slum Prevention through Conservation and Rehabilitation*, 4.

31. Paul M. Herron, "Easier Home Purchasing Urged by Ike," *WP*, January 26, 1954, 1. Explicitly to be avoided, by contrast, were policies that "would make our citizens increasingly dependent upon the Federal government to supply their housing needs." "Text of Message on Housing Aid," *LAT*, January 26, 1954, 13.

32. Robert Young, "Eisenhower Receives Report on U.S. Housing Goals," *CDT*, December 16, 1953, 13; Richard E. Mooney, "PHA and Realty Boards Warm Up Housing Battle," *WPTH*, October 17, 1954, R3; "Housing Proposals May Be Combined in One Law," *WPTH*, March 20, 1955, G9; Clayton Knowles, "Democrats Score Housing Program: Eisenhower Plan Falls 'Far Short' of One Taft Backed, Congressmen Declare," *NYT*, January 27, 1954, 18; "Aid of Democrats on Housing Urged," *NYT*, December 27, 1953, R1.

33. NAREB, *A Primer on Rehabilitation under Local Law Enforcement* (Washington, DC: NAREB, 1953), no page. Ruth Berman, "Neighborhood Conservation," *FHLBR* 12 (1946): 171–175, offers a lengthy explanation of the appeal and the limits of the focus on neighborhood associations. Other real estate industry support for conservation is expressed in James Follin, *Putting the Master Plan into Action* (Washington, DC: HHFA, 1955), VF NAC 1613.5 F; NAREB,

Build America Better Council, *Neighborhood Conservation* (Washington, DC: NAREB, 1953); NAHB, *A New Face for America* (Washington, DC: NAHB, 1953); "U.S. Opens Nation-Wide Home Improvement Drive," *CD*, January 28, 1956, 3; "Rehabilitation Projects' Benefits Are Disclosed," *LAT*, February 17, 1952, E5; NAREB, *Recommendations on Rehabilitation by the Committee on Rehabilitation* (Washington, DC: NAREB, 1952), VF NAC 1613 NAREB, 2, 5; ULI, *Neighborhood Reclamation* (Washington, DC: NAREB, 1950), VF NAC 1613 U; "Neighborhood Conservation as Proposed by the Build America Better Council, National Association of Real Estate Boards," and "Tentative Draft—Not for Publication, Proposed State Enabling Act for Creation of Municipal Conservation Authorities embodying the proposal for Rehabilitation and Elimination of Slums and Blight of the Build America Better Council," Box 769, Program Files, 1940–65, President's Advisory Committee on Housing Policies and Programs, 1953–54, Folder Presidential Advisory Commission—Subcommittee on Urban Redevelopment, Rehabilitation and Conservation, Material Prepared for Subcommittee (Folder 1 of 2), HUDR.

34. From their earliest studies, scholars had identified churches as important clues to the ecology of any city, a research theme that had endured for several decades to witness the challenge these institutions faced of remaining in place as the neighborhoods around them changed. Henry Allen Bullock, "The Urbanization of the Negro Church in Detroit" (PhD diss., University of Michigan, 1935); A. E. Holt, "The Ecological Approach to the Church," *AJS* 33, no. 1 (1927): 72–79; William Harlan, "An Ecological Study of Four Lincoln Churches" (MA thesis, University of Nebraska, 1940); Lauris B. Whitman, "An Ecological Study of the Rural Churches in Four Pennsylvania Counties" (PhD diss., Pennsylvania State University, 1953); Wilford E. Smith, *An Ecological Study of L.D.S. Orthodoxy* (Provo, UT: Brigham Young University, 1958); David Fosselman, *Transitions in the Development of a Downtown Parish: A Study of Adaptations to Ecological Change in St. Patrick's Parish, Washington, D.C.* (Washington, DC: Catholic University of America Press, 1952).

35. Action Inc., *Quaker 'Self-Help' Rehabilitation Program Philadelphia* (New York: Action Inc., 1956), VF NAC 1433.4 AC; Mary Roche, "An Antibiotic for the Slum," *NYT*, October 25, 1953, SM14. In Detroit, the Jewish Community Council of Detroit, Detroit Council of Churches, Archdiocese of Detroit, and National Council of Catholic Women were among the members of the Detroit Commission for Neighborhood Conservation.

36. NCCC, *The Pastor and the Conservation of his Parish* (Washington, DC: NCCC, 1952), 15; Testimony of Right Reverend Monsignor John O'Grady, Secretary, National Conference of Catholic Charities on 83-H1442-7, Monday, March 15, 1954, 645.

37. Remarks of Right Reverend Monsignor John O'Grady, Secretary of the National Conference of Catholic Charities. Cited in "Citizens in Urban Renewal" (Chicago: Renewal Division of Sears, Roebuck, and Company, 1959), VF NAC 1613.4 S. O'Grady had participated in the 1951 meeting in Detroit of the National Conference of Catholic Charities on conservation. Federal officials would actively encourage religious institutions' continued participation in city revitalization after the pasage of the 1954 Housing Act, noting how "churches located in the neighborhood can often play a key role in interpreting the program to neighborhood families with whom they have had long association." HHFA, *How Localities Can Develop a Workable Program for Urban Renewal* (Washington, DC: HHFA, December 1956), VF NAC 1613.5 U, 11. On other religious institutions and renewal, see Mel Ravitz, "The Church's Stake in Conserving Communities," *The City Church* (May–June 1961); Donald D. Frank, *Harlem Park Stewardship* (n. l: National Conference of Christians and Jews Inc., 1958), VF NAC 1434.4827 Bal; Flora Y. Hatcher, "A Promise to the People—Urban Renewal," *World Outlook*, 19, no. 6 (1959): 27–29; Mel Ravitz, "10 Ways to Help in Urban Renewal," *National Lutheran* (May–June 1958); George Younger, *The Church and Urban Renewal* (Philadelphia: J. B. Lippincott, 1965).

38. Certainly, Follin's predecessor Nathaniel Keith had expressed some interest in rehabilitation; for example, in Nathaniel Keith, *The Federal Picture in Slum Clearance and Urban Redevelopment—Present and Future* (Washington, DC: HHFA, 1953), VF NAC 1613 K, but having worked on reconditioning at the Federal Home Loan Bank Board in the 1930s, and at the Conservation Division of the Defense Production Administration in the 1940s, Follin was among the public officials who had campaigned to make rehabilitation and conservation more central in U.S. housing policy. In the lead up to the 1954 Housing Act, he lectured widely on integrated approaches to city improvement. Private communications underscored this commitment, for example, in a letter to the Subcommittee on Urban Redevelopment, Rehabilitation, and Conservation in which he called for a "Shift of Emphasis to Urban Conservation" that he suggested could easily be achieved with changes to extant urban policy. In short, "the Title I program should be modified to become and to be designated an 'Urban Conservation Program.'" See James Follin, address to National Association of Real Estate Boards meeting in Los Angeles, November 11, 1953, and James Follin, "Redevelopment and Rehabilitation: Complimentary Approaches to Better Housing and Better Neighborhoods" address to the NAHO meeting, in Milwaukee, October 16, 1953, both in Box 769, Program Files, 1940–65, President's Advisory Committee on Housing Policies and Programs, 1953–54, Folder Presidential Advisory Commission—Subcommittee on Urban Redevelopment, Rehabilitation and Conservation, Material Prepared for Subcommittee (Folder 2 of 2), HUDR; To James Rouse from J. W.

Follin, Re: Reducing the Cost of Urban Redevelopment Projects, November 19, 1953, Box 769, Program Files, 1940–65, President's Advisory Committee on Housing Policies and Programs, 1953–54, Folder Presidential Advisory Commission — Subcommittee on Urban Redevelopment, Rehabilitation and Conservation, Material Prepared for Subcommittee (Folder 1 of 2), 1, HUDR. The support among Follin's successors is documented in Richard L. Steiner, "Conservation of Middle-Aged Areas," in *Planning 1952*, ed. ASPO (Chicago: ASPO, 1952); Letter from Richard Steiner to Albert Cole, August 7, 1953, Box 762, Program Files, 1940–65, President's Advisory Committee on Housing Policies and Programs, 1953–54, Folder Presidential Advisory Commission — Material from Public Interest Groups, HUDR; David Walker, "Urban Redevelopment," *JH* (March 1947); and William L. Slayton, "Urban Redevelopment Short of Clearance," in Urban Redevelopment, ed. William Slayton (Chicago: University of Chicago Press, 1953).

39. This three-part scheme for urban lands was not a completely new proposition. One report from the Community Appraisal Study had used identical language in its assessment of Hyde Park–Kenwood, delimiting areas for "Redevelopment — in current need of replacement; Rehabilitation — major repairs, with planned life up to 20 years after rehabilitation; Conservation — planned life up to 20 years; Stable — planned life of 20 years or more." HPKCC, *A Report to the Community* (Chicago: HPKCC, 1951), 20. Even earlier, Robert Mitchell's original proposal for the Woodlawn conservation project had envisioned "rapid elimination of those structures which are harmful to their occupants, unprofitable to their owners, or by their condition a hazard to the desirability of the surrounding area; short-term steps to improve maintenance and management of others whose plan or condition would not justify a prolonged life; modernization and physical improvement of structures which are worth the added investment," although this plan for a comprehensive and continuous program of neighborhood rebuilding had been scaled back on implementation. Robert Mitchell, "Woodlawn: A Program of Neighborhood Conservation (Part 1: Objectives — Preliminary Draft)," 32. In Folder Mitchell, Bob, Box 12, Interoffice Correspondence, Regional Correspondence, Correspondence by Person Records of the Office of the Director, Records of the Urban Section, November 1941–June 1943, NRPBR.

40. The renewal program would be subsequently revised, but this legislation provided its foundation. Jon Teaford, "Urban Renewal and Its Aftermath," *HPD* 11:2 (2000).

41. Mark Gelfand, *A Nation of Cities: The Federal Government and Urban America, 1933–1965* (New York: Oxford University Press, 1975).

42. Miles Colean, *Renewing Our Cities* (New York: Twentieth Century Fund, 1953), 28n25. See Volker Welter and James Lawson, eds., *The City after Patrick Geddes* (Bern: Peter Lang AG, 2000); Charles Adams, "Patrick Geddes — Botanist and Human Ecologist," *Ecology* 26, no. 1 (1945): 103–104.

43. Lewis Mumford, *City Development: Studies in Disintegration and Renewal* (London: Secker & Warburg, 1946).

44. CPC, *Master Plan of Residential Land Use of Chicago* (Chicago: CPC, 1943), 67.

45. Homer Hoyt, "Urban Decentralization," *JLPUE*, 16, no. 3 (1940): 270, 274.

46. J. W. Toumey, "Re-shaping Our Forest Policy," *SM* 12, no. 1 (1921): 18–35; Henry S. Graves, "The Duty of Scientific Men in Conservation," *Science* 53, no. 1379 (1921): 505–509; W. J. McGee, "Current Progress in Conservation Work," *Science* 29, no. 743 (1909): 490–496; "Science and Service," *SM* 20, no. 2 (1925): 221–224; T. H. Frison, "Advances in the Natural Renewable Resources Program of Illinois," *Transactions of the Illinois State Academy of Science* 31 (1938): 19–34; Raphael Zon et al., *Conservation of Renewable Natural Resources* (Philadelphia: University of Pennsylvania Press, 1941); E. M. Dahlberg, *Conservation of Renewable Resources* (Appleton, WI: C. C. Nelson, 1939); William Van Dersal and Edward Graham, *The Land Renewed* (New York: Oxford University Press, 1946); U.S. Chamber of Commerce, *Our Renewable Resources Can Be Sustained, a Symposium* (Washington, DC: The Chamber, 1949); Charles Adams, "The Responsibilities of Governments for the Conservation of Renewable Natural Resources as a Phase of Human Ecology," in *Proceedings of the Inter-American Conference on the Conservation of Renewable Resources* (Washington, DC: U.S. State Department, 1949).

47. Colean, *Renewing Our Cities*, 42.

48. Ibid.

49. Ibid., 6, 69.

50. Ibid.

51. HHFA, *How Localities Can Develop a Workable Program for Urban Renewal* (Washington, DC: HHFA, October 1, 1954), VF NAC 1613.5 U, 1.

52. "Progress Report, Subcommittee on Urban Redevelopment, Rehabilitation and Conservation," October 29, 1953, Box 762, Program Files, 1940–65, President's Advisory Committee on Housing Policies and Programs, 1953–54, Folder Presidential Advisory Commission — Minutes of Meetings, 7, HUDR; S. Howard Evans, HHFA Division of Slum Clearance and Urban Redevelopment, "Record of First Meeting of Advisory Committee on Urban Renewal Held September 9, 1954, at U.S. Chamber of Commerce Building, Washington DC," 3, Box 625, Program Files, 1940–65, "HHFA Archives" 4-Program Planning and Administration 4/21/Title I (URA Program, General Information) — 4-21-1/UR Directories (1962–present), Folder 4–21, Title I (URA Program General Information), HUDR.

53. James Follin, *Slums and Blight — a Disease of Urban Life* (Washington, DC: URA, 1956), 3. Follin made his remarks in a 1955 lecture.

54. HHFA, *Aids to Your Community* (Washington: DC: HHFA, March 1958), VF NAC 1613.5 U, no page. The *Workable Program* was briefly referred to as a *Workable Plan*; before this, cities had to prepare "general community plans" for access to slum clearance funds.

55. HHFA, *How Localities Can Develop a Workable Program for Urban Renewal* (Washington, DC: HHFA, October 1, 1954), VF NAC 1613.5 U, 4.

56. In addition to many individual brochures were series, including "Rehabilitation Kits," "Local Public Agency Letters," and "URA Bulletins." Supplemental to the release of such formal guides were bibliographies with recommended readings for local public agencies and citizens groups. For example, an undated revision to HHFA, *Slum Clearance under the Housing Act of 1949: A Preliminary Explanatory Statement to American Cities* (originally published in 1949, the revision probably from 1959 or 1960) suggested readings, including Colean, *Renewing Our Cities*; Herbert Thelen, *Dynamics of Groups at Work* (Chicago: University of Chicago Press, 1954); Bettie Sarchet and Eugene Wheeler, "Behind Neighborhood Plans: Citizens at Work," *JAIP* 24, no. 3 (1958); William Brussat, *Citizens Organizations for Neighborhood Conservation* (Chicago: NAHRO, 1957); ACTION, *Citizen Organizations for Community Improvement* (New York: ACTION, 1957); Julia Abramson, *A Neighborhood Finds Itself* (New York: Harper, 1959); Martin Millspaugh and Gurney Breckenfeld, *The Human Side of Urban Renewal* (Baltimore: Fight-Blight, 1958). See Box 762, Program Files, 1940–65, "HHFA Archives," 4-Program Planning and Administration 4-20-6 Staff Studies and Surveys (URA) — 4-20-20 URA publications (general). Folder URA Publications (general), HUDR.

57. Colean, *Renewing Our Cities*, 5.

58. HHFA, *How Localities Can Develop a Workable Program for Urban Renewal* (Washington, DC: HHFA Agency, October 1, 1954), VF NAC 1613.5 U, 3.

59. The office's first head was Henry E. Price, former secretary of the Joint Committee on Appraisal and Mortgage Analysis; upon his departure to run the renewal program in Cleveland, he was replaced by Leonard Czarniecki, the former executive director of the Committee for Neighborhood Conservation in Detroit.

60. The Essay on Sources points to several cities' work in this area. Chicago officials were explicit that Hoyt's master residential plan provided the foundation of their *Workable Program* submission. See City of Chicago, Office of the Mayor, *Application for the Certification of the Workable Program of the City of Chicago, Prepared by the Housing and Redevelopment Coordinator* (Chicago: Office of the Mayor, 1954).

61. DHC, *The Detroit Plan: A Program for Blight Elimination* (Detroit: Office of the DHC, 1945). For a broader history of Detroit in this period, see

Thomas Sugrue, *The Origins of the Urban Crisis* (Princeton, NJ: Princeton University Press, 1996).

62. Memo to Executive Committee from Helen Fassett, Chairwoman, Planning Committee [of Detroit Committee for Neighborhood Conservation and Improved Housing], December 28, 1953, revised January 11, 1954, 1, Box 24, No Folder, DCPCR.

63. Parkins, *Neighborhood Conservation*, 1.

64. DCNC, *Notes on the Background and Objectives* (Detroit: DCNC, 1956), VF NAC 1613.1g27 Det, 2; DCNC, *Detroit's Neighborhood Conservation and Improved Housing Program* (Detroit: DCNC, 1955), VF NAC 1613.1g27 Det, 3.

65. Parkins, *Neighborhood Conservation*, 4; DCNC, *Detroit's Neighborhood Conservation and Improved Housing Program*, 1. The definition of the various areas were created by staff from the Department of Health, the Department of Buildings and Safety Engineering, and the Plan Commission, as well as the Conservation Committee. See Leonard J. Czarniecki, *The City of Detroit Program for the Elimination and Prevention of Blight* (Ann Arbor: University of Michigan, 1955) VF NAC 1613.1g27 Det.

66. DCPC, *A Ten-Year Investment and Program to Eliminate Deterioration and Prevent Blight and Slums in Detroit's 53 Middle-Aged Neighborhoods* (Detroit: DCPC, 1955), VF NAC 1613.1g27 Det.

67. Cited in Letter from Joseph Molner to the Honorable Common Council of Detroit, September 23, 1955, 2 attached to DCNC, *Reports Subsequent to Reviews to April and October 1955* (Detroit: DCNC, 1955–1956), VF NAC 1613.1g27 Det.

68. See, for example, Providence Redevelopment Agency, *The Scope of the Providence Housing Problem and the Role of Replanning, Urban Redevelopment, Rehabilitation and Neighborhood Conservation in Blight Elimination and Slum Prevention* (Providence, RI: Providence Redevelopment Agency, 1953); City Planning Board of Saint Paul, *New Life in Saint Paul* (St. Paul, MN: City Planning Board of Saint Paul, 1956), VF NAC 1613.5g27 St P, and City Planning Board of St. Paul, *Proposed Renewal Areas 1957* (St. Paul, MN: St. Paul City Planning Board, 1957), no page, VF NAC 1613.5g27 St P.

69. CPC, *Ten Square Miles of Chicago* (Chicago: CPC, 1948), 8.

70. CCB, "Checklist for Neighborhood Analysis" (Chicago: CCB, n.d.), 1, lp.lpccc.dur.0019 LPNDC.

71. Robert Mitchell, "National Objectives for Housing and Urban Renewal," Exhibit 1 in President's Advisory Committee on Government Housing Polices and Programs, "Appendix 2: Report of the Subcommittee on Urban Redevelopment, Rehabilitation, and Conservation," 134. " 'Blight' as designated in Detroit, At-

lanta, London and Bombay would not refer to identical conditions," planning staff in Detroit later observed, because "it is a relative matter." DCPC, *Renewal and Revenue: An Evaluation of the Urban Renewal Program in Detroit* (Detroit: DCPC, 1962), 8n1.

72. Philadelphia Office of the Development Coordinator, *Partnership for Renewal* (Philadelphia: Philadelphia Office of the Development Coordinator, 1960), VF NAC 1613.5g27 Phi; Portland City Planning Board, *Neighborhood Conservation Check List* (Portland: Portland City Planning Board, 1946), VF NAC 544g27 Por. Detroit was an especially interesting case insofar as city staff there employed multiple survey instruments, designing ones that were locally specific and also using more established scales. See Czarniecki, *The City of Detroit Program for the Elimination and Prevention of Blight*; DCNC, *Neighborhood Conservation — a Challenge to Better Living* (Detroit: DCNC, 1955) VF NAC 1613.1g27 Det, 3; Parkins, *Neighborhood Conservation*, 4.

73. Housing Authority of the City of Milwaukee, "Development and Redevelopment Studies, Quality of Dwelling Units, Health Department Field Survey 1948 to 1950, APHA Standards," in "Approaches to Urban Renewal in Several Cities" *Urban Renewal Bulletin 1*, February 1953, Box 623, HHFA Archives, Program Planning and Administration 40-20-20B/URA Bulletins — 4-20-20F/ URA Technical Memorandums, HUDR; Tacoma City Planning Commission, *Renewal Areas* (Tacoma, WA: Tacoma City Planning Commission, 1954), VF NAC 1613.5g27 Tac. Based on this work the Urban Renewal Administration would revise its own list of categories in later years. By the early 1960s, there were four and then five categories: "conservation areas needing no renewal treatment," "conservation areas needing little renewal treatment," "rehabilitation areas," and "clearance and redevelopment areas"; followed by "conservation areas needing no renewal treatment," "conservation areas needing little renewal treatment," "conservation areas needing extensive renewal treatment," "reconditioning areas," and "clearance and redevelopment areas." Box 624 Program Files, 1940–65, "HHFA Archives" 4/Program Planning and Administration / 4-20-20G URA Notes — 4-20-20J/Rehabilitation Kit. Folder Rehabilitation Kit 1, HUDR; HHFA, *Workable Program for Community Improvement* (Washington, DC: HHFA, 1961), VF NAC 1613.5 U.

74. Untitled document, Box 13, Records of the Federal Home Loan Bank System, Federal Home Building Service Plan, Program Subject Files, 1936–1942, Folder Pamphlets and Literature on Housing (Folder 1), 15, FHLBBR; HHFA, *How Localities Can Develop a Workable Program for Urban Renewal* (Washington, DC: HHFA, October 1954).

75. HHFA, *Urban Renewal: What It Is* (Washington: DC, HHFA, 1955), VF NAC 1613.5 U; HHFA, *Urban Renewal: What It Is . . . Teamwork by Citizens and Government to End Community Slums and Blight* (Washington: DC, HHFA,

1957), VF NAC 1613.5 U; HHFA, *An Introduction to Urban Renewal as Authorized by the Housing Act of 1954* (Washington: DC, HHFA, 1954), VF NAC 1613.5 U.

76. HHFA, *How Localities Can Develop a Workable Program for Urban Renewal* (Washington, DC: HHFA, December 1956), 10–11.

77. HHFA and URA, *Replacing Blight with Good Homes: FHA's Section 220 Mortgage Insurance for Urban Renewal* (Washington, DC: FHA, 1955), VF NAC 1613.5 U; HHFA Agency, *Urban Renewal: What It Is* (Washington, DC: US HHFA, multiple dates), VF NAC 1613.5 U; HHFA, *Aids to Your Community* (Washington, DC: HHFA, March 1958); URA, *Rehabilitation Kit No. 2* (Washington, DC: URA, 1959); URA, *Neighborhood Organization in Conservation Areas* (Washington, DC: URA, 1961).That citizen participation was anticipated to be a program without end is apparent in HHFA, *How Localities Can Develop a Workable Program for Urban Renewal* (Washington, DC: HHFA, October 1954); and URA, *Workable Program for Community Improvement: Answers on Citizen Participation*. Program Guide 7 (Washington, DC: URA, 1964).

78. HHFA, *The Workable Program: What It Is* (Washington, DC: HHFA, October 1957), VF NAC 1613.5 U 1957, 5.

79. San Francisco Planning and Housing Association, *A Report to the Community, Number 1* (San Francisco: San Francisco Planning and Housing Association, February 1957), 2–3.

80. *Summary Proceedings of Working Conference on Citizen Participation in Neighborhood Conservation and Rehabilitation* (Pittsburgh: ACTION-Housing, 1958), VF NAC 1613.5 Al, 20–21. Although a few local public agencies would argue that an understanding of the urban renewal program "cannot be communicated easily" to the broader public, most were confident that the basic outlines of the new approach to city improvement could in fact be taught. Richard H. Leach, "The Federal Urban Renewal Program: A Ten-Year Critique," *LCP* 25, no. 4 (1960): 780.

81. San Antonio Department of City Planning, *Urban Renewal for San Antonio, a New Way to End Slums and Blight* (San Antonio, TX: San Antonio Department of City Planning, 1957), VF NAC 1613.5g27 San A. This teamwork theme is widely found, for example, in J. Paul Holland, Speech delivered to Lincoln Park Conservation Association, November 24, 1958, Accession No. lp.lpnc .hst.0013, LPNDC; DCNC, *The Answer to Six Questions That Will Help You to "Live Better in Your Neighborhood"* (Detroit: DCNC, 1958), VF NAC 1613.1g27 Det, 1; St. Louis, Neighborhood Rehabilitation Program, *City . . . Citizens . . . Teamwork for a Better St. Louis* (St. Louis: St. Louis Neighborhood Rehabilitation Program, no date), VF NAC 1613.1g27 Sail S.

82. In New York City, however, citizens rather than local officials initiated

the creation of a conservation program. See Citizens Housing and Planning Council of New York, *Report and Recommendations of the Committee on Neighborhood Conservation* (New York: Citizens Housing and Planning Council, 1954), VF NAC 1613.1g27 NY.

83. DJBC, *Something Specific for Neighborhood Improvement* (Detroit: DJBC, 1955), VF NAC 1613.1g27 Det. This was not the city's first effort to make this point. A brochure prepared during World War II for property owners in one neighborhood had praised the quality of their community and yet warned it would not stay the same without active work. "Remember, if *nothing* is done, your neighborhood will not stay just as it is today. It will depreciate in value, value in terms of dollars and value in terms of health and a good environment for living. But if you and others work at it, this area can be made one of the really pleasant places to live—the changes can start quite soon." DCPC, *A Suggested Neighborhood Improvement* (Detroit: DCPC, 1944), VF NAC 1613g27 Det, 3.

84. On Oakland, Cleveland, and Wilmington, see NAHRO Redevelopment Section, *Rehabilitation and Conservation Aspects of Urban Renewal* (Chicago: NAHRO, 1960). On Baltimore, see ACTION, *Fight-Blight Fund, Inc., Baltimore* (New York: ACTION, 1956), VF NAC 1431.8g27 Balt AC. On Boston, see Boston College Seminar Research Bureau, *Home Improvements You Can Make with No Tax Increase* (Boston: Greater Boston Chamber of Commerce, 1959), VF NAC 1433.4 B. On Los Angeles, see Los Angeles Office of Urban Renewal Coordination, *Procedure for Obtaining Sec. 221, Relocation Housing* (Los Angeles: Los Angeles Office of Urban Renewal Coordination, 1959) VF NAC 1433.8g27 LosA.

85. Vallejo Redevelopment Agency, *Vallejo's Urban Renewal Plan for Marina Vista* (Vallejo, CA: Vallejo Redevelopment Agency, 1960), VF NAC 1613.5g27 Mar.

86. Metropolitan Center for Neighborhood Renewal, *Progress Report on Citizen Education and Participation in Community Conservation* (Chicago: Metropolitan Center for Neighborhood Renewal, 1959); Beryl Harold Levy and Shirley Adelson Siegel, *Toward City Conservation* (New York: League of West Side Organizations, 1959); NAHRO, *Films Helpful to Housing and Redevelopment Agencies* (Chicago: NAHRO, 1955), VF NAC 543 N; DCNC, *For a Better Home and Neighborhood: Help Us to Help You* (Detroit: DCNC, 1956), VF NAC 1613.1g27 Det; Howard Hallman, *Education to Forward Urban Renewal in Philadelphia, a Report* (Philadelphia: Philadelphia Housing Association, 1959), VF NAC 1613.5g27 Phi; Cleveland Department of Urban Renewal and Housing, *Neighborhood Rehabilitation and Conservation Program* (Cleveland: Cleveland Department of Urban Renewal and Housing, n.d.), VF NAC 1613.1g27 Cle; "Triple Drive Aims at Community, Slum Rehabilitation," *Electrical World*, January 9, 1956, 75–86; DCNC, *A Summary of Detroit's Neighborhood Conservation and*

Improved Housing Program (Detroit: DCNC, 1958) VF NAC 1613.1g27 Det; NAHRO, *Citizens Working Here! A List of References on Community Organization for Urban Renewal* (Chicago: NAHRO, 1959), VF Z-NAC 1613.1 N; "Urban Renewal: Problems of Eliminating and Preventing Urban Deterioration," *Harvard Law Review* 72, no. 3 (1959): 504–542.

87. DCNC, *Neighborhood Conservation—a Challenge to Better Living*, 9. Ravitz, who received his PhD degree in sociology from University of Michigan in 1955, subsequently joined the faculty of Wayne State University while still serving on the planning staff, and his courses frequently involved students in the conservation work.

88. Parkins, *Neighborhood Conservation*, 1. In Detroit, the slogan for conservation was "to help you and your neighbors live better where you are." DCNC, *Notes on the Background and Objectives*, 2. The Committee on Neighborhood Conservation offered an extensive series of brochures, speakers bureau, slides, films, photographs, meeting agendas, and technical aids with detailed instructions and an invitation to call on several city agencies for further help. DCNC, *The Answer to Six Questions That Will Help You to "Live Better in Your Neighborhood,"* VF NAC 1613.1g27 Det, 1.

89. Parkins, *Neighborhood Conservation*, 23. The neighborhood, "area 6E" in planning commission speak, was bounded by Mack, Concord, Gratiot, Warren, and Van Dyke.

90. DCNC, *Neighborhood Conservation—a Challenge to Better Living*, 6, 9, 6.

91. Ibid., 6; DCNC, *Proposed Neighborhood Improvement Plan Boulevard—Gratiot—Mack Pilot Area* (Detroit: DCNC, 1958), VF NAC 1613.1g27; Mel Ravitz and Adelaide Dinwoodie, "Detroit Social Workers Mobilize Citizen Aid for Urban Renewal."

92. DCNC, *Neighborhood Conservation—a Challenge to Better Living*, 6–7.

93. Parkins, *Neighborhood Conservation*, 41.

94. Ibid., 41.

95. ACTION, *Summary of Urban Renewal Research Program* (New York: ACTION, 1954), 1; *Remarks of Albert Cole, Chairman, to the Board of Directors Meeting, January 7, 1965* (New York: ACTION, 1965), VF NAC 22 A, 2. In its 1953 report to President Eisenhower, the Advisory Committee on Government Housing Policies and Programs had recommended the creation of a nongovernmental group to stimulate rehabilitation and conservation efforts nationwide.

96. "ACTION Granted $250,000 for Housing Research," *AC* 70 (December 1955): 22; "ACTION Clinic Offers Renewal Checklist," *AC* 71 (1956): 135; Charles Forrest Palmer, "ACTION," *Rotarian* 94 (April 1959): 28–29; "ACTION elects more board members," *AC* 72 (March 1957): 27; M. Hickey, "Toward bet-

ter housing," *Ladies' Home Journal* 73 (May 1956): 35; "ACTION vs. slums," *Colliers* 137 (April 13, 1956): 88–89; Frederick Augustus Irving, "It's time for ACTION," *Parents' Magazine and Family Home Guide* 31 (March 1956): 54; Henry Lee, "Is Your Neighborhood Ready for Action?" *Better Homes and Gardens* 33 (October 1955): 42; "ACTION to Fight Slums and Neighborhood Blight," *AC* 69 (December 1954): 23; Action Inc., *The Advertising Council campaign for ACTION* (New York: ACTION, 1955), VF NAC 540 A.

97. Action Inc., *This Is Action* (New York, ACTION, 1960), VF NAC 1430.02 A; Action Inc., *Officers, Directors & Associates as of October 17, 1960* (New York: ACTION, 1960), VF NAC 1430.02 A; William W. Nash, *Residential Rehabilitation: Private Profits and Public Purposes* (New York: McGraw-Hill, 1959); Albert M. Cole, *Address to the Opening Conference of the American Council to Improve Our Neighborhoods* (Washington, DC: HHFA, 1954), VF NAC 1432.5 C; Martin Meyerson, *Face of the Metropolis* (New York: Random House, 1963); Edward C. Banfield, "The Politics of Metropolitan Area Organization," *Midwest Journal of Political Science* 1, no. 1 (1957): 77–91; Action, *Urban Renewal Research Program* (New York: ACTION, 1954); Edward C. Banfield and Morton Grodzins, *Government and Housing in Metropolitan Areas* (New York, McGraw-Hill, 1958); Action Inc. Research Program, *Inventory of Research on Impediments in the Housing Market* (Philadelphia: ACTION: 1956), Z-NAC 1431 Am 35; Burnham Kelly et al., *Design and Production of Houses* (New York: McGraw-Hill, 1959). A number of the research studies were published through McGraw-Hill, which created the American Council to Improve Our Neighborhoods series in housing and community development. There were other connections to the earlier Chicago work, for example, former Illinois Institute of Technology president and land clearance commision head Henry Heald also served on the board.

98. On the growing importance of nongovernmental actors in the planning process see Frederick J. Adams, "Changing Concepts of Planning," *American Journal of Economics and Sociology* 15, no. 3 (1956): 245–252.

99. As federal officials noted of ACTION's 1954 debut, "It was indicated that some of those things covered by the Urban Renewal Service were closely related to the proposed activities of the new organization, 'Action, Inc.' " — but ACTION had its sights on more expansive work. S. Howard Evans, HHFA Division of Slum Clearance and Urban Redevelopment, "Record of First Meeting of Advisory Committee on Urban Renewal Held September 9, 1954, at U.S. Chamber of Commerce Building, Washington DC," Box 625, Program Files, 1940–65, "HHFA Archives" 4-Program Planning and Administration 4/21/Title I (URA Program, General Information) — 4-21-1/UR Directories (1962–present), Folder 4–21 Title I (URA Program General Information), HUDR. Indeed, this organiza-

tion took on all six functions Mitchell had proposed for a Federal Urban Conservation Service in 1941: to "(1) Awaken municipal and individual interest in the implications of urban blight, pointing out the place of neighborhood conservation as part of an integrated local housing program; (2) Stimulate necessary local planning as basic to a successful conservation program; (3) Provide a clearing house through which national conservation experience will be gathered and demonstrated; (4) Develop procedures and techniques for conservation and the relation of these to other housing measures; (5) Install and spot-supervise these procedures and techniques when requested to do so by local agencies; (6) Aid in preparing and promoting conservation legislation." Robert Mitchell, "Memorandum: Establishment of a Federal Urban Conservation Service," 2. Appended to letter from Robert Mitchell to Charles Ascher, June 6, 1941, Interoffice Correspondence, Regional Correspondence, Correspondence by Person, Records of the Office of the Director, Records of the Office of the Urban Section, November 1941–June 1943, NRPBR.

100. This was one of several series the organization produced, including *Report from ACTION; ACTION Research Memorandum; News for Action; News & Views about the American City from the ACTION Library.*

101. "Announcements and Reports," *Professional Geographer* 9, no. 5 (1957): 16; "Triple Drive Aims at Community, Slum Rehabilitation," *Electrical World; The Newark Conference on the ACTION Program for the American City* (New York: Action, 1959); "ACTION, Urban Renewal Clinic at Massachusetts Institute of Technology," *AC* 72 (February 1957): 113; ACTION, *Fight-Blight Fund, Inc. Baltimore*; ACTION, *Quaker "Self-Help" Rehabilitation Program: Philadelphia*; ACTION, *Organization of Block Groups for Neighborhood Improvement: The Hyde Park–Kenwood Community Conference* (New York: ACTION, 1956); Action Inc., *Housing Court: Baltimore* (New York: Action Inc., 1956); ACTION, *Summary Proceedings of Working Conference on Citizen Participation in Neighborhood Conservation and Rehabilitation*; ACTION Inc., *Housing Code Provisions: A Reference Guide for Citizen Organizations* (New York: ACTION, 1956); *Man of Action* (1955) Transfilm Productions; "New Film Spurs City Planning and Slum Elimination," *AC* 71 (November 1956): 18; ACTION Inc., *Business Takes Action for Urban Renewal, a Leadership Guide* (New York: ACTION, 1957), VF NAC 1613.5 Am 35a; James E. Lash, *New Faces on Your Customers* (New York: ACTION, 1961), VF NAC 1613.5 L; Action Inc., *URDOA: Urban Development Division of ACTION Council for Better Cities with the Cooperation of NAHB and NAHRO presents Redevelopment Professional School* (New York: ACTION, n.d.).

102. See the "Essay on Sources" for a sampling of other materials on urban renewal published by the federal government and by ACTION, Inc.

103. That renewal mobilized such a coalition was how the movement was understood at the time as well, for example, by David Wallace, "The Conceptualizing of Urban Renewal," *University of Toronto Law Journal* 18, no. 3 (1968): 248–258; and Peter H. Rossi and Robert A. Dentler, *The Politics of Urban Renewal: The Chicago Findings* (New York: Free Press of Glencoe, 1961).

104. "Conservation of Dwellings: The Prevention of Blight," *Indiana Law Journal* (1953): 113; Saint Paul City Planning Board, *Selecting Urban Renewal Projects* (St. Paul, MN: City Planning Board, 1956), VF NAC 1613.5g27 St. P, no page; Letter from Hyman Levine to VH Scher, October 12, 1955, Box 2, Folder Oct–Dec 1955, GLCCR.

105. "The Federal Home Loan Bank Board and Its Agencies," 1940, Box 13 Records of the Federal Home Loan Bank System, Federal Home Building Service Plan, Program Subject Files, 1936–1942, Folder Organization, 13, FHLBB; NAREB Committee on Rehabilitation, *A Primer on Rehabilitation Under Local Law Enforcement*, 15.

106. MHPC, *Conservation* (Chicago: MHPC), 3: 325.

107. Rossi and Dentler, *The Politics of Urban Renewal*, 54.

108. Richard M. Kovak, "Urban Renewal Controversies," *PAR* 32, no. 4 (1972): 359–372.

109. Colean, *Renewing Our Cities*, 84–85.

110. Jack Eisen, "Integration Expansion Urged," *WPTH*, September 11, 1957, B1; NAHRO, *Selected References on Family and Business Relocation Caused by Urban Renewal and Other Public Improvements* (Chicago: NAHRO, 1960); Detroit Municipal Reference Library, *Tenant Relocation: A Selected List of References* (Detroit: The Library, 1949), VF Z NAC 1433.8; HHFA, *Relocation in Urban Areas: A Selected List of Books and Articles, 1951–1961* (Washington: HHFA, 1961); Margery T. Ware, *Displaced Families—Where Will They Go?* (Washington, WUL, 1955), VF NAC 1433.8g27 Was.

111. Rossi and Dentler, *The Politics of Urban Renewal*; Ethel Payne, "Hyde Park Land Grab Gets OK; Residents Plan Appeal," *CDD*, November 27, 1956, 1.

112. "Urban Renewal," *Harvard Law Review*.

113. LACRA, *The Bunker Hill Area, Los Angeles, California, Determination of Blight* (Los Angeles: LACRA, 1951), VF NAC 1613g27 Los A; Henry A. Babcock, *Report on the Feasibility of Redeveloping the Bunker Hill Area, Los Angeles* (Los Angeles: LACRA, 1951); LACRA, *Bunker Hill Urban Renewal Project: History and Program for Completion* (Los Angeles: LACRA, 1982).

114. Parkins, *Neighborhood Conservation*, iii; Alice W. Scharrer et al., "Measuring Residential Maintenance and Improvement" (especially "Appendix Table: Indications of Having Made Specific Residential Improvement and Maintenance, as Reported by 130 Homeowners in Two Detroit Neighborhoods"), Box 24, Folder

"Measuring Residential Maintenance and Improvement in Two Detroit Neighborhoods: An Evaluation Study 1960," DCPCR; DCPC, *Renewal and Revenue*.

115. HHFA, *How Localities Can Develop a Workable Program for Urban Renewal* (Washington, DC: HHFA, October 1954), 6; *HPD* 11, no. 2 (2000): entire issue.

116. "History of Lincoln Park Conservation Association," LPNDC.

117. Scharrer et al., "Measuring Residential Maintenance and Improvement," 57. FHA had recognized this soon after the creation of Title I first enabled homeowners to take out loans for property improvement: "Modernization is a high unit sale, and requires much education and information before Mr. & Mrs. John Home Owner make up their minds to go down to the bank and go into debt." "FHA Plan for 1935," Box 7, Records of the Federal Home Loan Bank System, Federal Home Building Service Plan, Program Subject Files 1936–42, Folder Federal Housing Administration — General Correspondence and Press Releases, FHLBBR.

118. DJBC, *Brighten Detroit and Brighten Your Future!* (Detroit: DJBC, 1955), VF NAC 1613.1g27 Det; "Good Maintenance Does Not Raise Taxes" (Neighborhood Conservation Information Bulletin, 1959), Box 8, Folder Neighborhood Conservation Projects (1) Mack-Concord (2) Eight Mile-Wyoming, DCPCR; NAHRO, *Rehabilitation and Conservation Aspects of Urban Renewal* (Chicago: NAHRO, 1960); John Ihlder, *Memorandum on Baltimore's Slum Improvement: Exaggerations Compel Response* (Washington, DC: U.S. National Capital Housing Authority, 1948); *Home Improvements You Can Make with No Tax Increase*; Buffalo Board of Redevelopment, *Neighborhood Conservation, Buffalo's Pilot Project* (Buffalo: Buffalo Board of Redevelopment, 1960), 6, VF NAC 1613.1g27 Buf.

119. Morton Neighborhood Council, *A Neighborhood Acts* (New York: National Federation of Settlements, 1957), VF NAC 1613.1g27 Phi.

120. See Box 1, Folder 1951–1952; Folder 1953, GLCCR. Indeed, the group had been so enthusiastic about conservation it had invited the three candidates for alderman to debate "How we can conserve our community." North Lawndale Citizens Council Newsletter, December 14, 1953, Box 1, Folder 1953, GLCCR.

121. Scharrer et al., "Measuring Residential Maintenance and Improvement," 58. These frustrations endured even though both cities (like others that included St. Louis and Portland) set aside extra monies to support this provision of the renewal program beyond what matching funds for federal assistance required.

122. Martin Anderson, *The Federal Bulldozer* (Cambridge, MA: MIT Press, 1964); John Staples, "Urban Renewal: A Comparative Study of Twenty-Two Cities, 1950–1906," *Western Political Quarterly* 23, no. 2 (1970): 294–304.

123. Alexander von Hoffman, "A Study in Contradictions: The Origins and

Legacy of the Housing Act of 1949," *HPD* 11, no. 2 (2000): 299–326. In Chicago, a related critique was offered—that the city was not sequencing redevelopment according to areas of most urgent need. See "Killing 3 Birds with One Stone," *CST* (July 21, 1960).

124. Leach, "The Federal Urban Renewal Program." Not only did such concurrent programs make it difficult for local public agencies to fit their *Workable Programs* into an overall city plan, but they also made it difficult to execute the *Workable Program* itself. In Chicago, for example, more citizens were to be relocated through highway construction than clearance projects, challenging the city's ability to rehouse those displaced by clearance for redevelopment alone, as noted by OHRC, *Relocation Reports* (Chicago: OHRC, Various Dates), VF NAC 1433.8g27 Chi.

125. Leach, "The Federal Urban Renewal Program," 783; "A Method of Estimating the Total Cost of Detroit's Neighborhood Conservation Program," 1959, 1, Box 8, Folder Method: "A Method of Estimating the Total Cost of Detroit's Neighborhood Conservation Program," 1959, DCPCR; Parkins, *Neighborhood Conservation*, 96–97.

126. von Hoffman, "A Study in Contradictions," 317.

127. Teaford, "Urban Renewal and Its Aftermath," 448.

128. Although the federal government provided funds to individual homeowners, "not eligible" for federal funding were "costs for community-wide citizen organizations." Untitled Undated Memo, Box 762, Program Files, 1940–65, "HHFA Archives" 4-Program Planning and Administration 4-20-3 Branch Memos, Division Memos, Slum Clearance Memos, Operations Memos, Commission Memos — 4-20-5/ Relocation (Publications), Folder 4-20-4 Conservation and Rehabilitation, HUDR. The Housing and Home Finance Agency's commissioned research on voluntary organizations "sponsored by the American Friends Service Committee, the Church of the Brethren, Saul Alinsky's various neighborhood organizations, the Peace Corps trainees in New York, and others" are documented in "Proposal for the Study of the Use of Volunteers in Rehabilitation," undated, Box 762, Program Files, 1940–65, "HHFA Archives" 4-Program Planning and Administration 4–20–6 Staff Studies and Surveys (URA) — 4-20-20 URA publications (general), Folder Staff Studies and Surveys (URA), HUDR.

129. OHRC, *Statement on the Enforcement and Conservation Program in the City of Chicago* (Chicago: OHRC, December 1953), 15, VF NAC 1613.5g27 Chi.

130. Siegel and Brooks, *Slum Prevention through Conservation and Rehabilitation*, 4; "Cole Cites New Markets for Homebuilders," *WP*, January 3, 1954, F13; Abe Gottlieb, "The Gray Areas of American Cities," *Commentary* 33 (April

4, 1962); H. N. Osgood and A. H. Zwerner, "Rehabilitation and Conservation," *LCP* 25, no. 4 (1960): 705–731; NAHRO Redevelopment Section, *Rehabilitation and Conservation Aspects of Urban Renewal*. This despite the longevity of Title I. Some local offices did keep files on "reputable firms" to assist as noted in "U.S. Opens Nation-Wide Home Improvement Drive," *CD*, January 28, 1956, 3. Even in Detroit, where ample assistance was available from local and federal sources, studies measuring participation in conservation found few people using 220 loans. In one conservation area in Detroit, for example, residents were "unaware as to what steps to take in obtaining bids from reliable contractors" for modernization projects. Surveyors also discovered an unexpected impediment, that "the use of credit is frowned upon by many residents." DHC, *The Vernor Springwells Conservation Project* (Detroit: DHC, 1962), VF NAC 1613.1g27 Det.

131. NAREB Build America Better Committee, *Conservation-Centered Renewal in the University City Area of West Philadelphia* (Washington, DC: NAREB Build America Better Committee, 1960), VF NAC 1613.5g27 Phi, 23. See also Leonard Czarniecki, "Rehabilitation and Conservation: Urban Renewal's Ugly Ducklings," *AJ* 34 (July 1966): 427–435.

132. Cited in D. E. Mackelmann, "Community Organization in Conservation: Speech to the Monthly Meeting of the Associated Clubs of Woodlawn, Inc. October 19, 1960, at the Woodlawn Regional Library," lp.lpccc.dur.0020 LPNDC.

133. Repeated in many U.S. cities, such outcomes are widely described in assessments of urban renewal both from the time and from later historians; see the "Essay on Sources."

134. "U.S. Opens Nation-Wide Home improvement Drive," 3.

135. Anderson, *The Federal Bulldozer*, 6.

136. James Scheuer, "To Renew Cities, Renew Their People," in *Planning 1964*, ed. American Society of Planning Officials (Chicago: ASPO, 1964), 21.

137. "Metropolis in a Mess: Special Report on New York City," *Newsweek*, no. 54, July 27, 1959, 29–31; American Institute of Architects, Georgia Chapter, et al., *The Mess We Live In* (Atlanta: Authors, 1963); Roger Starr, *The Living End: The City and Its Critics* (New York: Coward-McCann, 1966).

138. Gerald Lloyd, "Redirecting Government's Role in Urban Affairs," in *Planning 1964*, 11.

139. Jack Meltzer, "Urban Renewal: Its Potentialities and Its Limitations," in *Proceedings of Renewing Chicago in the Sixties* (Chicago: University of Chicago University College, 1961), 16.

140. "Back of the Yards Neighborhood Council," *Encyclopedia of Chicago* (Chicago: Newberry Library, 2005).

141. Charles Barresi, "Racial Transition in an Urban Neighborhood," *Growth and Change* 3, no. 3 (1972): 16–22.

142. Buffalo Board of Redevelopment, *Neighborhood Conservation*, 6.

143. John A. McDermott, *Eight Observations about Neighborhood Stabilization* (Philadelphia: PCHR, 1957), VF NAC 1613.1g27 Phi.

144. DCPC, *Report on Social Attitude Study of Conservation, Pilot Neighborhood 6E* (Detroit: DCNC, 1955), VF NAC 1613.1g27 Det, 2; Czarniecki, *The City of Detroit Program for the Elimination and Prevention of Blight*; George Henderson, *The Block Club Movement within the Detroit Tenth Police Precinct* (Detroit: DUL Community Services Department, 1962), VF NAC 1613.1g27 Det; Scharrer et al., "Measuring Residential Maintenance and Improvement."

145. DCPC, *Report on Social Attitude Study of Conservation, Pilot Neighborhood, 6E*, 3.

146. Henderson, *The Block Club Movement within the Detroit Tenth Police Precinct*, 4–5; Scharrer et al., "Measuring Residential Maintenance and Improvement," 63.

147. Scharrer et al., "Measuring Residential Maintenance and Improvement," 59.

148. PCHR, *Urban Renewal and Intergroup Relations* (Philadelphia: PCHR, 1959), 16, VF NAC 1613.5g27 Phi.

149. Although there were a few positive portrayals of clearance even in the African-American press (see for example, Albert Barnett, "Chicago Fights Its Slums: Puts New Meaning into Word 'Home,'" *CD,* July 28, 1951, 7), negative response to the 1947 Illinois redevelopment act and concerns about housing for minorities more generally prompted the creation of this bureau, described in To Raymond M. Foley From Frank S. Horne "Subject: Bi-weekly report—Racial Relations Service," November 26, 1948, Box 7, Subject Correspondence Files, Raymond M. Foley Administrator HHFA, 1947–53, Folder Report of Activities, Racial Relations Service (1 of 2), HUDR.

150. "Sengstacke Aids Fix-Up Campaign," *CDD*, February 22, 1956, 5.

151. Thomas Philpott, *The Slum and the Ghetto* (New York: Oxford University Press); Frankie V. Adams, "The Community-Wide Stake of Citizens in Urban Renewal," *PQ* 19, no. 1 (1958): 92–96.

152. "HANA Backs Weaver's Rap at Laxity in Civic Betterment," *NPC*, June 3, 1961, 3; "Morningside Area Slated for First Neighborhood Conservation District in City; Everett Reid Is Chief," *NPC*, June 3, 1961, 15; Thomas Johnson, "Mass Rally Protests Jamaica Urban Renewal Pilot Project," *NPC*, July 14, 1962, 1; Thomas Johnson, "Long Island . . . Inside Out," *NPC*, August 4, 1962, 5; St. Louis Land Clearance for Redevelopment Authority, *Facts about Urban Renewal in St. Louis* (St. Louis, MO: St. Louis Land Clearance for Redevelopment

Authority, 1964), VF NAC 1613.5g27 St.Lou 1963, no pages; Interview with Monroe F. Brewer, Chief Engineer, Housing Rehabilitation Project, St Louis, Missouri, December 13, 1956, cited in Parkins, *Neighborhood Conservation*, 55n3.

153. As Jack Siegel and C. William Brooks described in their report to the Presidential Advisory Subcommittee, the Urban League's "block beautiful" campaigns were "precursors of present block conservation efforts." (Siegel and Brooks, *Slum Prevention through Conservation and Rehabilitation*, 68.) Evidence of the support for slum prevention at other Urban League chapters is broadly apparent. In Warren, Ohio, for example, the Urban League directed a project on housing rehabilitation. In Manhattanville, New York, the local conservation association was headed by Reginald Johnson, associate director of the National Urban League. In Detroit, the city's Urban League helped to expand citizen participation in conservation by providing training to neighborhood groups. *Slum Clearance and Related Housing Problems: Hearings before the Subcommittee on Housing of the Committee on Banking and Currency, House of Representatives*, 85th Cong., 2nd sess. January 7, 8, and 9, 1958 (Washington, DC: Government Printing Office, 1958), 311; "Week-Long 'Clean-Up' Campaign a Success at Hamilton Grange," *NPC*, July 11, 1964, 11; Letter from Joseph Molner to the Honorable Common Council of Detroit, April 19, 1956, VF NAC 1613.1g27 Det, appended to DCNC, *Reports Subsequent to Reviews to April and October 1955*. Lester Granger, executive director of the National Urban League, served on the board of ACTION Inc., suggesting support echoed at the national level. As these League chapters promoted property conservation, its research department was expressing some new interest in human conservation, commissioning a study from Eli Ginzberg's Columbia group. Ginzberg, *The Negro Potential*.

154. Henderson, *The Block Club Movement within the Detroit Tenth Police Precinct*, 14–16. Henderson had received undergraduate and master's degrees in sociology from Wayne State University and worked on the conservation program in Detroit.

155. Ibid., 14–16.

156. Compare race-based statistics on relocation in URA, *Relocation from Urban Renewal Project Areas through December 1961* (Washington, DC: URA, 1961), with imagery in PCPC, *West Poplar Redevelopment Area Plan* (Philadelphia: PCPC, 1953), VF NAC 1613.5g27 Phil P 1953; PCPC, *For a Better Life: The Story of the East Poplar Housing Plan* (Philadelphia: PRA, 1953), VF NAC 1613g27 Phi; Los Angeles Office of Urban Renewal Coordination, *Procedure for Obtaining Sec. 221 Relocation Housing*, VF NAC 1433.8g27 LosA; Sacramento Redevelopment Agency, *Redevelopment Facts for Property Owners and Occupants of Project 3* (Sacramento, CA: Sacramento Redevelopment Agency, 1959), VF NAC 1613.5g27 Sac; PRA, *Information Booklet for Those Who Will Have to*

Move to Make Way for Improvement and Redevelopment of the East Poplar Area, Unit No. 3 (Philadelphia: PRA, 1953), VF NAC 1433.8g27 Phi; PRA, *The next steps in Morton* (Philadelphia: PRA, 1959), VF NAC 1613.5g27 Phi; PRA, *The next steps in Eastwick* (Philadelphia: PRA, 1959), VF NAC 1613.5g27; Pittsburgh Urban Renewal Authority, *What You Should Know about East Liberty Renewal* (Pittsburgh: Pittsburgh Urban Renewal Authority, 1960), VF NAC 1613.5g27 EasL; Citizens' Council on City Planning, *A Guide to Eastwick Redevelopment* (Philadelphia: Citizens' Council on City Planning, 1956), VF NAC 1613.5g27 Phi; Vallejo Redevelopment Agency, *Vallejo's Urban Renewal Plan for Marina Vista* (Vallejo, CA: Vallejo Redevelopment Agency, 1960), VF NAC 1613.5g27 Mar.

157. Siegel and Brooks, *Slum Prevention through Conservation and Rehabilitation*, 6; Parkins, *Neighborhood Conservation*, 96; HPKCC, *A Report to the Community*, 8; Guy Stuart, *Discriminating Risk* (Ithaca: Cornell University Press, 2003).

158. Letter from Charles Abrams to Albert Cole, August 5, 1953, Box 762, Program Files, 1940–65, President's Advisory Committee on Housing Policies and Programs, 1953–54, Folder Material from Citizens' Housing Groups, HUDR.

159. CCB, "Checklist for Neighborhood Analysis," 2, lp.lpccc.dur.0019 LPNDC.

160. National headquarters to Sears, Lawndale had been described as a middle-aged area a decade earlier by the *Chicago Land Use Survey*. It was one of the city's centers for Jewish residents during this period, with the survey estimating African American occupancy at a half percent. As the HOLC reported, "the poorer type Jewish family is concentrated in a few fairly-well defined sections of Chicago; in the Woodlawn section, area D-78, about 40% are families of this type. The entire Lawndale area is mostly made up of Jewish families of the class described and this area is now spilling over into the Albany Park, Rogers Park and West Rogers Park Sections of Chicago." HOLC Division of Research and Statistics, "Confidential Report of a Re-Survey of Metropolitan Chicago, IL," June 1940, Box 85, Records of the Federal Home Loan Bank Board, Home Owners Loan Corporation, Records Relating to the City Survey File, 1935–40, Illinois, Folder Chicago Re-Survey Report 2, 1 (5): 35, FHLBBR. This profile changed rapidly during the 1940s as the area faced "infiltration" from African American neighbors. Interracial integration was on the agenda from the early 1950s, as a summary of a meeting of the North Lawndale Citizens Council laid out. President Turner Trimble, who would be appointed president of the city's Association of Community Councils later that year, "defined the 'basic purpose' as the solution of the problems arising out the changing community. He stated the hope that we could become an interracial community able to maintain harmonious living and decent

standards. He felt it was not too much to believe that we could become a 'pilot community' of an interracial makeup. The problem was to stabilize the population changes." Memo M. Hyman to I. Gold, June 26, 1951, Box 1, Folder 1951–1952, GLCCR. An offshoot of this organization became the Greater Lawndale Conservation Commission, with a diversity of interests represented on its board: the presidents of several local banks, a reverend from a local Baptist church, an editor of the *Chicago Defender*, a rabbi from local Hebrew theological college, a writer from the *Jewish Daily Forward*, and a representative from Sears. "GLCCR Minutes of Meeting of February 23, 1954, prepared by E. Vrana," Box 1, Folder 1953, GLCCR. Further discussion of the history of this neighborhood and its experience with renewal is documented in Amanda Seligman, *Block by Block* (Chicago: University of Chicago Press, 2005).

161. As paraphrased by meeting minutes. "GLCCR Minutes of Meeting of February 23, 1954, prepared by E. Vrana."

162. Jessamine Cobb, "Community Area No 29 — North Lawndale," October 1955, Box 2, Folder Oct–Dec 1955, GLCCR; "GLCCR Minutes of Meeting of February 23, 1954, prepared by E. Vrana."

163. Ragland was well-known in real estate circles in Chicago and nationally. Letter from Robert Mitchell to Howard Haylett, March 29, 1943, In Folder National Association of Real Estate Boards, Box 20, Alphabetical Correspondence G-N, Records of the Office of the Director, Records of the Urban Section, November 1941–June 1943, NRPBR.

164. "Tenants and Property Owners Unite against This Hoax," July 1955, Box 2, Folder Jan–March 1955 GLCCR. (Note this document is misfiled.) As a reflection of the generally good race relations in the neighborhood, in both cases, the turmoil would be temporary. Because Cox and Dooley recognized that "The strong sentiment expressed here [by the whole group] was that the groups in opposition to the Conservation program" did not represent the larger community, they soon were convinced to take on roles as active participants in the program "GLCCR Minutes of Meeting of February 23, 1954, prepared by E. Vrana"; "Seat Officers for Greater Lawndale Unit," *CDT*, August 1, 1954, WA3. And consistent with the Greater Lawndale Conservation Commission's belief that the leafleters "did not represent the feelings of the Negro residents and property owners in the community, and that the community would support such a program and the people heading it up," efforts to incite protest backfired. "GLCCR Minutes of Meeting of February 23, 1954, prepared by E. Vrana." Vigilant representatives of Chicago's NAACP immediately sent a telegram to the Lawndale group informing them of the hoax: "Please be advised and inform persons attending this evening's meeting that the NAACP is NOT sponsoring tonight's meeting . . . Such approval can only be given by the executive committee of the Chicago branch NAACP . . . all NAACP is

seriously disturbed by the unauthorized use of this name and identification on either side of present controversy." Telegram from Cora Batton and Willoughby Abner NAACP to LC branch, Box 1, Folder Jan–November 1954, GLCCR. Yet, both episodes indicate the continued controversies of intergroup relations in cities across the United States as the renewal program unfolded.

165. McDermott, *Eight Observations about Neighborhood Stabilization*, 2.

166. Arnold Hirsch has documented the race relations service at HHFA assigned few staff to renewal projects. See Hirsch, "Searching for a 'Sound Negro Policy'" *HPD* (2001): 393–441; Leonard Blumberg, "Urban Rehabilitation and Problems of Human Relations," *PQ* 19, no. 1 (1958): 97–105.

167. NAHRO Redevelopment Section, *Rehabilitation and Conservation Aspects of Urban Renewal*; von Hoffman, "A Study in Contradictions," 318; Hugh O. Nourse, "The Economics of Urban Renewal," *LE* 42, no. 1 (1966) 65–74; William Widnall, "The Question Of the Soundness of the Present Federal Urban Renewal Program," *Congressional Digest*, 43, no. 4 (1964): 109–113; William Alonso, "The Historic and the Structural Theories of Urban Form: Their Implications for Urban Renewal," *LE* 40:2 (1964), 227–231; Jack Meltzer, "Urban Renewal"; Letter from T. V. Houser March 5, 1955, to W. Frederic Mosel, 3, Box 2, Folder Jan.–March 1955, GLCCR.

168. On positive press coverage, see the *Time* article "In Defense of Urban Renewal" and discussion in *Remarks of Albert Cole, Chairman, to the Board of Directors Meeting, January 7, 1965* (New York: ACTION, 1965), VF NAC 22 A.

169. NAHRO Redevelopment Section, *Rehabilitation and Conservation Aspects of Urban Renewal*, 22.

170. David Walker, "A New Pattern for Urban Renewal," *LCP* 25, no. 4 (1960): 633. "The urban renewal process will be moving into the suburbs before we are through with the central city," Walker predicted. "And then, unless we plan to abandon our cities, certainly before we are through with the suburbs, we will be back to renewing the central city, if, indeed, we have ever discontinued there." Ibid., 634.

171. Rossi and Dentler, *The Politics of Urban Renewal*; Adam Cohen and Elizabeth Taylor, *American Pharaoh* (Boston: Little, Brown, 2000); Audrey Probst, "The Community That Saved Itself," *Nation*, March 5, 1955; United States General Accounting Office, *Ineffective Administration Contributing to Unsatisfactory Progress in Rehabilitating the Hyde Park–Kenwood Urban Renewal Area, Chicago, Illinois* (Washington, DC: U.S. General Accounting Office, 1964). Residents of Chicago's Woodlawn neighborhood grew especially frustrated, organizing the Woodlawn Organization in response. See John Hall Fish, *Black Power / White Control* (Princeton, NJ: Princeton University Press, 1973).

172. John Sengstacke, "Are We Telling the Urban Renewal Story II," in *Pro-*

ceedings of Renewing Chicago in the Sixties; Anderson, *The Federal Bulldozer*; Charles Haar, *Between the Idea and the Reality* (Boston: Little, Brown, 1975), 16–17; John Dyckman, "National Planning in Urban Renewal: The Paper Moon in the Cardboard Sky," *JAIP* 26, no. 1 (1960): 49–59; NAHRO Redevelopment Section, *Rehabilitation and Conservation Aspects of Urban Renewal*, 26; CUL Community Services Department, *Urban Renewal and the Negro in Chicago* (Chicago: CUL, 1958); Martin Meyerson, "The Conscience of the City," *Daedalus* 97, no. 4 (1968).

Conclusion: From Ecology to System

1. See the "Essay on Sources" for an overview of the importance of "scientific" approaches within the urban professions.

2. National Resources Committee, *Our Cities: Their Role in the National Economy* (Washington, DC: NRC, 1937), 73.

3. Untitled Document, Box 13, Records of the Federal Home Loan Bank System, Federal Home Building Service Plan, Program Subject Files, 1936–1942, Folder Pamphlets and Literature on Housing (Folder 1), 13, FHLBB.

4. FHLBB, *Waverly: A Study in Neighborhood Conservation* (Washington, DC: FHLBB, 1940), viii.

5. FHA, *The Structure and Growth of Residential Neighborhoods in American Cities* (Washington, DC: Government Printing Office, 1939).

6. CPC, *Master Plan of Residential Land Use of Chicago* (Chicago: CPC, 1943), 67.

7. On the new environmentalism, see Carroll Pursell, ed., *From Conservation to Ecology* (New York: Crowell, 1973); Samuel Hays, *A History of Environmental Politics since 1945* (Pittsburgh, PA: University of Pittsburgh Press, 2000); Stephen Fox, *The American Conservation Movement: John Muir and His Legacy* (Madison: University of Wisconsin Press, 1986); Robert Gottlieb, *Forcing the Spring: The Transformation of the American Environmental Movement* (Washington, DC: Island Press, 1993); Alfred Crosby, "The Past and Present of Environmental History," *AHR* 100, no. 4 (1995): 1177–1189; Hal Rothman, *The Greening of a Nation? Environmentalism in the United States since 1945* (Fort Worth: Harcourt Brace College, 1998). However, Rothman sees conservation lasting far longer than his colleagues.

8. Scholars of paradigms often note a final burst of enthusiasm before one widely accepted worldview gives way to another. Texts such as Miles Colean, *Renewing Our Cities* (New York: Twentieth Century Fund, 1953); Jack Siegel and C. William Brooks, *Slum Prevention through Conservation and Rehabilitation* (Washington, DC: SURRC, 1953); and NAREB, *A Primer on Rehabilitation under*

Local Law Enforcement (Washington, DC: NAREB, 1953) can be interpreted in this light.

9. Maurice Parkins, *Neighborhood Conservation: A Pilot Study* (Detroit: DCPC, 1958), 52–53.

10. "To cure a disease, both effect and cause must be treated. If a person gets a skin rash from poison ivy in the back yard, it is not enough to treat the rash. The poison ivy must be removed." CCB, *The Chicago Conservation Program: A New Plan Whereby Private Citizens and Public Officials Work Together to Make Chicago a Better Place in Which to Live* (Chicago: CCB, 1956), VF NAC 1613.1g27 Chi, 7. Other Chicagoans similarly followed this rhetoric. Prominent among them was T. V. Houser, chief of Chicago-based Sears. As he explained in a 1957 book, "Neighborhood deterioration represents the same kind of need in the case of the cities that poor livestock breeds and poor crop yields represented a generation ago on the farms. Unquestionably, it is one of today's most urgent national problems, and our interests as businessmen and citizens coincide as they do in many urban areas." Houser, *Big Business and Human Values* (New York: McGraw-Hill, 1957), 2.

11. Cleveland Department of Urban Renewal and Housing, *Neighborhood Rehabilitation and Conservation Program, Cleveland, Ohio* (Cleveland, OH: Cleveland Department of Urban Renewal and Housing, n.d.), VF NAC 1613.1g27 Cle. As the Federal Home Loan Bank Board had put it: "The initial point of infection is a single neglected property. Like the one rotten apple in the barrel which may infect all the others, from such a property neighborhood blight may spread gradually throughout an area until decay is complete and only total demolition remains as the final solution." "Waverly," *FHLBR* 6 (1940): 330.

12. Cited in San Francisco Planning and Housing Association, *A Report to the Community, Number 1* (San Francisco: San Francisco Planning and Housing Association, February 1957), 2.

13. J. Martin Klotsche, *The Urban University and the Future of Our Cities* (New York: Harper and Row, 1966), 52; H. N. Osgood and A. H. Zwerner, "Rehabilitation and Conservation," *LCP* 25, no. 4 (1960): 73on 83; "Rutgers to Set Up Urban Aid Center," *CST*, August 9, 1960, 5; "Research Project Backed by Ford Gift of $750,000 — Other Grants Listed," *NYT*, July 22, 1959, 22; Virginia M. Esposito, ed., *Conscience and Community: The legacy of Paul Ylvisaker* (New York: Lang, 1999).

14. Ylvisaker's personal familiarity with natural resource conservation programs is detailed in Paul Ylvisaker, *Intergovernmental Relations at the Grass Roots: A Study of Blue Earth County, Minnesota, to 1946* (Minneapolis: University of Minnesota Press, 1956), and Paul Ylvisaker and Paul Bedard, *The Flagstaff Federal Sustained Yield Unit, Inter-university Case Program 37* (Syracuse, NY: Inter-University Case Program, 1955).

15. The universities were all land-grant institutions — Rutgers, Berkeley, Delaware, Illinois, Missouri, Oklahoma, Purdue, and Wisconsin. See Ford Foundation, *Urban Extension: A Report on Experimental Programs Assisted by the Ford Foundation* (New York: Ford Foundation 1966); *Proceedings of the Pittsburgh Urban Extension Conference* (Pittsburgh, PA: ACTION-Housing, 1962); William Pendleton, *Urban Studies and the University — The Ford Foundation Experience* (New York: Ford Foundation, 1974); National 4-H Foundation, *4-H in Urban Areas* (Washington, DC: National 4-H Foundation, 1964); Allegheny Council to Improve Our Neighborhoods-Housing, *A Report on the Pilot Program for Neighborhood Urban Extension* (Pittsburgh, PA: ACTION-Housing, 1964); ACTION-Housing, Inc., *Plan of Operations, Neighborhood Urban Extension* (Pittsburgh, PA: ACTION-Housing Inc., 1963); Allegheny Council to Improve Our Neighborhoods, *Neighborhood Urban Extension: A Report on the First Year, 1963* (Pittsburgh, PA: ACTION-Housing Inc., 1964); *Neighborhood Urban Extension — The Fifth Year and Summary* (Pittsburgh, PA: ACTION-Housing, 1970); Clifford Cham Jr., *The Neighborhood Church in Urban Extension* (Pittsburgh, PA: ACTION-Housing, 1964); "This month's feature: proposal for a youth conservation corps," *Congressional Digest* 39 (December 1960): 295–301; Osgood and Zwerner, "Rehabilitation and Conservation"; and Robert Weaver, "Civic Invention and Urban Change: Address at George Washington University, January 12, 1965," in George Washington University, *Man and Metro* (Washington, DC: George Washington University, 1965), 8. The federal Higher Education Act of 1965, which enabled new funds to flow to community service programs, enabled the expansion of such programs — but not on the scope of the Morrill Act.

16. That Clements's most sustained influences in urban contexts postdate the acceptance of his theories in natural resources planning was not unique. Henrika Kucklick explores a similar story of outdated ecological ideas influencing Homer Hoyt's work in Henrika Kuklick, "Chicago Sociology and Urban Planning Policy: Sociological Theory as Occupational Ideology," *Theory and Society* 9 (1980): 821–845.

17. On areas for rehabilitation and conservation as "grey areas," see Richard L. Steiner, *Homebuilders and Urban Renewal* (Washington, DC: URA, 1957), VF NAC 1613.5.

18. Examples of scrutiny were extremely rare and included (just before the passage of the Housing Act of 1954) Colean's comments on the problem of ecological analogies, and Isaacs and Siegel's reflections on how, despite many similarities, soil conservation might not be an ideal model for urban conservation. Reflections from the Ford Foundation and Robert Weaver on the ways in which a nationwide urban extension program would need to differ from the earlier agricultural one followed the Ford experiments.

19. I refer to work by Sarah Phillips, Randal Beeman, James Pritchard, and others; see the "Essay on Sources" for a survey of studies looking back at the conservation movement in theory versus implementation.

20. Compare, for example, C. W. Thornthwaite, "The Relation of Geography to Human Ecology," *EM* 10, no. 3 (1940): 343–348, and idem, "Operations Research in Agriculture," *Journal of the Operations Research Society of America* 1, no. 2 (1953): 33–38. On ecology's postwar transformations, see Peter Taylor, "Technocratic Optimism, H. T. Odum, and the Partial Transformation of Ecological Metaphor after World War 2," *JHB* 21 (1988): 213–244; and Sharon E. Kingsland, *The Evolution of American Ecology, 1890–2000* (Baltimore: Johns Hopkins University Press, 2005). Among urbanists the work of William Slayton, Harvey Perloff, Richard Meier, Paul Ylvisaker (and the Ford Foundation), and even President John F. Kennedy reflected this shift. See William Slayton, "Impact of the Community Renewal Program on Urban Renewal," Address to the 1964 Conference of the American Institute of Planners, Newark, New Jersey, August 18, 1964; Harvey Perloff et al., *Lessons from Urban Renewal (for NTITS)* (Los Angeles: University of California, 1972); Richard Meier, "A General Systems Party," *General Systems* 13 (1968): 209–212; Harry B. Wolfe and Martin L. Ernst, "Simulation Models and Urban Planning," in *Operations Research for Public Systems*, ed. Philip Morse (Cambridge, MA: MIT Press, 1967), 49–81; David Jardini, "Out of the Blue Yonder: The RAND Corporation's Diversification into Social Welfare Research, 1946–1968" (PhD diss., Carnegie Mellon University), 1996.

21. John Dyckman, "The Technological Obsolescence of Planning Practice," *JAIP* 27, no. 3 (1961): 244.

22. Leland M. Swanson and Glenn O. Johnson, eds., *The Cybernetic Approach to Urban Analysis* (Los Angeles: University of Southern California Graduate Program in City and Regional Planning, 1964); Richard Meier, "Significance of Research Developments, Urban Growth in a Communications Framework," in *Proceedings of the 1960 Annual Conference*, ed. AIP (Chicago: AIP, 1960); Donald F. Blumberg, "The City as a System," *Simulation* 17, no. 4 (1971): 155–167; Richard D. Duke, *Gaming-Simulation in Urban Research* (East Lansing: Michigan State University Institute for Community Development and Services, 1964).

23. Robert Wood, "To Government Science: The Good Urban Witch," in *Cybernetics and the Management of Large-Scale Systems*, ed. Edmond Dewan (New York: Spartan Books, 1969), 129; Emanuel Savas, "Cybernetics in City Hall," *Science* 168 (May 29, 1970): 1066–1071; Jennifer Light, *From Warfare to Welfare: Defense Intellectuals and Urban Problems in Cold War America* (Baltimore: Johns Hopkins University Press, 2003); Jardini, "Out of the Blue Yonder"; Charles Haar, *Between the Idea and the Reality* (Boston: Little, Brown, 1975); Abt

Associates, *A Study and Provision of Technical Assistance through Simulation for More Effective Citizen Participation in the Model Cities Program: Report on Tasks I and II: Game Development* (Cambridge, MA: Abt Associates, 1970). The renewal program persisted into the 1970s.

24. Historians of environmental science have described how systems ecology reframed some of Frederic Clements's theories; Eugene Odum's concept of "climax ecosystem" is a notable example. A similar phenomenon is apparent in urban theory and practice, where the idea of the urban life cycle, and indeed ecological concepts more generally, were refashioned within the systems tradition. On the persistence of life cycle ideas in urban theory, see Edgar Hoover and Raymond Vernon, *Anatomy of a Metropolis* (Cambridge, MA: Harvard University Press, 1959); Leo Schnore, "The City as a Social Organism," *Urban Affairs Review* 1, no. 1 (1966): 59–69. Discussions of the relevance of urban life cycles to program implementation for Community Renewal in Denver and for Model Cities in Lowell and Seattle are documented in Denver Planning Office, *Denver 1985: A Comprehensive Plan for Community Excellence* (Denver: Denver Planning Office, 1967); Marshall Kaplan, Gans and Kahn, *The Model Cities Program: A History and Analysis of the Planning Process in Three Cities: Atlanta, Georgia; Seattle, Washington; Dayton, Ohio* (Washington, DC: U.S. Department of Housing and Urban Development, 1969); Walter Schroeder with John Strongman, "Comparing Urban Dynamics with Lowell, Massachusetts." In *Urban Simulation: Models for Public Policy Analysis*, ed. Marshall Whithed and Robert Sarly (Leiden: Sijthoff, 1974). Certainly, there is evidence that a few individuals and institutions, especially in real estate, held fast to older interpretations of urban processes even in the face of these disciplinary trends. See, for example, Arthur Ring, *The Valuation of Real Estate* (Englewood Cliffs: Prentice-Hall, 1963), esp. 65–67; and subsequent editions of real estate texts from the American Institute of Real Estate Appraisers (e.g., pp. 192–193 of the 1966 printing of *The Appraisal of Real Estate* [Chicago: AIREA]) and from Homer Hoyt and Arthur Weimer (e.g., p. 308 of the 1966 printing of *Real Estate* [New York: Ronald Press]).

25. St. Louis, Neighborhood Rehabilitation Program, *City . . . Citizens . . . Teamwork for a Better St. Louis* (St. Louis: St. Louis Neighborhood Rehabilitation Program, n.d.), VF NAC 1613.1g27 Sail S, 10; Chesley Manly, "How Chicago Is Winning War against Slums," *CDT*, December 19, 1554, 20; HPKCC, *A Report to the Community* (Chicago: HPKCC, 1951), 32; Bettie Belk Sarchet and Herbert Thelen, *Block Groups and Community Change* (Chicago: Human Dynamics Laboratory, University of Chicago, 1955); Thomas Furlong, "Chicago Declares War on Blight," *CDT*, November 24, 1946, B8; Baltimore Citizens Planning and Housing Association, *The Battle of the Slums* (Baltimore: The Association, 1950), VF

NAC 1613g27 Bal; Untitled Document, Box 13, Records of the Federal Home Loan Bank System, Federal Home Building Service Plan, Program Subject Files, 1936–1942, Folder Pamphlets and Literature on Housing (Folder 1), 12–13, FHLBBR.

26. During the 1960s and 1970s, the inspiration that urban professionals found in systems science, with its belief in universal principles ordering structure and function in plants, animals, humans, machines and organizations, and confidence in the possibilities for modeling and monitoring such systems, was nurtured by the many biologists, physicists, ecologists, and engineers eager to collaborate on problem solving in the urban realm. Scientific commentaries on the relevance of systems science to the analysis and management of cities include Jay Forrester, *Urban Dynamics* (Cambridge, MA: MIT Press, 1969); Volta Torrey, *Science and the City* (Washington, DC: U.S. Government Printing Office, 1967); C. H. Waddington, "Progressive Self-Stabilizing Systems in Biology and Social Affairs," *Ekistics* 22, no. 133 (1966), 402–405; Anthony Catanese and Alan Steiss, "Systemic Planning for Very Complex Systems," *Planning Outlook* 5 (1968): 7–27; Michael Woldenberg, *Hierarchical Systems: Cities, Rivers, Alpine Glaciers, Bovine Livers and Trees* (Cambridge, MA: Harvard Graduate School of Design, 1968); Leo Kadanoff, "Simulations, Urban Studies, and Social Systems [7 papers]." In Leo Kadanoff, ed., *From Order to Chaos II; Essays: Critical, Chaotic, and Otherwise* (Singapore: World Scientific, 1999); Savas, "Cybernetics in City Hall." Magoroh Maruyama, "Human Futuristics and Urban Planning," *JAIP* 39, no. 5 (1973), 346–358, and contributions from Howard Odum, Stafford Beer, and Gordon Pask in a special issue of *Architectural Design* 42, no. 10 (1972).

27. Social development aspects of Cold War military programs and their influence on the U.S. design of domestic urban programs are documented in Light, *From Warfare to Welfare*.

28. Indeed, as Carl Brauer notes in passing, one of the competing monikers for the "war on poverty" was a program of "human conservation and development." See Carl Brauer, "Kennedy, Johnson, and the War on Poverty," *JAH* 69, no. 1 (1982): 98–119. Eli Ginzberg, still working on the Columbia project and later a presidential manpower adviser, continued to be a prominent advocate for the conservation of human resources.

Essay on Sources

Primary Materials

A central argument of this book is that, while ample attention has focused on the University of Chicago Sociology Department as headquarters to human ecology in the 1920s and 1930s, the importance of ecological thinking about cities was far broader and endured for much longer than has been conventionally understood. Five categories of primary sources help to make this case.

First, educational materials from sociologists reveal how, from the origins of the field and continuing into the postwar period, scholars in the discipline credited with creating the human ecology tradition saw it as a cross-disciplinary endeavor that spanned plant and animal science, geography, economics, and planning in addition to sociology. Course readers, including Roderick McKenzie, *Human Ecology* (n.l.: n.p., 1929); Roderick D. McKenzie, ed., *Readings in Human Ecology* (Ann Arbor, MI: George Wahr, 1934); and Jesse Steiner, *Readings in Human Ecology* (Seattle: University of Washington Bookstore, 1939), are among the texts that illustrate this idea. These readings also document how, while Clemensian theories of succession to climax dominated the understanding of growth, decay, and change in cities (especially after social scientific theories of an urban life cycle were appropriated by the city planning and real estate communities), a long list of other plant and animal scientists such as Eugenius Warming, William McDougall, Charles Darwin, J. Arthur Thompson, Warder Allee, Friedrich Alveredes, James Baker, Josias Braun-Blanquet, J. Richard Carpenter, Royal N. Chapman, Charles Elton, Richard Hesse, Arthur Sperry Pearse, Edward Salisbury, Karl Schmidt, John E. Weaver, and F. A. E. Crew influenced these scholars' work. E. Gordon Ericksen, *Introduction to Human Ecology* (Los Angeles: University of California, Los Angeles, 1949), suggests the continuity of sociologists' cross-disciplinary aspirations after World War II.

Unpublished theses and dissertations are a second category of materials that document the enduring popularity of ecological studies after 1940 as a research subject. At Marquette University; University of Nebraska, Omaha; University of Washington; University of Michigan; University of California, Berkeley; University of Texas; and Washington University in St. Louis, these sources are especially numerous. A small sampling includes Merle D. Hibbert, "An Ecological Study of Mental Deficiency in Milwaukee Based on the Records of Private Training Schools" (MA thesis, Marquette University, 1941); Mona Brown, "The Jewish Community in Racine, Wisconsin: A Study of Human Ecology" (MA thesis, Marquette University, 1942); Murphy Cleophas Williams, "An Ecological Study of the Negro in Ward Seven" (MA thesis, University of Nebraska, Omaha, 1947); James R. Mead, "An Ecological Study of the Second Ward of Omaha" (MA thesis, University of Nebraska at Omaha, 1953); Heinz John Graalfs, "Demographic and Ecological Correlates of the Changing Structure of American Cities" (PhD diss., University of Washington, 1955); Mary Carras, "Everett, Washington: A Demographic and Ecological Analysis" (MA thesis, University of Washington, 1954); Robert L. Carneiro, "Subsistence and Social Structure: An Ecological Study of the Kiukuru Indians" (PhD diss., University of Michigan, 1957); Thelma F. Batten, "Functional Organization in Metropolitan Areas: An Ecological Analysis of Standard Metropolitan Areas' Employed Labor Force Distribution Among Various Industrial Groups, 1950" (PhD diss., University of Michigan, 1955); Donald Q. Innes, "Human Ecology in Jamaica with a Detailed Study of Peasant Agriculture in the Mollison District of Northern Manchester" (PhD diss., University of California, Berkeley, 1959); Marilyn Esther McCurtain, "Political Ecology of Three Metropolitan Areas of California: San Francisco, Los Angeles, San Diego, 1850 to 1950" (MA thesis, University of California, Berkeley, 1955); Donald Anthony Petesch, "Mexican Urban Ecology" (MA thesis, University of Texas, 1960); George Daniel White, "Status Indices and Ecological Areas: Austin, Tex., 1950" (MEd thesis, University of Texas, 1950); William Mary Bates, "The Ecology of Juvenile Delinquency in St. Louis" (PhD diss., Washington University, 1959); Sarah Lee Boggs, "The Ecology of Crime Occurrence in St. Louis: A Reconceptualization of Crime Rates and Patterns" (PhD diss., Washington University, 1964).

A third set of sources includes the educational brochures on urban renewal produced by the U.S. Housing and Home Finance Agency and Urban Renewal Administration for local public agencies and citizens. Many of these readings on redevelopment, rehabilitation, and conservation suggest how the life cycle models of cities that had captivated professionals in real estate and planning became the intellectual basis for this city improvement program. Among them are HHFA, *An Introduction to Urban Renewal as Authorized by the Housing Act of 1954* (Washington, DC: HHFA, 1954); HHFA, *Urban Renewal: What It Is* (Washington, DC: HHFA, 1957); HHFA, *Aids to Your Community: Programs of the Housing and*

Home Finance Agency (Washington, DC: HHFA, March 1958); HHFA, *How Localities Can Develop a Workable Program for Urban Renewal* (Washington, DC: HHFA, multiple editions); URA, *Implementing Conservation* (Washington, DC: URA, 1960); URA, *Selecting Areas for Conservation* (Washington, DC: URA, 1960); URA, *Residential Property Surveys in Conservation Areas* (Washington, DC: URA, 1961). Juxtaposing such documentary evidence with earlier publicity about scientific conservation from several federal agencies reveals a variety of common themes. See, for example, SCS, *The Fight against Erosion, Greatest Enemy of the Soil* (Washington, DC: SCS, 1934); Wilbur R. Mattoon, *Stop Gullies — Save Your Farm: United States Department of Agriculture with Contribution from Forest Service* (Washington, DC: USDA, 1934); USRA, *Helping the Farmer Help Himself* (Washington, DC: Government Printing Office, 1936); USRA, *Better Land for Better Living* (Washington, DC: USRA, 1936); and *Effective Operations of the Soil Conservation Districts* (Washington, DC: SCS, 1941).

Complementing these federally produced materials are publications from ACTION Inc., a private organization based in New York whose affiliates included the many urban professionals who had long pressed for a national city improvement program on the model of natural resource conservation. These materials, which addressed the diverse stakeholders in urban renewal, reveal how, after 1954, an ecological understanding of the evolution of cities endured even as explicit rhetoric linking the urban agenda to the nation's conservation program declined. See, for example, brochures that include ACTION Inc.: *What Is ACTION?* (1954); *A Program of Research in Urban Renewal for the American Council to Improve Our Neighborhoods* (1954); *Why ACTION: Because It Can Help You Help Yourself to Better Living Conditions: A Few Questions and Answers Will Introduce You to ACTION* (1955); *We All Need ACTION to Conserve and Improve America's Housing* (1955); *Time for Action* (1955); *ACTION: Its Objectives, Its Methods, Its Program* (1955); *A to Eliminate, C and Prevent Slums, T to Conserve, I Sound Dwellings, O and Neighborhoods N* (1955); *100 Selected References: Problems of Community Development* (1955); *Urban Renewal Outline* (1956); *Our Living Future: Citizen ACTION for Urban Renewal* (1957); *ACTION Says Why Don't You Have an Urban Renewal Conference* (Chicago: ACTION, 1958); *Organizations in Renewal* (1958); *Publications and Program Materials* (1961); *Agenda for a Good City* (1963); *ACTION to Date* (1963). A central theme in several of these readings, as well as those from the federal government cited earlier, is with proper treatment cities might be renewed while being used, an idea whose conceptual links to the conservation movement are apparent when read alongside documents about renewable natural resources.

Publications for a spectrum of readers produced by local public agencies during the 1950s and early 1960s provide a fifth category of sources documenting how the initial blueprint for renewal in many cities took the form of a program for the scientific management of the urban life cycle. That this theme—reflected in most cities' *Workable Program* submissions for Urban Renewal Administration approval—appears in other publications suggests that taking action to slow, to halt, and to reverse blight was more than simply rhetoric to garner federal funds. See San Diego Urban Renewal Commission, *Blight Abatement Yardstick* (San Diego, CA: San Diego Urban Renewal Commission, 1961); idem, *This Is Your City* (San Diego, CA: San Diego City Planning Department, July 1960); Providence Redevelopment Agency, *The Scope of the Providence Housing Problem and the Role of Replanning, Urban Redevelopment, Rehabilitation and Neighborhood Conservation in Blight Elimination and Slum Prevention* (Providence, RI: Providence Redevelopment Agency, 1953); San Antonio Department of City Planning, *Urban Renewal for San Antonio: A New Way to End Slums and Blight* (San Antonio, TX: San Antonio Department of City Planning, 1957); Oakland Redevelopment Agency, *A General Neighborhood Renewal Plan for West Oakland* (Oakland, CA: Oakland Redevelopment Agency, 1959); Honolulu Office of the Urban Renewal Coordinator, *Your Neighborhood Today* (Honolulu, HI: Honolulu Office of the Urban Renewal Coordinator, 1958); Houston City Planning Commission, *Background for Plan: 1c Population, Land Use and Growth* (Houston: Houston City Planning Commission, 1959); Portland City Planning Board, *Portland 1957: Conservation, Renewal, Redevelopment* (Portland, ME: Portland City Planning Board, 1957); Los Angeles Office of Urban Renewal Coordination, *What Is Urban Renewal?* (Los Angeles: Los Angeles Office of Urban Renewal Coordination, 1961); Vancouver City Planning Department, *Workable Program, A Blueprint for 1957: Urban Renewal* (Vancouver, WA: City Planning Department, 1957); Houston City Planning Commission, 2a: *Condition of Residential Structures Survey* (Houston, TX: Houston City Planning Commission, 1959).

Unpublished and Ephemeral Materials

Six archival and library collections proved especially valuable for tracing how notions of "ecological" communities and national "resources" shaped the evolution of urban policy in the United States.

This book's argument about the breadth of ecology's influences in urban studies does not negate the centrality of the University of Chicago Sociology Department to popularizing this interpretive tradition, making collections there of special interest. Although the university's archives contain syllabi and research produced by department faculty, the theses and dissertations that they supervised offer especially valuable insights into how, in the 1920s and 1930s, ecological approaches to understanding urban phenomena first diffused, and how this mode of interpretation continued to attract new students to the program in the 1940s and 1950s. The majority of these student studies are on the shelves of the Regenstein Library; a few are held in Special Collections.

Within the federal government, America's urban improvement efforts largely concentrated on housing, making files of federal housing agencies held at the U.S. National Archives and Records Administration an important source. To understand the Home Owners Loan Corporation's early work on reconditioning and later conservation, the records of its parent agency, the Federal Home Loan Bank Board (Record Group 195) are the place to look, with information describing personnel and programs in the 1930s and 1940s. These include an early draft of the Waverly report and assorted memoranda, pamphlets, press releases, and newspaper clippings on modernization efforts at the corporation and the Federal Housing Administration. The well-known security surveys that provided the basic research for the conservation programs in Waverly and Woodlawn, organized by city, are contained in this collection as well. Homer Hoyt's contributions to risk rating standards at the Federal Housing Administration are split between files of that agency (Record Group 31) and those of the Housing and Home Finance Agency (Record Group 207), with cities' mortgage risk district maps available in separate cartographic collections. Evidence of Hoyt's efforts to build on the city surveys already completed by the Home Owners Loan Corporation in his work to create the housing agency's Economic Data System is also available in these record groups. Information on the period between the passage of the Housing Acts of 1949 and 1954, when federal officials were considering the importance of neighborhood conservation at the national level, are found in files of the Presidential Subcommittee on Redevelopment, Rehabilitation, and Conservation, and records of Housing and Home Finance Agency Administrator Raymond Foley (both part of Record Group 207). Also in this record group are the agency's general files, which contain ample materials on federal efforts to acquaint cities and citizens with renewal after 1954, including through the newly created Urban Renewal Administration. Documents include memoranda, staff studies, and publicity materials, as well as some information on city improvement efforts around the United States.

Held separately from the files of the nation's housing agencies at the National Archives are the records of the National Resources Planning Board Research Committee on Urbanism and later Urban Conservation and Development Section. Papers produced by these organizations are available in the files of the board's Central Office, subsequently called Office of the Director (Record Group 187). These include drafts of *Our Cities*, the original proposal for community conservation in Woodlawn (found here rather than in Record Group 195), and much correspondence about postwar planning, including the possibilities of establishing a federal urban department modeled after the U.S. Department of Agriculture.

Three local history repositories provided essential details about the workings of city improvement programs on the ground. The Chicago Historical Society holds a number of collections detailing the work of neighborhood improvement associations and other community groups in the city. Records of the Greater Lawndale Conservation Commission, active in 1950s conservation efforts, gather meeting minutes, association newsletters, correspondence with city officials, and copies of materials the city of Chicago distributed to local community groups. The Lincoln Park Neighborhood Collection at DePaul University contains records of the Lincoln Park Neighborhood Conservation Association from its founding in the early 1950s. These records, which detail the life of the organization, in addition gather materials from conservation and renewal planning in other neighborhoods around Chicago. Some items have been digitized and placed online. The Detroit Public Library maintains the archive of the city's plan commission, where some remaining records of Detroit's conservation program are found. These include files from the Committee for Neighborhood Conservation and Improved Housing as well as the plan commission's two divisions involved in conservation activities. Materials include meeting minutes, newsletters, and publicity materials for neighborhood improvement efforts, as well as several folders devoted to research on program efficacy.

The Harvard Graduate School of Design Vertical Files collection is an outstanding repository of both national and local perspectives on the history of urban redevelopment; it brings together unpublished documents with ephemera such as leaflets, brochures, and other published materials no longer widely available. Documents on the New Deal conservation movement are well represented, and their presence at a design school supports this project's central argument. A range of items related to conservation education, including pamphlets, speeches, bibliographies, and other documents from the U.S. Department of Agriculture, including the Soil Conservation Service and Resettlement Administration, are especially numerous. This collection also offers insights into the evolution of theory and practice in city planning and real estate during the period between 1920 and 1960, with course syllabi from several institutions; reports and publicity materials from professional organizations in planning and real estate; and studies commissioned by local and federal governments. Because of Harvard planning professor Reginald Isaacs's leadership role in the Community Appraisal Study and Chicago's later conservation program, many materials on conservation in U.S. cities are part of this collection, with holdings for Chicago and Detroit especially strong. Following the passage of the 1954 Housing Act, both the federal government and ACTION

Inc. geared up for a massive educational campaign and documents connected to their publicity efforts are assembled here. Also available are several cities' *Workable Program* submissions, blight rating studies, and local education efforts, including brochures advising citizens of the impending demolition of their homes.

Secondary Sources

With the urban professionals described here seeking to refute longstanding assumptions about rural-urban divides, studies of American antiurbanism as well as the evolving meanings of nature provide essential background to this book. Two classics on nature in the American psyche are Leo Marx, *The Machine in the Garden: Technology and the Pastoral Ideal in America* (New York: Oxford University Press, 1964), and Roderick Nash, *Wilderness and the American Mind* (New Haven, CT: Yale University Press, 1967). William Cronon, ed., *Uncommon Ground: Toward Reinventing Nature* (New York: W. W. Norton, 1995), provides a more recent set of meditations on this topic. Whether the subject is the Garden City movement, early advocacy for environmental protection, or migration as a search for the pastoral ideal, literature from urban and environmental historians has emphasized an enduring conceptual separation of "nature" and "the city," even as it attends to efforts to deliver to property owners in an urbanizing society closer relations with the natural world. Examples from this rich field include Kermit C. Parsons and David Schuyler, eds., *From Garden City to Green City: The Legacy of Ebenezer Howard* (Baltimore: Johns Hopkins University Press, 2002); David Stradling, *Smokestacks and Progressives: Environmentalists, Engineers, and Air Quality in America, 1881–1951* (Baltimore: Johns Hopkins University Press, 1999); Virginia Jenkins, *The Lawn: A History of an American Obsession* (Washington, DC: Smithsonian Books, 1994); and Kenneth Jackson, *Crabgrass Frontier* (New York: Oxford University Press, 1985).

The historiography of twentieth-century American environmental science features Frederic Clements's ideas about succession and climax in a central position for their influences on both ecological theory and resource planning practice. Works by Ronald Tobey and Joel Hagen, including Ronald Tobey, *Saving the Prairies: The Life Cycle of the Founding School of American Plant Ecology, 1895–1955* (Berkeley: University of California Press, 1981); Joel B. Hagen, "Organism and Environment: Frederic Clements's Vision of a Unified Physiological Ecology," in *American Development of Biology*, ed. Ronald Rainger et al. (Philadelphia: University of Pennsylvania Press, 1988); and Joel Hagen, "Clementsian Ecologists: The Internal Dynamics of a Research School," *Osiris* 8 (1993): 178–195, emphasize the overarching dominance of the Clementsian shool, albeit not its influences on urban theory and practice. Sharon E. Kingsland's *The Evolution of American*

Ecology, 1890–2000 (Baltimore: Johns Hopkins University Press, 2005), and *Modeling Nature: Episodes in the History of Population Ecology* (Chicago: University of Chicago Press, 1985) set these developments in the broader context of American environmental science during the twentieth century and ecology's quest for disciplinary legitimacy. The influences of ecological theory on resource planning practice in forest, farm, and range management are detailed in Randal S. Beeman and James A. Pritchard, *A Green and Permanent Land: Ecology and Agriculture in the Twentieth Century* (Lawrence: University Press of Kansas, 2001); Donald Worster, *Nature's Economy* (San Francisco: Sierra Club Books, 1977); Donald Worster, *Dust Bowl: The Southern Plains in the 1930s* (Oxford: Oxford University Press, 2004); and Christopher Masutti, "Frederic Clements, Climatology and Conservation in the 1930s," *Historical Studies in the Physical and Biological Sciences* 37:1 (2006): 27–48; Ashley L. Schiff, "Innovation and Administrative Decision Making: The Conservation of Land Resources," *Administrative Science Quarterly* 11, no. 1 (1966): 1–30; and Emily Brock, "The Challenge of Reforestation: Ecological Experiments in the Douglas Fir Forest, 1920–1940," *Environmental History* 9, no. 1 (2004): 57–79, as well as several of the sources that treat the dominance of the Clementsian school. Many of these scholars, in tracking the rise and fall of a paradigm in the history of environmental science, also observe the critical response to Clements's theories throughout his career, with special attention to his claims about the organismic character of plant communities and the deterministic aspects of succession.

Several of the aforementioned studies, especially the work of historians of science, treat connections between scientific knowledge of nature and scientific management of nature as the central theme in American conservation history. A broader literature on the rise and fall of the conservation movement in the United States addresses a range of other thematic concerns. Many U.S. and environmental historians, for example, focus on politics and policies at national, state, and local levels, with accounts in this genre, including Carroll Pursell, ed., *From Conservation to Ecology* (New York: Crowell, 1973); Kendrick Clements, *Hoover, Conservation, and Consumerism: Engineering the Good Life* (Lawrence: University Press of Kansas, 2000); Donald Pisani, "The Many Faces of Conservation: Natural Resources and the American State, 1900–1940," in *Taking Stock: American Government in the Twentieth Century*, ed. Morton Keller and R. Shep Melnick (Cambridge: Cambridge University Press, 1999); Henry L. Henderson and David B. Woolner, eds., *FDR and the Environment* (New York: Palgrave Macmillan, 2005); Elmo R. Richardson, *Dams, Parks, and Politics: Resource Development and Preservation in the Truman-Eisenhower Era* (Lexington: University of Kentucky Press, 1973), combining to trace the evolution of conservation policy over several decades. The American population's broad familiarity with conservation is another common theme, detailed in sources that touch on the multimedia campaign to enlist public participation in the movement. These include Finis Dunaway, *Natural Visions: The Power of Images in American Environmental Reform* (Chicago: University of Chicago Press, 2005); Phoebe Cutler, *The Public Landscape of the New Deal* (New Haven, CT: Yale University Press, 1985); Cara Finnegan, *Picturing Poverty: Print Culture and FSA Photographs* (Washington, DC: Smithsonian Books, 2003); Sidney Baldwin, *Poverty and Politics: The Rise and Decline of the Farm Security Administration* (Chapel Hill: University of North Carolina Press, 1968), as well as several of the previous readings. In these and other studies, such as Samuel P. Hays, *Conservation and the Gospel of Efficiency: The Progressive Conservation Movement, 1890–1920* (Cambridge, MA: Harvard University Press, 1959); J. Leonard Bates, "Fulfilling American Democracy: The Conservation Movement, 1907 to 1921," *Mississippi Valley Historical Review* 44, no. 1 (1957): 29–57; and Hal Rothman, *The Greening of a Nation? Environmentalism in the United States since 1945* (Fort Worth, TX: Harcourt Brace College, 1998), the flexible and multiple meanings of the term "conservation" are central. Disparities between short-term and long-term assessments of conservation policy is another subject of concern, treated by many of these historians as well as Marion Clawson, *New Deal Planning: The National Resources Planning Board* (Baltimore: Published for Resources for the Future by Johns Hopkins University Press, 1981); Samuel Hays, *A History of Environmental Politics since 1945* (Pittsburgh, PA: University of Pittsburgh Press, 2000); and Karl Brooks, "A Legacy in Concrete: The Truman Presidency Transforms America's Environment," in *Harry's Farewell: Interpreting and Teaching the Truman Presidency*, ed. Richard Kirkendall (Columbia: University of Missouri Press, 2004); Wayne Rasmussen, "The New Deal Farm Programs: What They Were and Why They Survived," *American Journal of Agricultural Economics* 65, no. 5 (1983): 1158–1162; and Philip Selznick, *TVA and the Grassroots* (Berkeley: University of California Berkeley Press, 1984). Sarah Phillips, *This Land, This Nation: Conservation, Rural America, and the New Deal* (New York: Cambridge University Press, 2007), which attends to how the movement around this shape-shifting ideal mobilized a coalition of interest groups for policy planning that fell apart as the program was implemented, provides an especially relevant parallel to the urban renewal story presented here.

Despite the recognition among historians of environmental science and policy of the broad importance of ecological theory and conservation practice in American life, this discipline has paid less attention to developments in human ecology and their significant influences on theory and practice in the urban professions. A few exceptions include Eugene Cittadino, "The Failed Promise of Human Ecology," in *Science and Nature: Essays in the History of the Environmental Sciences*, ed. Michael Shortland (London: British History of Science Society, 1993); Kingsland, *The Evolution of American Ecology, 1890–2000*; and Gregg Mitman, *The State of Nature: Ecology, Community, and American Social Thought, 1900–1950* (Chicago: University of Chicago Press, 1992), which focus on the ties or lack thereof between studies of plants, animals, and humans; links between conservation of natural resources and conservation of cities are not part of these accounts. Sociologists and urban historians have generated the majority of studies that attend to the influences of ecological thinking in the social science, real estate, and city planning professions. The 1920s and 1930s is the primary subject of Emanuel Gaziano, "Ecological Metaphors as Scientific Boundary Work: Innovation and Authority in Interwar Sociology and Biology," *American Journal of Sociology* 101, no. 4 (1996): 874–907; Marlene Shore, *The Science of Social Redemption* (Toronto: University of Toronto Press, 1987); Martin Bulmer, *The Chicago School of Sociology* (Chicago: University of Chicago Press, 1984); Robert Faris, *Chicago Sociology, 1920–1932* (San Francisco: Chandler, 1967); Richard Helmes-Hayes, "'A Dualistic Vision': Robert Ezra Park and the Classical Ecological Theory of Social Inequality," *Sociological Quarterly* 28, no. 3 (1987): 387–409; Amy Hillier, "Residential Security Maps and Neighborhood Appraisals: The Home Owners' Loan Corporation and the Case of Philadelphia," *Social Science History* 29, no. 2 (2005): 207–233; Marc Weiss, "Richard T. Ely and the Contribution of Economic Research to National Housing Policy, 1920–1940," *Urban Studies* 26, no. 1 (1989): 115–126; Kenneth Jackson, "Race, Ethnicity, and Real Estate Appraisal: The Home Owners' Loan Corporation and the Federal Housing Administration," *Journal of Urban History* 6, no. 4 (1980): 419–52; Amy E. Hillier, "Redlining and the Home Owners' Loan Corporation," *Journal of Urban History* 29, no. 4 (2003): 394–420; and Henrika Kuklick, "Chicago Sociology and Urban Planning Policy: Sociological Theory as Occupational Ideology," *Theory and Society* 9 (1980): 821–45. Studies that take the story past World War II include Marc Weiss and John Metzger, "The American Real Estate Industry and the Origins of Neighborhood Conservation," in *Proceedings of the Fifth National Conference on American Planning History*, ed. Laurence C. Gerckens (Hilliard, OH: Society for American City and Regional Planning History, 1994); Rick Cohen, "Neighborhood Planning and Political Capacity," *Urban Affairs Quarterly* 14, no. 3 (1979): 337–362; and John T. Metzger, "Planned Abandonment: The Neighborhood Life-Cycle Theory and National Urban Policy," *Housing Policy Debate* 11, no. 1 (2000): 7–40.

Questions about the sociology of knowledge are not a central concern in accounts that touch on ecological thinking in the American urban professions, but historians of urban studies, city planning, and real estate provide a broader context for understanding the appeal of a vision of cities as "ecological" communities and national "resources" in a range of studies examining the quest for professional legitimacy among figures in these fields in the four decades after the U.S. Census Bureau recognized America as an officially urban nation. General analyses that touch on participants' scientific aspirations are Alice O'Connor, *Poverty Knowledge: Social Science, Social Policy, and the Poor in Twentieth-Century U.S. History* (Princeton, NJ: Princeton University Press, 2001); M. Christine Boyer, *Dreaming the Rational City: The Myth of American City Planning* (Cambridge, MA: MIT Press, 1983); and Jeffrey Hornstein, *A Nation of Realtors* (Durham, NC: Duke University Press, 2005). Scientific management is the subject of John Jordan, *Machine-Age Ideology: Social Engineering and American Liberalism, 1911-1939* (Chapel Hill: University of North Carolina Press, 1995) and Otis Graham, *Toward a Planned Society* (New York: Oxford University Press, 1976). The historiography of scientific management in the urban professions does not treat in depth the history of slum prevention activities characterized as conservation by actors in this book. Instead, the movement for rehabilitation receives its most detailed discussion in accounts of the history of home improvement and property protection that attend not to the work of urban professionals but rather to the efforts of homeowners and homeowner associations. See Carolyn Goldstein, *Do It Yourself: Home Improvement in 20th-Century America* (New York: Princeton Architectural Press, 1998); Steven Gelber, *Hobbies: Leisure and the Culture of Work in America* (New York: Columbia University Press, 1999), and Thomas Philpott, *The Slum and the Ghetto: Neighborhood Deterioration and Middle Class Reform, Chicago, 1880–1930* (New York: Oxford University Press, 1978).

Analyses of urban renewal range from overview studies of the policy planning process to close readings of its implementation in specific cities and neighborhoods based on interviews or archival research. Emphasizing clearance aspects of the program, much of the scholarship takes a dark view of renewal's effects, for example, Jane Jacobs, *The Death and Life of Great American Cities* (New York: Vintage, 1961); Marc Fried and Peggy Gleicher, "Some Sources of Residential Satisfaction in an Urban 'Slum,'" *Journal of the American Institute of Planners* 27 (1961): 305–315; C. Gruen, "Urban Renewal's Role in the Genesis of Tomorrow's Slums," *Land Economics* 39 (1963): 285–291; Martin Anderson, *The Federal Bulldozer* (Cambridge, MA: MIT Press, 1964); James Q. Wilson, ed., *Urban Renewal: The Record and the Controversy* (Cambridge, MA: MIT Press, 1966); John Staples, "Urban Renewal: A Comparative Study of Twenty-Two Cities, 1950–1960," *Western Political Quarterly* 23, no. 2 (1970): 294–304; Robert Kessler and Chester Hartman, "The Illusion and Reality of Urban Renewal: A Case Study of San Francisco's Yerba Buena Center," *Land Economics* 49, no. 4 (1973): 440–453; John Bauman, *Public Housing, Race, and Renewal* (Philadelphia: Temple University Press, 1987); and Jon C. Teaford, *The Rough Road to Renaissance: Urban Revitalization in America, 1940–1985* (Baltimore: Johns Hopkins University Press, 1990). Amanda Seligman's *Block by Block: Neighborhoods and Public Policy on Chicago's West Side* (Chicago: University of Chicago Press, 2005), and June Manning Thomas, *Redevelopment and Race: Planning a Finer City in Postwar Detroit* (Baltimore: Johns Hopkins University Press, 1997), both of which touch on the conservation and rehabilitation components rather than clearance alone, report a more positive reception for this federal program in specific communities, at least in the short term. Although the influences of ecological theory and conservation policy are not a central theme, that this program was the culmination of several decades of work to make city problems a national priority is widely acknowledged in this body of research. For more specific reference to how activities in Chicago shaped the direction of federal policies, see discussions in Peter H. Rossi and Robert A. Dentler, *The Politics of Urban Renewal: The Chicago Findings* (New York: Free Press of Glencoe, 1961); and Arnold Hirsch, *Making the Second Ghetto: Race and Housing in Chicago, 1940–1960* (Chicago: University of Chicago Press, 1998).

Finally, in treating an episode in American urban history in which analogical thinking served to construct a scientific understanding of urban problems and solutions, this project also draws on studies of the rhetoric of science and social science, analyses of changing "images of the city" in the urban professions, and scholarship on the construction and reconstruction of public problems. Helpful readings in these areas include George Lakoff and Mark Johnson, *Metaphors We Live By* (Chicago: University of Chicago Press, 1980); Sabine Maasen, Everett Mendelsohn, and Peter Weingart, eds., *Biology as Society, Society as Biology: Metaphors* (Boston: Kluwer Academic, 1995); Lloyd Rodwin and Robert M. Hollister, eds., *Cities of the Mind: Images and Themes of the City in the Social Sciences* (New York: Plenum Press, 1984); Lawrence J. Vale and Sam Bass Warner Jr., eds., *Imaging the City: Continuing Struggles and New Directions* (New Brunswick, NJ: Center for Urban Policy Research, 2001); Joseph Gusfield, *The Culture of Public Problems: Drinking—Driving and the Symbolic Order* (Chicago: University of Chicago Press, 1981), and John Kingdon, *Agendas, Alternatives, and Public Policies* (Boston: Little, Brown, 1984.) For more specific accounts of scientific images of cities and city problems, see Christopher Hamlin, "The City as a Chemical System? The Chemist as Urban Environmental Professional in France and Britain, 1780–1880," *Journal of Urban History* 33, no. 5 (2007): 702–728; Anthony Vidler, "The Scenes of the Street: Transformations in Ideal and Reality, 1750–1871," in *On Streets*, ed. Stanford Anderson (Cambridge, MA: MIT Press, 1979); Antoine Picon and Alessandra Ponte, eds., *Architecture and the Sciences: Exchanging Metaphors* (New York: Princeton Architectural Press, 2003); and Jennifer Light, *From Warfare to Welfare: Defense Intellectuals and Urban Problems in Cold War America* (Baltimore: Johns Hopkins University Press, 2003).

Index

Page numbers in italics indicate figures.

Abrams, Charles, 8, 68, 70
ACTION Inc., 148–51, 160, 167–68, 270n97, 270n99
Adams, Charles: as conservationist, 8, 39, 50, 198n19; and human ecology, 9, 32–34, 107, 193n115, 201n48, 238n36
African Americans: in Baltimore, 61; in Buffalo, 156; in Chicago; in conservation, 113–14, 130, 152, 156–60; in Detroit, 147, 156–57; in New York, 157; in Philadelphia, 157; in Richmond, 29, 225n75; 62, 87, 96, 105, 113–14, 153, 159, 231n125, 276n149, 278n160, 279n164; in St. Louis, 157; in Washington, DC, 12
agriculture: importance in United States, 8, 47–48, 129, 235n24; as model for urban professions, 1, 16, 49–50, 67–68, 107, 119, 122, 133–34, 150, 167, 171, 204n63, 211n107, 212n112, 242n49, 283n18; problems in depression, 38, 65, 72, 88; scientific approaches to, 8, 38–39, 42, 109–10, 195n6, 197n15, 257n16; and social science, 40, 42, 205n73. See also rural areas
Alinsky, Saul, 114, 126, 156, 274n128
Allee, Warder, 106, 110, 241n42
Almack, John C., 19
analogical reasoning: in city planning, 37, 44, 50, 79, 95, 125, 161, 163, 166,

169; in real estate, 51–52, 69, 161, 163; in human ecology, 3, 5–7, 15, 19–22, 26, 30, 32, 106, 161–62; limits to, 32, 137, 168, 240n40, 283n18; and urban renewal, 136–38
Anderson, Nels, 12, 20, 27, 44, 187n52, 228n98
antiurbanism, 2–5, 161. See also rural areas
appraisal, of real estate, 3–4, 52, 62, 64, 71, 73, 78, 102, 144, 206n80, 207n82, 210n102, 211n111, 219n22, 252n122. See also valuation
Ascher, Charles, 67, 81, 95, 132, 242n49, 245n68

Babcock, Frederick, 30, 31, 53, 55–56, 71, 76, 205n74, 214n133
Back of the Yards, in Chicago, 156
Back of the Yards Council, 126
Baker, Oliver, 204n63, 205n73
Baltimore, 58, 60–62, 63, 83, 95, 132–33, 135, 138, 145–46, 149, 153, 158, 163–64, 166–67, 231n121, 252n127. See also Waverly
Baltimore Housing Authority, 61
Banfield, Edward, 104, 108–10, 149, 244n64
Barrows, Harlan, 18
Bartholomew, Harland, 44
Bates, Marston, 240n40
Becker, Howard, 104

299

Behavior Research Fund, 22, 189n73
Bennett, Hugh, 40, 65, 130, 163, 197n16, 198n19, 215n135, 256n10
Berman, Ruth, 96, 231n123, 259n33
Better Homes in America, 190n90, 209n101
Bews, John William, 33, 110
Bingham, Robert, 44
blight, urban, 44–47, 45, 54–55, 59, 62, 66–67, 73, 86–7, 89–92, 92, 94, 119–21, 123, 133, 135–36, 139–40, 143–44, 150–52, 155, 158, 166, 170, 214n131, 225n76, 226n78, 226n81, 234n15, 234n19, 250n102, 251n120, 257n19, 265n71, 271n99, 282n11. *See also* Illinois Blighted Areas Redevelopment Act
block groups, 62, 93, 113, 115, 119–20, 134, 138, 146–47, 150, 153–54, 157–58, 160, 259n33, 277n153. *See also* homeowners associations
Blucher, Walter, 51, 56–57, 109, 122, 132, 149, 155, 160, 243n56, 249n92, 251n121
Blumer, Herbert, 104
Bodfish, Morton, 103, 235n23
Bogardus, Emory, 19
Bogue, Donald, 104, 105, 107
Boston, 72, 83, 146–47, 153
Boulevard-Gratiot-Mack, in Detroit, 147–48
Breese, Gerald, 86, 105
Broadbent, Elizabeth, 86
Brooks, C. William, 132–34, 277n153
Brown, Oscar, Sr., 119
Brownlow, Louis, 49
Brownlow Committee, 49, 203n59
Buffalo, NY, 154, 156
Burgess, Ernest, 10, 17, 99, 104–5, 117, 190n82, 207n82; and city planning, 44, 52, 107, 254n149; Hoyt and, 71, 75, 80, 87, 91, 111; and scientific community, 33, 107, 189n73; zonal model of, 22–28, 23, 29–30, 51, 53, 75, 80–81, 87, 91, 105, 111, 188n70, 192n100, 192n104, 193n111, 236n33, 252n122
Burroughs, Roy, 29
Burton, John, 71

Butler, Walker, 124
"Butler Bill." *See* Illinois Urban Community Conservation Act
Buttenheim, Harold, 56

Catholic Church, 114, 135
Chandler, Albert, 19
Chaney, Ralph, 39, 88
Chicago: as geographical focus, 4–5; Hyde Park-Kenwood neighborhood in, 116, 118–27, 152, 160, 262n39; Lawndale neighborhood in, 80, 151, 153, 157, 159, 255n153, 273n120, 278n160, 279n164; Lincoln Park neighborhood in, 114, 126–27, 222n49, 255n153; sector model of, 75; Woodlawn neighborhood in, 5, 58, 62–66, 68, 93–94, 214n131; zonal model of, 23;
Chicago Community Conservation Board, 124, 126–27, 143, 166
Chicago Community Inventory, 104–8, 112, 115, 117, 235n29, 226n31, 242n50, 249n95, 250n102
Chicago Defender, 114, 157, 279n160
Chicago Department of Buildings, 127, 153
Chicago Dwellings Association, 121
Chicago Housing Authority, 77, 94, 108, 121, 255n156
Chicago Land Clearance Commission, 101–4, 111, 121, 124, 126, 234n20, 251n121, 270n97
Chicago Land Use Survey, 76–81, 88, 91, 102, 107, 220n36, 222n42, 235n23
Chicago Plan Commission, 5, 64, 68–69, 76–81, 84–85, 87–89, 93–94, 96, 102–5, 108, 111, 114, 121, 127, 164, 220n36, 221n38, 222n39, 226n78, 229n111, 254n147, 254n151, 230n115, 253n138
Child, Charles M., 7, 13, 20, 22, 27, 186n49, 194n117
Cincinnati, 146
cities: Cold War and, 169–71; as frontier, 1, 10, 50, 96, 134; as gardens, 69; in Great Depression, 29–30; as laboratory, 10, 183n19; as spider webs, 22,

2236n33; as trees, 27; World War II and, 81–84, 163–64. *See also* multiple nuclei model; sector theory; zonal structure

city planning, 2–5, 12, 35–37, 43–49, 51–60, 64–71, 76–79, 81–82, 84–89, 91, 93–105, 107–12, 115–27, 134, 136, 139–40, 143, 145, 147, 149–52, 155, 160, 162–67, 169–70, 179n4, 190n90, 200n41, 201n48, 202n49, 203n50, 203n59, 204n65, 210n102, 214n129, 217n3, 220n37, 221nn37–38, 223n54, 226n80, 229n108, 230n115, 231n121, 232n130, 234n17, 241n48, 242n50, 242n54, 243n55, 243nn57–58, 244n59, 245nn66–68, 248n92, 249n94, 251n120, 254n149, 265n71, 269nn87–89, 270n98

Civilian Conservation Corps, 40, 49, 83, 113, 198n24, 199n27

Civil Works Administration, 28, 72, 218n13

Clawson, Marion, 202n50

Clements, Frederic, 3–4, 7, 9, 15, 20, 26, 33–34, 39, 88, 130, 256n14, 285n24

Clements, Kendrick, 197n18

Cleveland, 29, 78, 96, 114, 139, 143, 146, 166–67, 231n122, 235n21, 264n59

climax: in ecology and natural resources planning, 4, 11, 26–27, 39, 42, 51, 129–30, 162, 169, 190n84, 227n92, 257n14, 285n24; in urban theory and practice, 17, 27, 65, 136, 190n82, 227n92, 240n39

Colby, Charles, 30–31, 44, 108, 202n49, 226nn78–79, 236n33, 242n54

Cold Spring Harbor Symposium on Quantitative Biology, 107

Cold War, 99–100, 165, 169–71

Cole, Albert, 131, 149

Colean, Miles, 54, 58, 132, 136–37, 139, 149, 210n102, 217n155, 283n18

Cole Committee, 131–34

Comey, Arthur, 48, 202n49

Committee on Government Housing Policies and Programs. *See* Cole Committee

Committee on Housing and Blighted Areas, 54

Community Appraisal Study, 5, 116–24, 147, 149, 158, 249n94, 250n102, 252n124, 253n127, 262n39

Community Conservation Agreements, 112–13, 118, 247n80, 249n101

concentric arrangement: of cities, 22, 23, 27–30, 31, 57, 74–75, 80, 86–87, 105, 108, 193, 111, 207n82, 226n79, 226n81, 236n122, 252n122; of ecological communities, 23–25, 24–25, 189n74. *see also* zonal structure

conservation: calls for city departments of, 36, 67, 124–25, 253n138; in Chicago, 112–15, 121–27, 159, 254n149, 255n153, 255n154, 262n39, 277n153, 279n160, 279n164; as city improvement tool, 48–51, 58–60, 64–68, 112–15, 117–20, 202n49, 203nn58–59, 204n65, 210n102, 212n113, 214n132, 215n135, 216n147, 216n153, 224n65, 229n111, 229n114, 231n122, 243n58, 251n121, 252n124, 252n127, 253n129, 258n22, 259n23, 261nn37–38, 268n82, 269n95, 271n99, 283nn17–18; in Detroit, 140, 142, 147–48, 152–54, 156–57, 260n35, 264n59, 269n84, 269n88, 275n130, 277nn153–54; ecology and, 3–4, 8, 37–40, 41, 42, 83, 88, 99, 130, 162–66, 168, 195n6, 196nn11–12, 197n17, 198n19, 199n26, 204n63, 211nn110–11, 215n135, 218n17, 226n77, 256n13, 257n164; in economics, 196n11, 211n109; Eisenhower and, 130–31; in Hyde Park–Kenwood neighborhood, 118–21; in *Master Plan of Residential Land Use of Chicago*, 87–89, 92, 91–94; in New Deal, 38–43, 163, 204n65, 254n147, 259n28; Truman and, 99–100; in Washington, DC, 82–83; in Waverly neighborhood, 60–62, 63, 65–66, 229n109, 231n121; in Woodlawn neighborhood, 62–66, 93–94, 229n109, 262n39; World War II and, 81–84, 163–64. *See also* human resources; national resources; natural resources

Conservation Reserve Program, 131
Conservation Service, Home Owners Loan Corporation, 67, 82, 95, 231n120
Cooley, George, 117
Country Life Commission, 49
Cowles, Henry, 15, 18, 23–26, 186n49, 187n54
Crane, Jacob, 210n102, 217n155
Cunfer, Geoff, 197n15
Czarniecki, Leonard, 264n59

Daley, Richard J., 153
Darwin, Charles, 13–14, 238n36
Davis, William, 187n53
Dayton, OH, 146
Delano, Frederic, 47–48, 201n47
Dentler, Robert, 152, 255n156
Department of Agriculture, 39–40, 42, 49–50, 67–68, 72, 107, 109–10, 119, 150, 167, 171, 204n63, 235n24, 242n50, 244n64
Detroit, 75, 114, 128, 135, 139–40, 142, 143–44, 146–48, 152–54, 156–57, 166, 219n18, 260n35, 261n37, 264n59, 264n61, 265n71, 268n83, 269n88, 275n130, 276n153, 277n154
Dewey, Richard, 125
Dice, Lee, 240n40
Dickinson, Robert, 34, 236n33, 240n39
do-it-yourself, 65, 146, 211n111
Dorau, Herbert, 13, 21, 28, 53, 207n87, 211n109
Douglass, Harlan Paul, 13, 15
Downs, James, 122, 125–26, 149, 250n102, 252n124
Downs-Mohl Real Estate, 122, 125, 217n1, 250n102
Drake, St. Clair, 119, 152
Draper and Kramer Real Estate, 121, 251n121
Dryer, Charles Redway, 10–11, 16, 19, 190n82
Dudziak, Mary, 233n13
Duncan, Beverly Davis, 105
Duncan, Otis Dudley, 104, 105, 107, 236n31, 238n37
Dunham, H. Warren, 128
Dust Bowl, 130, 163, 196n15

Dyckman, John, 111, 120, 149, 158, 160, 169–70, 249n94

Ecological Society of America, 8–9, 18, 39
ecology: conservation movement and, 3–4, 8, 37–40, 41, 42, 83, 88, 99, 130, 162–66, 168, 196nn11–12, 197n17, 198n19, 199n26, 204n63, 215n135, 257n164; and human ecology, 9–10, 32–34, 182n16, 183nn21–22, 185nn28–29, 188n56, 193n112, 193n115, 194n117, 195n6, 200n35, 201n48, 204n65, 224n62, 237n36, 240n40, 241n41; research traditions in, 8, 10–11, 13–16, 20, 23–25, 27, 31, 37–40, 42, 79, 83, 105, 130, 136, 169–70, 180n3, 180n9, 187n54, 188n66, 189n79, 237n36 257n14, 284n20, 285n24, 285n26; systems, 169, 285n24. *See also* human ecology
Ecology (journal), 187n52
education: of citizens on urban renewal, 144–51; in city planning, 44, 108–12, 242n54, 243n58, 245n67; on conservation, 36, 40–42, 61–62, 66–67, 83, 93, 95–96, 114, 119–20, 125–27; in human ecology, 17–19, 104–11; of local public agencies, 138–44, 148–51; in real estate, 4, 53, 90, 91, 285n24
Eisenhower, Dwight, 127, 130–32, 134–35, 150, 165, 257n17, 258n20, 269n95
Eisenhower Center for the Conservation of Human Resources, 257n17
Ekblaw, W. Elmer, 32
Eliot, Charles, 203n50
Elton, Charles, 14, 110
Ely, Richard, 12, 28, 35, 52, 59, 205nn73–74
Emerson, Alfred, 33, 106–7, 110
Emerson, Ralph Waldo, 2–3
eminent domain, 45–46, 54–58, 88, 94–95, 100, 123, 154, 207n84, 210n102, 229n114
Ericksen, E. Gordon, 34, 105–6
eugenics, 7, 181n9, 199n24
Europe, 181n6
Evans, Francis, 107

302   INDEX

Fahey, John, 61, 95
Fair Deal, 100
Faris, Robert, 236n30
Farm Credit Bureau, 210n102
farming, 1, 16, 38, 43, 47, 57, 110, 235n24; conservation and, 3, 37, 39–42, 41, 49, 65, 73, 83, 88, 99–100, 110, 130–31, 134, 163, 167, 196n15, 201n44, 210n102, 215nn134–35, 218n17, 229n114, 257n14, 282n10
Farm Mortgage Bankers Association, 205n78
Farm Security Administration. *See* Resettlement Administration
Farr, Newton, 66
Federal Home Building Service Plan, 209n101
Federal Home Loan Bank Board, 54, 60–61, 95, 119, 167, 261n38, 282n11. *See also* Home Owners Loan Corporation
Federal Home Loan Bank Review (journal), 66, 96, 207n82
Federal Housing Administration, 4–5, 53–55, 58, 60–61, 68, 71–72, 74, 76, 78, 80, 85, 87, 92–93, 118, 136, 144, 149
Filley, Robert, 77
filtering, of populations, 74, 219n22, 235n23
Financial Surveys of Cities. *See* Real Property Inventories
Fine, Gary, 104
Fisher, Ernest, 28, 44, 53–54, 71, 75, 208n89, 208n96, 219n26, 220n27
Foley, Donald, 77, 79, 86, 93, 226n78
Foley, Raymond, 113–14
Follin, James, 135, 138, 143, 154, 261n38
Ford Foundation, 149, 167–68, 283n18, 284n10
forests, 3, 9–11, 16–17, 21, 37–40, 43, 47–48, 57, 59, 64, 73, 109–10, 130, 163, 182n16, 196n12, 201n44, 210n102, 218n17, 229n114, 244n64, 257n14
Frank, Glenn, 59

Gallery, John Ireland, 114
garden, city as, 69–70

Garden Cities, 2, 22, 161
Geddes, Patrick, 136, 204n65
Ginzberg, Eli, 257n17, 277n153, 286n28
Gluck, Eleanor, 31
Goffman, Erving, 104
Goode, J. Paul, 18, 187n53
Goodwillie, Arthur, 58, 61, 67, 82–83, 210n102, 217n155, 223n58, 224n65, 231n120
Graham, H. W., 24, 25
Granadosin, Isaac, 19
Gras, Norman Scott Brien, 16, 22, 220n27, 236n33
Gray, L. C., 39, 205n73
Great Depression, 6, 28–35, 37–38, 43–45, 48, 53, 58, 72, 83, 88, 99, 130, 162–63, 206n70, 252n124
Greater Lawndale Conservation Commission, 151, 157, 159, 279n160, 279n164
Great Society, 170
Gropius, Walter, 115–16, 248n92
Grunsfeld, Ernest, 241n48
Gusfield, Joseph, 104

Hagen, Joel, 257n14
Haig, Robert Murray, 14–15, 73
Hanson, Herbert, 8, 39, 42, 196n12, 199n26
Harlan, Howard, 29
Harris, Chauncy, 105, 220n27, 247n33
Hartley, William, 198n24
Harvard University, 15, 20, 31, 98, 115–17, 122, 125, 165, 187n53, 202n49, 248n92
Haskell, Edward, 33
Hauser, Philip, 104–5, 117, 119, 244n66, 248n92
Heald, Henry, 103, 270n97
Henderson, George, 158, 277n154
Henry, L. K., 24, 25
Hilbersheimer, Ludwig, 117, 248n92
Hillier, Amy, 180n8, 195n4, 206n81, 212n113
Hinman, Albert, 13, 21, 33, 207n84, 211n109
Hirsch, Arnold, 234n17, 280n166
Hirsch, Morris, 121
Hollingshead, A. B., 42

Holsman, Henry, 46
homeowners associations, 54–55, 57, 60, 62, 65, 87, 113, 170, 209n101, 229n111. *See also* block groups
Home Owners Loan Corporation, 5, 36, 53–55, 58, 61–62, 64, 66–67, 73, 81–82, 84, 87, 92–93, 95, 101, 144, 149, 151, 207n83, 210n102, 212n113, 218n17, 227n92
Hoover, Herbert, 28, 38, 190n90, 196n9, 197n18, 201n101
Horne, Frank, 113–14
Housing Action Committee, 101
Housing Act of 1934, 66, 71, 209n101
Housing Act of 1937, 230n114
Housing Act of 1949, 103, 113, 120, 121–32, 134–35, 235n24
Housing Act of 1954, 134–35, 138, 148, 152, 166, 261nn37–38, 283n18
Housing and Home Finance Agency, 103, 113, 131, 135, 138–40, 148–50, 154, 166, 257n19, 274n128. *See also* Urban Renewal Administration
Howard, Ebenezer, 22
Hoyt, Homer, 68–80, 84–90, 91, 93–94, 99, 101–5, 115–16, 122–23, 132–33, 136–37, 140, 158, 164–65, 168, 194n122, 217n5, 218n10, 218n17, 219n22, 219n25, 220n36, 221n37–38, 222n39, 223n54, 225n75, 225n76, 226n79, 227n92, 228n94, 235n23, 237n34, 253n129, 253n138, 264n60, 283n16, 285n24
Hughes, Everett, 31, 107, 119, 125, 208n88, 238n37
Human Dynamics Laboratory, 118–19, 121
human ecology: analogy in, 6–7, 10–17, 19–28, 23, 31–34, 106; and city planning, 37–28, 43–46, 48, 50–51, 56–57, 65, 69–70, 76, 79, 85, 94, 96–98, 108–12, 116, 128, 136–37, 161, 163–64, 167, 170, 195n4, 201n48, 204n65, 220n37, 222n39, 232n130, 242n54, 244n58, 244n66; and economics, 4, 7, 13, 15–16, 26, 30, 34, 51, 196n11, 218n9, 219nn25–26, 239n39; and geography, 4, 7, 10–11, 15–16, 18–19, 26, 30–32, 34, 107, 185n34, 187n55,

239n39; informal appropriations of science, 7, 19, 20, 32–34, 107–8, 160, 168, 183n24, 200n37, 241n41, 283n16, 283n18; and real estate, 4, 37–38, 43–44, 46, 52–57, 60–61, 65–66, 69–70, 75–76, 87, 89, 136, 161, 163, 164, 167, 170, 195n4 206nn80–81, 207n83, 212n113, 252n122; and sociology, 3–4, 7, 12–15, 17–18, 20, 22, 25–27, 29, 33–34, 52, 91, 105, 107, 128, 180n3, 182n111, 187n52, 187n55, 239n39; and urban renewal, 138, 144, 160, 168; and urban studies, 4, 7–10, 14, 18–19, 21, 28, 43, 66, 73, 76, 79, 85, 98–99, 102, 104–7, 112, 117, 136, 144, 161, 167, 169–70, 183n19, 184n25, 184n27, 186n49, 188n57, 188n66, 236n30, 237n33, 239n39, 260n34. *See also* conservation
human resources, 42, 49–50, 131, 163, 164, 171, 198n24, 199n25, 199n27, 257n17, 286n28
Huntington, Ellsworth, 9, 181n9
Hurd, Richard, 30, 192n104, 218n9
Hyde Park–Kenwood, in Chicago, 116, 119, 125, 152, 160, 246n79, 262n39
Hyde Park–Kenwood Community Conference, 118–21, 150, 152, 249n95

Ickes, Harold, 42, 47–49, 113, 171
Ihlder, John, 210n102, 217n155
Illinois Blighted Areas Redevelopment Act, 98, 101–2, 104, 116–17, 123–24, 126, 131, 135, 143, 157, 276n149
Illinois Emergency Relief Commission, 88
Illinois Institute of Technology, 103, 109, 116–17, 248n92, 270n97
Illinois Neighborhood Redevelopment Corporation Law, 94–95, 255n156
Illinois Postwar Planning Commission, 107
Illinois Urban Community Conservation Act, 98–99, 124, 135, 255n156
Innis, Donald, 245n66
Institute for Research in Land Economics and Public Utilities, 4, 52–53, 206n81, 214n129, 221n38, 235n23, 236n33

Insured Mortgage Portfolio (journal), 73–74, 76, 206n81, 207n82, 208n89
Interim Commission on Conservation, 122, 125, 254n151
Introduction to the Science of Sociology (Park & Burgess), 17, 19, 187n49
invasion: in cities, 15, 17, 25, 37, 42, 53, 73–74, 105, 136, 170, 190n82, 204n82, 235n22, 240n39; in nature, 15, 26, 37, 42
Isaacs, Reginald, 98, 110–11, 115–17, 121–25, 132–33, 149, 165, 168, 264n147, 283n18

Johnson, Charles, 28
Johnson, Earl, 71
Johnson, Lyndon, 170
Johnson, Reginald, 277n153
Jones, William, 12
Joyce, L. A., 257n14

Kelly, Edward, 83
Kennedy, John F., 1, 129, 167, 170, 179n2, 284n20
Kennelly, Martin, 101, 112, 122
Kincaid, H. Evert, 87
Kingsland, Sharon, 185n28, 237n36
Kligman, Miriam, 97
Klove, Robert, 77, 86, 221n38, 226n78
Klüver, Heinrich, 189n73
Korean War, 99, 165
Kramer, Ferd, 132, 149, 220n32, 251n121
Kuklick, Henricka, 180n8, 195n4, 283n16

laboratory, city as, 10, 183n19
Landis, Paul, 42
land use capability, map of. *See* maps
Lanham Act, 82, 83
Lawndale, in Chicago, 92, 151, 153, 157, 159, 255n153, 273n120, 278n160, 279n164
Levi, Julian, 255n156
Levine, Hyman, 151
Lewis, Sinclair, 52
life cycle, of cities: in Burgess, 26–27, 34–35, 105, 189n82; in Hoyt, 68, 70–71, 233n12, 235n23, 251n120, 253n138;
interventions in, 5, 36, 39, 52–53, 55, 65–66, 69, 88, 99, 100, 102, 112, 116, 127, 212n113, 216n147; *Master Plan of Residential Land Use* and, 84–88, 90; in Metropolitan Housing and Planning Council Study, 123–24; renewal and, 1, 33, 99, 128–29, 136–38, 151, 155, 158, 166, 170; spreading awareness of, 138–44, 189n79, 234n20, 240n39, 252n112, 285n24
"life history" research techniques, 20–21, 80, 188n66
Lincoln Park, in Chicago, 114, 222n49
Lincoln Park Conservation Association, 126–27, 255n153
Lind, Andrew, 20
Lindeman, Eduard, 20, 27, 44, 51, 122, 187n52, 199n25
Local Community Research Committee, 104–5, 221n38
Lohmann, Karl, 44
Long Beach, CA, 29–30
Longmoor, Elsa Schneider, 29–30
Los Angeles, 11, 30, 72, 139, 146, 152, 207n82, 268n84
Ludlow, William, 109, 243n56
Lynd, Helen, 32
Lynd, Robert, 32

MacChesney, Nathan William, 54, 57
MacCornack, Walter, 96, 230n128
Maizlish, Mae Schiffman, 86, 226n79
Male, Charles, 35, 195n5
Manhattanville, NY, 277n153
Man of Action (film), 150
maps: in ecology and natural resources planning, 10, 39–40, 41, 73, 87, 218n17, 238n36, 256n13; in mortgage risk evaluation, 39–41, 53, 60–51, 67, 72–73, 76, 78, 87, 102, 170, 206n81, 207n82, 212n113, 218n11, 218n15, 218n17, 222n42, 225n75; in urban renewal, 140, 141, 142, 144, 158; in urban theory and practice, 10, 18, 22, 27–28, 43–44, 53, 70, 72–74, 79–80, 86, 91, 102–3, 117, 119, 188n70, 189n70, 190n86, 193n111, 194n122, 205n76, 225n76, 226nn78–79, 242n50, 252n122, 253n129, 254n151

INDEX 305

Marshall, Robert, 32
Master Plan of Residential Land Use of Chicago (Chicago Plan Commission), 5, 81, 84–87, 89–91, 92, 93–94, 102–3, 115, 136, 140, 158, 165, 225n75, 226n78, 227n88, 233n23, 234n19, 248n91, 253n129
Masutti, Christopher, 180n9, 196n12, 197n17, 257n14
Maxey, Alva, 113
Mayer, Harold, 86, 94, 104, 107, 111, 117, 119, 248n92
McDermott, John, 159–60
McDougall, William, 31, 34, 182n16, 241n41
McFall, Robert, 34, 80
McKenzie, Roderick, 13, 17–19, 23, 27–28, 33–34, 52, 71, 187n55, 190n82, 194n121–22, 219n26, 220n27, 238n37
McMichael, Stanley, 44
McNeal, Donald, 36, 54, 60, 67, 216n147
Mears, Eliot, 19, 31
Meier, Richard, 117, 243n56, 243n58, 248n92, 249n96,
Meltzer, Jack, 155, 160
Merriam, Charles, 108, 203n50
Merriam, Robert, 125
Mertzke, Arthur, 205n74
Metropolitan Housing and Planning Council, 121, 123, 124, 133, 251n121, 255n156
Metropolitan Housing Council, 94, 107, 113, 220n32
Meyerson, Martin, 104, 109–11, 116–17, 119, 149, 160, 244n66, 248n92, 249n96
Michael Reese Hospital, 103, 115–16, 121, 165
Michigan Academy of Science, 110
Millspaugh, Martin, 217n3
Mitchell, Robert, 64, 66, 85, 95, 111, 124, 132, 143, 149, 214nn132–33, 220n32, 242n49, 262n39, 271n99
Mitchell, Wesley Clair, 71, 203n50
Mitman, Gregg, 183n24, 237n36
Monchow, Helen, 221n38
Moore, Loring, 117

Morehouse, Edward, 12, 35
Mowbray, John McC., 60
Mowrer, Ernest, 236n30
multiple nuclei model, 220n27, 237n33
Mumford, Lewis, 136
Mumford, Milton, 234n17

National Association of Housing Officials, 64, 67, 160, 180n10
National Association of Real Estate Boards, 4–5, 52, 53–57, 60, 66, 122, 134, 145, 151, 160, 167, 205n78, 217n1, 230n115
National Conference of Catholic Charities, 114
National Council of Catholic Charities, 135
National Housing Agency, 69, 92, 96, 103, 115, 125
nationality groups, 12–13, 18, 31, 51–52, 57, 61–62, 74, 78, 96, 106, 110–11, 118, 158, 181n9, 190n86, 205n77, 222n49, 225n75, 246n68
National Opinion Research Center, 117, 236n32, 248n92
National Planning Board. *See* National Resources Planning Board
national resources, 2, 3, 48–50, 59, 69, 99–100, 108–10, 112, 128, 130–31, 134, 138, 161–64, 167–69, 203n50, 244n54, 244n66. *See also* conservation; natural resources
National Resources Committee. *See* National Resources Planning Board
National Resources Planning Board: 28, 37, 48, 50, 54, 58–59, 67, 70, 72, 80–81, 89, 95, 97, 107, 116, 119, 122, 133, 137, 149, 171, 191n92, 210n102, 217n155, 229n114, 242n40, 243n56, 245n68. *See also* Research Committee on Urbanism
national security, 68, 82–84, 94, 99–100, 120, 130–31, 163–65, 224n62
natural areas, 11–15, 26, 51, 85
natural resources, 4, 35, 37–39, 42–43, 46–48, 50–51, 57–60, 64–65, 68, 73, 76, 82, 85–87, 91, 93–94, 96, 98, 100, 109–10, 123, 125, 128–32, 136, 161–64, 166, 168–69, 198n24, 201n48,

203n50, 206n79, 210n102, 229n114, 257n14, 282n14, 283n16. *See* also conservation; national resources
Neighborhood Improvement Act, 54–58, 64, 67, 209n100, 230n115
neighborhood unit, 56, 65, 89, 110–11, 118
Nelson, Herbert, 28, 52, 53, 55, 57, 205n76, 208n96, 217n1
Nelson, Otto, 122, 251n121
Nelson, Richard, 69, 74, 114, 118, 125, 217n1
Newcomb, Charles, 28, 222n48, 226n79
New Deal, 3, 38–43, 48, 59, 65, 81, 88, 107, 109, 113, 130–31, 169, 196n11, 254n147, 257n116
New York City, 11–12, 14, 20, 31, 72, 78, 94, 110, 138, 140, 146, 148, 153–55, 157, 219n18, 222n39, 267n82, 274n118, 277n153
Nichols, Jesse Clyde "J.C.," 52, 211n107
North Lawndale, in Chicago, 80, 153, 255n143, 273n120, 278n160
Northwestern University, 4, 21, 30, 52–54, 64, 71, 77, 101, 184n25, 205n74, 206n81, 220n32. *See also* Institute for Research in Land Economics and Public Utilities

Oakland-Kenwood Planning Association, 121
Odum, Eugene, 240n40, 285n24
Odum, Howard, 28, 286n26
Ogburn, William, 108
O'Grady, John, 135, 247n85, 261n37
Olmsted, Frederick Law, 2, 202n49
ontography, 18, 187n53

PACE Associates Architects, 121
Pacific Coast Race Relations Survey, 19
Palmer, Vivien, 17, 20–21, 187n51, 188n66
Park, Robert, 6, 12, 14, 17–19, 21, 28–29, 71, 99, 106, 113
Park, Thomas, 107
Parkins, Almon Ernest, 43
Parkins, Maurice, 148, 166
Perloff, Harvey, 104, 108, 119, 245n66, 252n127, 284n20

Pettibone, Holman, 234n17
Philadelphia, 27–28, 72, 83, 111, 133, 135, 139, 143, 146–47, 150, 153–54, 157–60, 248n92, 252n127
Phillips, Sarah, 99, 130, 215nn134–35, 284n19
Philpott, Thomas, 62, 157, 208n97, 220n36, 239n38
Price, Henry E., 264n59
Program of Education and Research in Planning, 108–12, 115
public purpose, 68, 94, 103, 123, 207n84, 210n102, 229n114. *See also* eminent domain
Public Works Administration, 42, 47, 54, 64

Quakers, 135, 150
Queen, Stuart, 30, 192n100, 193n111

race, 7, 12–13, 19, 28–29, 31, 37, 57, 61–62, 65, 72, 74, 78–79, 82, 95–96, 100, 105–6, 110–14, 118, 122, 135, 155–56, 158–59, 181n9, 185n27, 190n86, 205n77, 219n22, 222n50, 223n58, 225n75, 233n10, 246n79, 248n92, 249n92, 249n100, 250n102, 277n156, 278n160
Ragland, John, 159, 279n163
Ratcliff, Richard, 50, 53, 232n130, 240n39
Ravitz, Mel, 128, 147
real estate, 2–5, 35–37, 43, 51–60, 62, 64–65, 68–71, 73, 75–78, 81–82, 89, 93–97, 100–103, 114, 118, 121–22, 125–26, 134, 139, 149–50, 155–56, 159–60, 162–65, 167, 170, 179n4, 195n5, 201n43, 205n74, 205n76, 206nn80–81, 207n82, 208n89, 208n97, 210n102, 211n111, 212n113, 214n129, 218n15, 220n32, 223n54, 228n102, 229n108, 230nn114–15, 235nn22–23, 249n100, 250n102, 252n112, 252n124, 258n24, 259n33, 279n163, 285n24
Real Property Inventories, 28, 72–74, 76–78, 218n13, 218n15, 222n39
Reconstruction Finance Corporation, 64

Redick, Richard, 105
regional planning, 30, 48–49, 59, 94, 103, 115, 201n48, 210n102, 242n54, 243n58, 245n66
Regional Planning Association of America (Regional Plan Association) 202n48, 208n90, 240n39
renewable resources, 4, 42–43, 91, 136–37, 166, 196n11. *See* also urban renewal
Renner, George, 7, 30, 32, 46, 181n9, 186n43, 198n24, 202n49, 240n39
Research Committee on Urbanism, 28–29, 37, 48–50, 54, 56, 59, 65, 95, 107, 116, 119, 137, 202n49, 204n63
Resettlement Administration, 40, 42, 50, 109, 198n24, 244n64
resources. *See* national resources; natural resources
Riesman, David, 104
Robinson, Ira, 111, 117
Roosevelt, Franklin Delano, 28, 38, 40, 42, 47–48, 53, 81–82, 99–100, 109, 130, 163–64, 196n9, 198n24, 199n24, 203n50
Roosevelt, Teddy, 49, 199n24
Rosenberg, Albert, 145
Ross, D. Reid, 117
Rossi, Peter, 152, 247n80, 255n156
Rothman, Hal, 82, 281n7
Rouse, James, 132, 149
Rugh, Charles E., 19
rural areas, continuities with urban areas, 16, 22, 27, 38, 47–49, 50, 52, 59, 109, 113, 129, 138, 161–62, 164, 167, 201n43, 202n49, 204n63, 237n33, 244n58; versus urban areas, 1–2, 8, 29, 35, 47–49, 60, 72, 94, 129, 161, 210n102, 230n114. *See also* agriculture

San Antonio, 138, 145
San Francisco, 12, 72, 140, *141*, 145
Sanger, Margaret, 7
Sasaki, Hideo, 149, 248n92
Sauer, Carl, 183n22, 238n37, 240n39, 245n66
Schiff, Ashley, 257n14
Schilling, Walter, 111

Schmid, Calvin, 28, 194n122, 238n37
Schmoll, Hazel, 24
Sears, Paul, 33, 194n117, 215n135
Seattle, 20, 28, 285n24
sector theory, 74–75, 75, 87, 235n23
Segoe, Ladislas, 48
segregation, 13–14, 29, 37, 105–6, 110–11, 114, 131, 155–56, 185n32, 205n77, 239n38, 251n120
Sengstacke, John, 157
sequent occupance, 15–16, 20–21, 29, 31, 186n43, 252n122
Shelford, Victor, 33, 39, 50, 198n19
Shelley v. Kramer, 106, 113, 118
Shideler, Ernest, 12, 186n46, 220n27
Siegel, Jack, 98, 123, 125, 133–34, 149, 254n147, 277n153, 283n18
Simpson, Herbert, 71
Slayton, William, 125, 254n147, 284n20
slum clearance, 57, 69–70, 76, 87–88, 94, 101, 104, 112, 114, 121, 122–27, 129, 131, 135, 140, 146, 152, 154, 157–58, 164, 210n102, 228n98, 228n101, 230n114, 247n85, 251n120, 256nn3–4, 257n19, 258n24, 264n54, 266n73, 274n124, 276n149
slum prevention, 87, 115, 118, 123, 126–27, 132–33, 135, 144–45, 148, 151, 153–54, 156–57, 159, 166, 168, 258n22, 259n25
Smith, William, 19
social science, 2–3, 7–12, 15, 17–20, 26, 29, 32–34, 37–38, 42–44, 48, 51, 53, 69, 71, 74, 76–78, 80, 85, 96–97, 100, 102, 104, 106–11, 116, 118, 122, 127, 162–63, 167, 170, 179n4, 182n17, 183n24, 187n51, 190n82, 200n37, 201n48, 205n76, 207n82, 218n9, 219n25, 221n37, 223n54, 226n79, 236nn31–32, 236n33, 238n36, 241nn41–42, 241n43, 243n55. *See also* human ecology
Social Science Research Committee, 104–5, 227n87
Society for Social Research, 17, 49, 106, 186
soil conservation, 68, 123, 126, 130, 134, 162, 196n12, 211n111, 216n153, 254n152, 259n28, 283n18

Soil Conservation Act, 98
Soil Conservation Service, 39–40, *41*, 130, 211n10, 227n92
South Side Industrial Committee, 111
South Side Planning Board, 103, 111, 115, 121, 248n92, 251n120
Spencer, Herbert, 188n63
spider webs, cities as, 22
Steiner, Jesse, 28, 33–34, 189n81, 194n121, 238n37
Steiner, Richard, 160
St. Louis, 15, 29, 146, 157, 167, 193n111, 273n121
Structure and Growth of Residential Neighborhoods in American Cities (Hoyt), 73, 75, 76, 80, 87, 225nn75–76
Subcommittee on Urban Redevelopment, Rehabilitation, and Conservation, 132–34, 149, 258n22, 261n38, 277n153
succession: in cities, 11, 15–17, 21, 25–26, 29, 30, 34, 37, 51, 53, 65, 74, 105, 129, 136, 162, 169, 182n16, 190m82, 207n82, 227n92, 239n22, 248n92; in nature, 4, 11, 15–17, 24–26, *24*, *25*, 34, 37, 39, 51, 129–30, 162, 169, 182n16, 190n84, 199n26, 257n14
Sullenger, T. Earl, 29
systems ecology, 169, 257n14, 285n24

Tansley, Arthur, 10, 30, 33, 238n36
Taylor, Carl, 42, 199n26
Taylor, Griffith, 181n9
Taylorism, 110, 119n32, 244n61
Teaford, John, 262n40
Tennessee Valley Authority, 125–26, 199n27, 254n148
Thelen, Herbert, 118–21, 156, 248n92
Thomas, Lewis, 30, 192n100, 193n111, 194n122
Thompson, J. Arthur, 17, 27
Thompson, Warren, 29
Thrasher, Frederic, 20
Transeau, Edgar Nelson, 24, *24*
trees, cities as, 27
Trimble, Turner, 278n160
Truman, Harry, 99–104, 120, 130, 165, 233n10, 233n13, 257n19

Tugwell, Rexford: and agricultural planning, 40, 42, 65, 109–10, 129, 163, 215n135, 244n61; and urban planning, 104, 108–10, 115, 243n54, 244n61, 244n63

Ullman, Edward, 105, 220n27, 237n33
Underwriting Manual (Federal Housing Administration), 53, 72, 76, 207n82
University of Chicago, 4, 10–12, 14, 17–19, 22–23, 30, 49, 52, 54, 62, 64, 70, 77, 79–80, 86, 99, 104–12, 115–17, 119, 121–22, 149, 152, 155, 180n10, 188n70, 190n86, 191n94, 203n50, 205n74, 220n32, 221n38, 214n131, 248n92
Urban Community Conservation Act, 98–99, 124, 135, 255n156
Urban Conservation and Development unit, National Resources Planning Board, 67, 81, 97, 245n68
Urban League, 113, 144, 156–58, 160, 277n153
Urban Redevelopment Corporations Act, 216n151
urban renewal, 1, 99, 127–29, 134–40, *141*, *142*, *143*–60, 162, 165–71, 231n122, 256n3, 262n37, 262n40, 264n59, 266n97, 267n80, 271n102, 272n103, 273n121, 275n133, 279n160, 280n164, 280n166, 280n170, 285n23
Urban Renewal Administration, 135, 138–39, 143–44, 148, 153–54, 160, 266n73

Vallejo, CA, 146
valuation, 36, 55, 62, 118, 146, 156, 159, 211n111, 250n102, 285n24
Vance, Rupert, 33
van der Rohe, Mies, 116, 248n92
von Humboldt, Alexander, 10
von Thünen, Johann, 22, 181n6

Walker, David, 160, 280n170
Walker, Mabel, 44, *45*, 51
Walther, Herman, 64, 101
Warming, Eugenius, 7, 10, 12, 17, 27, 106

Warren, OH, 277n153
Washington, DC, 12, 53, 70, 72, 75, 77, 82–83, 85, 127, 138, 140, 144, 152, 219n18, 231n120
Watt, Richard, 233n11, 233n14
Waverly, in Baltimore, 60–62, 63, 65–66, 82, 124–25, 215n140
Weaver, John, 39, 196n12
Weaver, Robert, 171, 283n18
Weaver, W. Wallace, 21, 27–28
Weber, Ada, 181n6
Weimer, Arthur, 53, 90–91, 91, 285n24
Weiss, Marc, 60, 206n81, 215n140
Wheeler, William Morton, 14, 17, 32
Whitaker, Joe Russell, 43
White, C. Langdon, 32, 185n34, 186n43
White, Gilbert, 108, 119
Whittlesey, Derwent, 15–16, 20, 31
Wieboldt Foundation, 105, 118, 236n30
Wiley, Clarence, 86
Wilson, M. L., 50
Wirth, Louis, 14, 29, 32, 44, 48, 71, 77, 79, 104–5, 107–8, 115–17, 121, 222n48, 241n48, 242n49, 244n66, 248n92
Woodbury, Coleman, 77, 227n87, 246n73
Woodlawn, in Chicago, 5, 60, 62, 64–66, 68, 83–85, 88, 89, 93–95, 101, 112, 116, 124, 145, 149, 156, 212n112, 213n125, 213n127, 214nn131–32, 229n109, 229n111, 246n79, 254n147, 262n39, 278n160, 280n171
Woodlawn Plan, 64, 94
Workable Program for Urban Renewal, 138–40, 143–45, 154, 263n54, 264n60, 274n124
Works Projects (Progress) Administration, 61, 64, 76, 78, 218n15
World War II, 38, 68–69, 81–85, 94–96, 99, 113, 163–65, 223n61, 224n62, 231n122, 233n11, 268n83
Wright, Henry, 45

Ylvisaker, Paul, 149, 167–68, 282n14, 284n20
Young, Erle Fiske, 18, 30
Young, Hugh, 77, 93, 222n39

zonal structure: in cities, 21–22, 23, 25–27, 29–30, 50–51, 75, 80, 91, 188n70, 192n100; in nature, 23, 25, 189n74
zoning, 12, 43, 56, 61, 65, 76, 91, 160, 190n90, 211n109, 216n156, 226n80
Zorbaugh, Harvey, 13, 43, 117
Zueblin, Charles, 242n54